Sensors and Probes for Bioimaging

Sensors and Probes for Bioimaging

Young-Tae Chang
Nam-Young Kang

WILEY VCH

Authors

Prof. Young-Tae Chang
Department of Chemistry
POSTECH, Pohang
South Korea

Dr. Nam-Young Kang
Department of Convergence
IT Engineering
POSTECH, Pohang
South Korea

Cover Image: Courtesy of Young-Tae Chang

Library of Congress Card No.: applied for

British Library Cataloguing-in-Publication Data
A catalogue record for this book is available from the British Library.

Bibliographic information published by the Deutsche Nationalbibliothek
The Deutsche Nationalbibliothek lists this publication in the Deutsche Nationalbibliografie; detailed bibliographic data are available on the Internet at <http://dnb.d-nb.de>.

Print ISBN: 978-3-527-34960-9
ePDF ISBN: 978-3-527-83434-1
ePub ISBN: 978-3-527-83435-8
oBook ISBN: 978-3-527-83436-5

Typesetting Straive, Chennai, India
Printing and Binding CPI Group (UK) Ltd, Croydon, CR0 4YY

C116210_300523

Contents

1

Introduction to Bioimaging

Bioimaging can be defined as visualization of a biological object. The most basic bioimaging may be just "seeing" the living object using our own eyes. This function is called "vision" and the procedure is mediated by visible light. The visible light is a part of electromagnetic wave in the wavelength range of 400–700 nm, and the image information is generated by the interaction between light and object, such as reflection, scattering, and diffraction. The generated information-rich light package travels and reaches our eyes. The focused light through lens would be projected to the screen as in a camera. The retina in our eye is the screen of the image, which is composed of the two-dimensional array of optic nerves. The photon in light signal (containing the information of the object) reaches the retina and activates optical neurons, and the signal is transferred to the brain and is reconstructed into the image of the object by neuronal processing. Even though the screen is two-dimensional, the processed images via two retinas provide three-dimensional information about the shape and distance of the object. Visible light travels at a so-called speed of light (3×10^8 m/s), so the information transfer in the vision could be almost instantaneous. If there is a possible delay, it may be from the signal transition step from the optical nerve to the brain and the information processing time in the brain.

Bats live in the dark environment without enough environmental light for vision. Instead, they use ultrasound for bioimaging platform. If other conditions are same, the light vision could be million times faster than ultrasonic sensing (340 m/s) (Figure 1.1). Among all the sensors, light vision is the fastest and most information-rich system. Therefore, the invention of eye (in more general term, photoreceptor) is one of the most dramatic events in the evolution of life. Due to the high quality and also huge quantity of information, vision is the most important sense, easily accounting for more than 90% of information we receive through all other senses, including hearing, taste, smell, and touch.

1.1 Color

Visible light sensing not only generates black-and-white images, but also can provide color information. The visible light is composed of a spectrum of electromagnetic wave in the range of 400–700 nm. Human eye has three color photoreceptors,

Sensors and Probes for Bioimaging, First Edition. Young-Tae Chang and Nam-Young Kang.

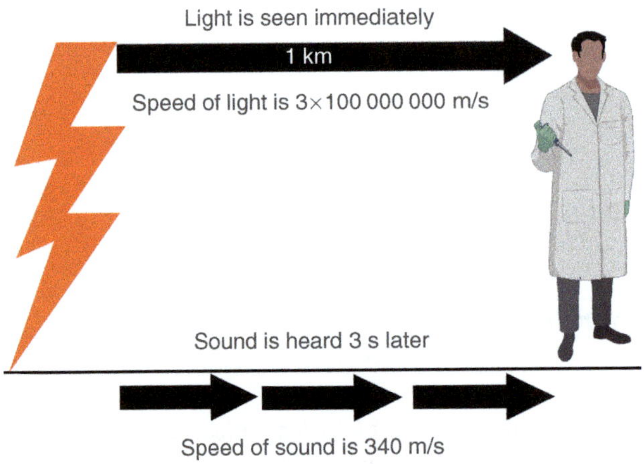

Figure 1.1 Vision through light and sound in different speed.

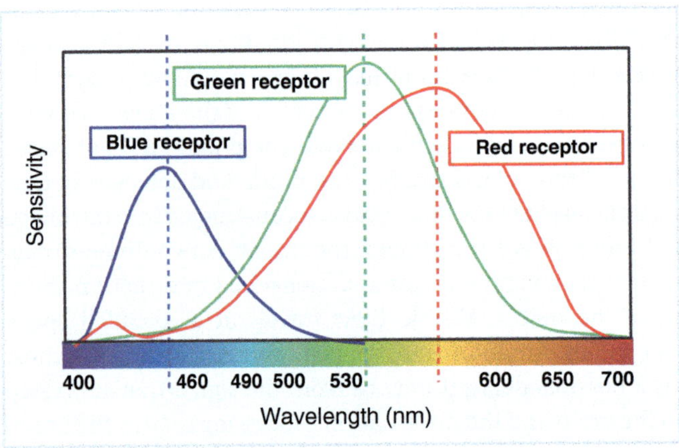

Figure 1.2 Three color receptors and their sensitivity to different wavelengths.

of which the maximum sensitivity is for blue (445 nm), green (535 nm), and red (575 nm) (Figure 1.2). For example, when we receive 445 nm light, we sense it as a blue color, and 575 nm light as a red color. Therefore, color recognition is the ability of sensing different wavelengths of light. And, the term spectroscopy is derived from spectrum, i.e. spectroscopy is the study of the wavelength-dependent interaction between the light and the object.

If we have three color receptors, then do we recognize only three colors? No, it is not. At least, we give seven names of color to the rainbow! Our color sensors have the maximum sensitivity to a specific wavelength, but the sensing wavelengths are broad and overlap with each other. If the eye receives 560 nm light, both green and red receptors are activated, and we sense it as a yellow color. The light with 590 nm will more strongly activate red receptors and less strongly green receptors, making the color as orange. That means our color sense is determined by the ratio of the three

Figure 1.3 Comparison between black-and-white picture and colored picture.

Figure 1.4 The color spectrum of dogs and humans.

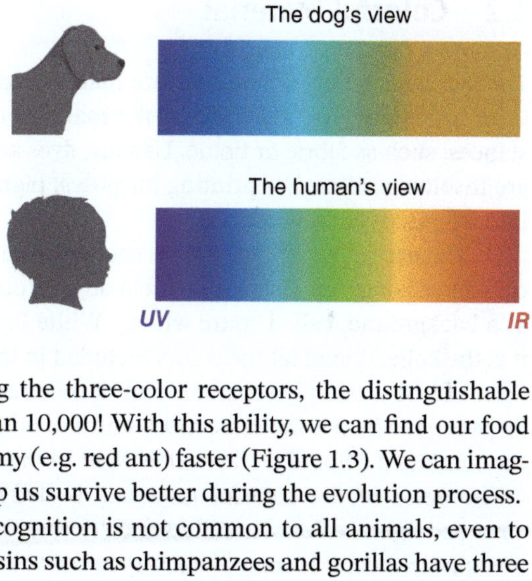

receptors' activation degree. Using the three-color receptors, the distinguishable colors by human eyes are more than 10,000! With this ability, we can find our food (e.g. red apple) better, also our enemy (e.g. red ant) faster (Figure 1.3). We can imagine how useful this ability is to help us survive better during the evolution process.

Interestingly, this three-color recognition is not common to all animals, even to mammals. Only our very close cousins such as chimpanzees and gorillas have three color receptors, but even then not all the monkeys do. Including our remote cousins, dogs and cows have only two-color receptors. It sounds trivial whether it is two or three. But, with two color receptors, the distinguishable colors are narrowed down only to the level of 100! It is the difference between two- and three-dimensional combination power (Figure 1.4). Therefore, the visions of cows and dogs would be much more boring than the colorful flowers and spectrum of rainbow we see. This is why many sensors are designed for color change to achieve maximum effect to the naked eyes. Our eyes are a wonderful color sensory system!

There is a funny story in the bull fight. The fighters use red cloth to stimulate bulls, as red color may be related to the image of blood. Funny thing here is the bull may see red color more like dark gray rather than bloody red. The red cloth is to stimulate the audience, not the bulls at all!

Color blindness arises when part of the color receptor is defunctionalized. In humans, most common type is green-red blindness, which occurs when either green or red receptor has problem. If you look at the receptor property carefully, you may realize that the maximum wavelengths of green (535 nm) and red (575 nm) receptors are rather closer, compared to that of blue (445 nm). We call the receptors green and red, but they are more like yellow and orange. To maximize the combination power in color contrast, this design may not be the optimum choice. If we design the color pixel of a computer screen, we may choose more even distribution of the colors, such as 465, 525, and 630 nm [1]. Not surprisingly, the green and red receptors are structurally closer to each other, implying that they evolved from the common ancestor. So, we can imagine, a long time ago, we also had two color receptors similar to dogs or bulls (blue and yellow), and the yellow receptor diverged to two receptors, green and red. Without this evolution of color receptors, we might not be able to enjoy the beautiful sunset!

1.2 Colorful Material

The synthetic colorful materials are mainly organic dyes and inorganic pigments. Conventionally, dye is defined as the material that imparts its color to other substances, such as fabric or tissue. Usually, dyes are soluble in solvents, but pigments are insoluble solids. For printing purposes, pigment powder needs to be dispersed into a liquid binder before use.

On the earth, the strongest light source is the sun. To minimize the background of light sensing, our visionary system adjusts our sensors to recognize the sunlight as a background, called “pure white.” White light is not the status of no color, but it is the collection of all the colors included in the sunlight. The colors of the white light can be manually separated into a spectrum by a prism through a process of dispersion, which is the same mechanism of rainbow formation. Therefore, white is the combined color of all the visible light in the rainbow.

The color of the colorful materials is determined by the wavelength of the absorbed light, i.e. leftover reflected color after absorption of white light. Therefore, the appeared color is complementary to the absorbed color. The concept of complementary color has been known for a long time and is widely used in painting art for vivid color contrast. Even though the wavelength of visible light is in linear scale (400–700 nm, violet to red), our color receptors deceive our color recognition due to the tiring of receptors. The relationship of the complementary color in our color sensing system is described in a color wheel (Figure 1.5).

A chromophore is the part of the molecule which is responsible for the color. The chromophore of inorganic pigments is usually is transition metal, which has a visible light range of electron excitation energy. The chromophore of organic dyes

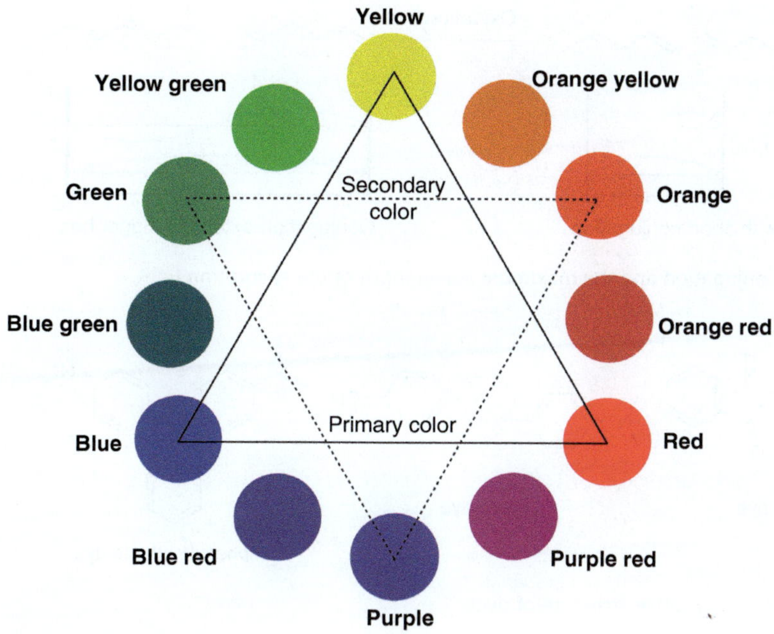

Figure 1.5 Color wheel.

is a long-conjugated double bond system. The light absorption had been modeled in early quantum mechanics era through a particle-in-a-box model, which later led to Schrödinger equation for atomic structure of electrons. Interestingly, the organic conjugation system could be described as a particle-in-a-box model, where the σ bond electrons define the size of the box and π electrons are the particles in the box. As the box size becomes bigger, the wavelength of absorption light gets longer, through the narrowing electron transition gap. When the absorption maximum reaches the boundary of visible light (violet color), the appeared color of the material would be the complementary color of violet, yellow (colors in opposite direction in the color wheel). If the conjugation gets longer, the absorption maximum moves from violet to blue and then green. Accordingly, the appeared color changes from yellow to orange and then brown. You may recall the old books turn into the yellow color first, and then change into reddish tone. This is the result of the extension of conjugation in the lignin component in the paper pulp and is one of the examples of the organic dye model of particle in a box (Figure 1.6).

However, if the conjugation is too long, any oxidation or reduction reaction in the middle of the chromophore will break the conjugation bridge, which is the principle of bleaching agents (either oxidants or reductants). That means the dye with a long conjugation system is weak for the chemical damages, and usually the color of naturally occurring organic dyes easily diminish over time. As a result, the color of simple carbon-conjugated systems is not vivid, as shown in the paper of old books. In the nineteenth century, German organic chemists opened the way to synthetic dyes to replace the natural dyes. Adding electron-donating and -accepting motifs at

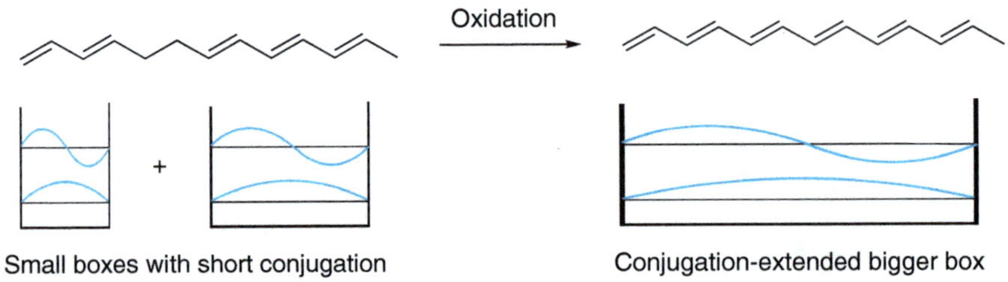

Figure 1.6 Conjugation and the maximum wavelength of the absorption light.

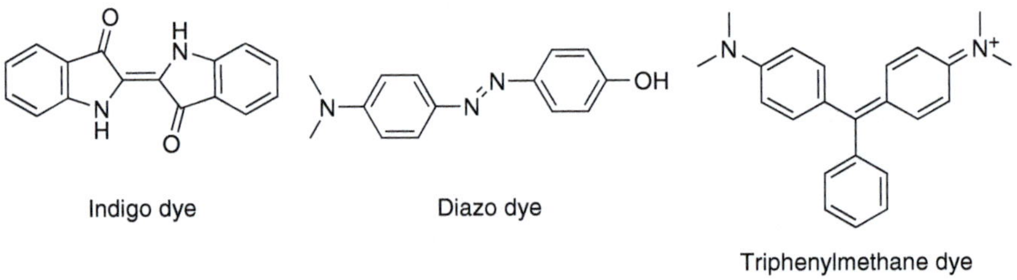

Figure 1.7 Representative structure of dyes.

each end of the conjugation system provides a stronger dipole, which makes the conjugation effect longer and also makes the absorption stronger to give a vivid color. So, most of the synthetic dyes are composed of the conjugation system with electron-donating and -accepting motifs at each end (Figure 1.7). The wavelength of absorption of the organic dyes can be predicted by molecular orbital calculations, and Pariser–Parr–Pople (PPP) method is one of the best-known models [2].

1.3 Light Source of Bioimaging

If bioimaging is visualizing a biological object, which part of the body is the object? If we take the daylight photography as an example, we can use a casual camera catching the visible light to visualize the surface of our body. The surface imaging is easy, and the damage by light exposure to skin is trivial. However, what if we want to visualize the inside of our body? Usually we use catheters composed of a metal tube with a light source and camera on the tip, and insert them through the mouth or anus to visualize the gastrointestinal tract (Figure 1.8). While we call this technique as an endoscopy, is this really inside of our body? Well, compared to the exposed skin surface, the gastrointestinal tract seems more like a hidden part of our body. But, topologically speaking, if we consider our body to be donut shaped, the surface of the gastrointestinal tract should be considered as part of the outside, not the real inside.

In contrast, if the camera visualizes the beneath skin area, we may consider it as a real inner part of our body imaging. For this real endoscopy, one possible way may be to put the camera penetrating the skin to reach the target tissue. However,

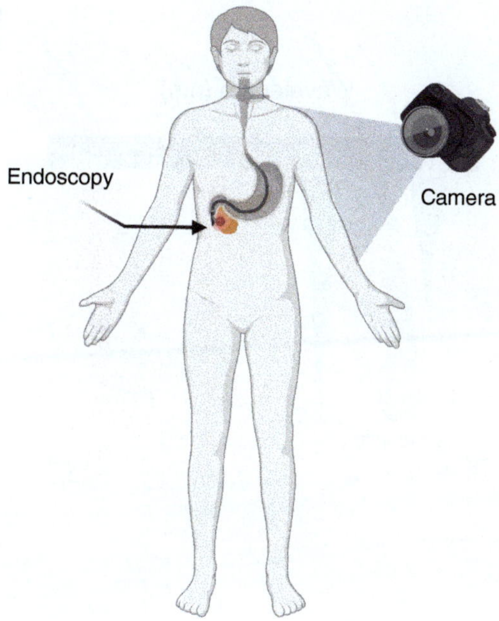

Figure 1.8 Regular camera and endoscopy.

if it is not really necessary for treatment purposes, such an invasive approach may not be desirable for humans or any live animals. If it is not physical penetration of the camera, how about light penetration? If light penetrates the skin reaching the inner space and returns back with the information of the target area, it would be much less damaging than physical insertion of the camera. In this case, the penetration depth of light would be an important factor. If our skin is like a transparent jelly fish, the inner world imaging may be straightforward. The term "transparency" itself implies free penetration of visible light. Unfortunately, our body skin is not so transparent, and most visible light can penetrate at most several millimeters of depth under the skin. Then, how can we increase the penetration depth of light through the tissue?

Visible light lies in the 400–700 nm range, and there are other light sources outside of visible light. The shorter wavelength of visible light comprises ultraviolet (UV), X-ray, and γ-ray, etc. The longer wavelength makes infrared (IR), terahertz light, microwave, and radio wave. Coming back to tissue imaging, the penetration of electromagnetic waves is diminished by mainly scattering and absorption of the light source in the tissue. Absorption wavelength is dependent on the tissue composition, but the scattering is usually higher in shorter wavelength light. So, longer wavelength light tends to penetrate better than the shorter wavelength light, by reduced loss by scattering. For this reason, near-infrared (NIR: longer than 700 nm light up to 1000 nm) light is a popular optical imaging source for noninvasive tissue imaging or whole-body imaging of small animals such as mice. Recently, even longer wavelength of 1000–1700 nm is popularly used for bioimaging as the second NIR window or NIR-II [3]. With further reduced scattering and negligible autofluorescence, NIR-II may provide a higher signal-to-noise ratio and deeper tissue

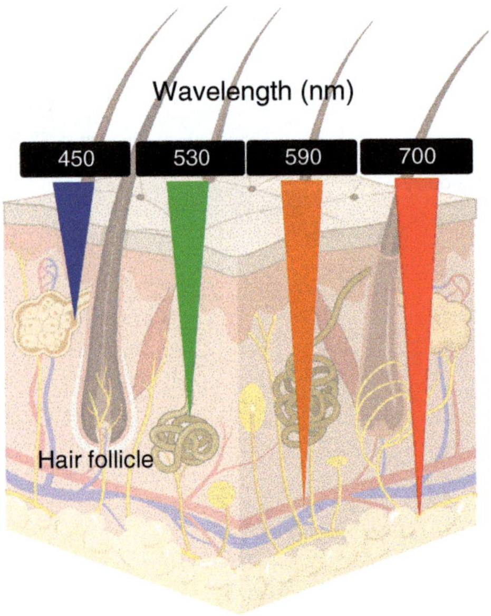

Figure 1.9 Penetration of light into the skin.

penetration than the conventional NIR imaging (Figure 1.9). As our eyes cannot sense NIR directly, the detected NIR should be converted to artificial visual light image as in the night vision goggle of battle field. The green color in the night vision goggle image is a processed artificial color. Green is the usual choice of color due to its best sensitivity to our eyes. NIR is better than visible light for the penetration depth, but still it is difficult to proceed further than a centimeter into the tissue.

The even longer wavelength light sources, such as microwaves or radio waves, have better penetration through the whole body and have been used in magnetic resonance imaging (MRI). MRI requires a high magnetic field to separate the nuclear spin energy status of protons in the body. The separated energy gap of nucleus absorbs microwaves and MRI detects the signal of the relaxation of the absorbed microwave light. Protons in different environments (such as water or lipid) generate distinguishable signals and through a computed tomography (CT), three-dimensional sectional images could be constructed. MRI is a noninvasive CT technique, especially useful for soft tissue (which contains protons) imaging.

If we go to the other direction of shorter wavelength light, there are still possibly different applications in bioimaging. In the X-ray range, the wavelength of light is a hundred times shorter than visible light, and the photon of X-ray would be small and rigid. If visible light photon is like a tennis ball, X-ray is like a needle and can easily penetrate soft matter. Thus, X-ray images mainly show the rigid bone structure, through which X-ray cannot penetrate (Figure 1.10). By adopting CT techniques, X-ray three-dimensional imaging has been well developed even before MRI is introduced. Due to the order of historical development, conventionally the term “CT” is used for “X-ray CT”, unless other description is provided.

Although most of the X-ray light does not interact with soft part of the body, in a molecular level, there could be a small amount, but strong damage to

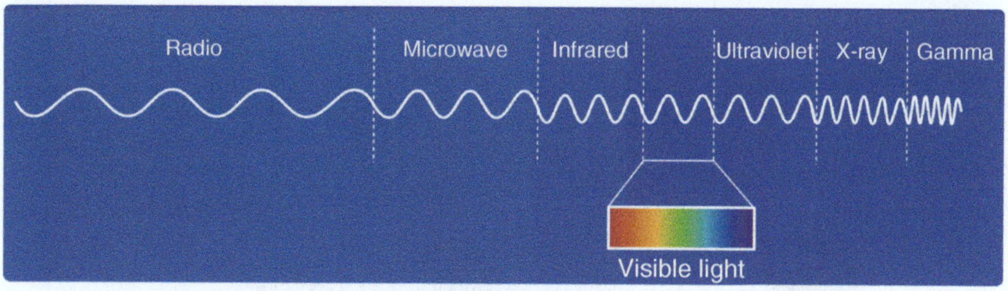

Figure 1.10 Electromagnetic waves as the light source of bioimaging.

the biomolecules can occur by breaking the covalent bonds or ionization. The accumulated damages in DNA can cause mutations of cells, resulting in cancer in somatic cells and mutagenesis in fetus. So, excess amount of exposure to X-ray is not recommended due to health concerns, especially for pregnant women.

γ-Ray has a shorter wavelength than X-ray and the higher energy allows its penetration even through bones, the hardest part of our body. The intensive γ-ray can be used for tumor treatment, which is called as γ-knife technique. To minimize the damage to normal tissue, multiple sources of γ-ray are used from different directions, and only the tumor site is focused to accumulate a high density of γ-ray. In principle, γ-ray also can be used for bioimaging in a similar way of X-ray imaging, but it is not common in practice. Instead, γ-ray-generating radioactive materials are used as imaging agents. In this case, the γ-ray is not provided from outside as in the X-ray method, but is irradiated from inside of the body, through an administered imaging probe into the target site of the body. The position of the isotope could be imaged through a γ-camera similar to an X-ray film. When CT technique is combined with γ-ray-generating radioactive isotope, a three-dimensional single-photon emission computed tomography (SEPCT) imaging is also possible. A similar, but higher performance technique is positron emission tomography (PET). In PET, instead of a direct γ-ray-generating isotope, a positron-generating isotope is used. Positron is a positively changed electron, a kind of anti-particle of electron. When a positron meets an electron, they are annihilated, generating one pair of γ-ray photons. As the two γ-ray photons travel in direct 180° providing richer information for the original position of positron, usually the spatial resolution of PET is better than SPECT. It is noteworthy that X-ray uses an external light source for the imaging, but SPECT and PET use endogenous γ-ray generated from an isotope-labeled probe (Figure 1.11). That is why SPECT and PET are called molecular imaging techniques, in contrast to X-ray imaging.

As shown earlier, electromagnetic waves with different wavelengths from visible light also can be used as the light source of various imaging techniques, when coupled with a proper detector or camera system. The different wavelengths of light render different modes of interaction with matters, and each can generate unique information for the target object. Therefore, unexplored areas of electromagnetic waves would provide novel chance of new imaging technology or modality. Terahertz light is such an emerging new source of light.

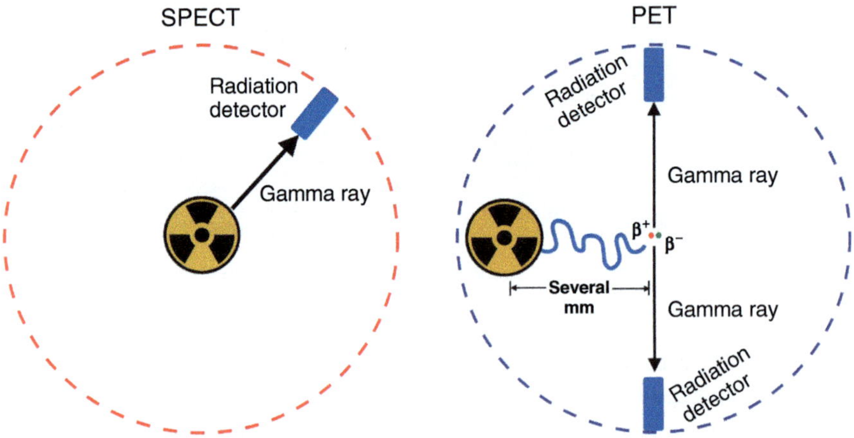

Figure 1.11 Endogenous γ-ray imaging in SPECT and PET.

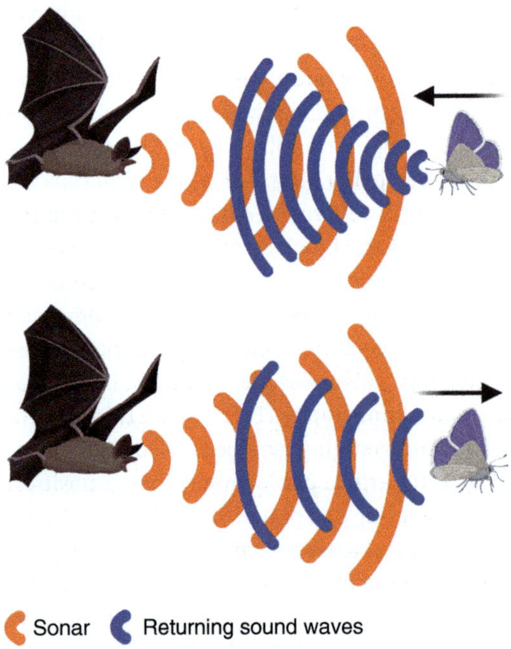

Figure 1.12 Sound imaging of bats.

Not only electromagnetic waves, sound waves or seismic waves also can generate processed images through interaction with matters. The bat's vision through ultrasonic waves would be a good example. Combining electromagnetic waves and sound waves for improved or unique imaging technique, such as photoacoustic imaging, is also a powerful visualization technique (Figure 1.12).

Electron beam is another source to provide ultrahigh-resolution imaging of materials. There are several modes of electron microscopy, such as transmission electron microscopy (TEM) and scanning electron microscopy (SEM). In TEM, an ultrathin sample is irradiated with an electron beam and the transmitted electrons

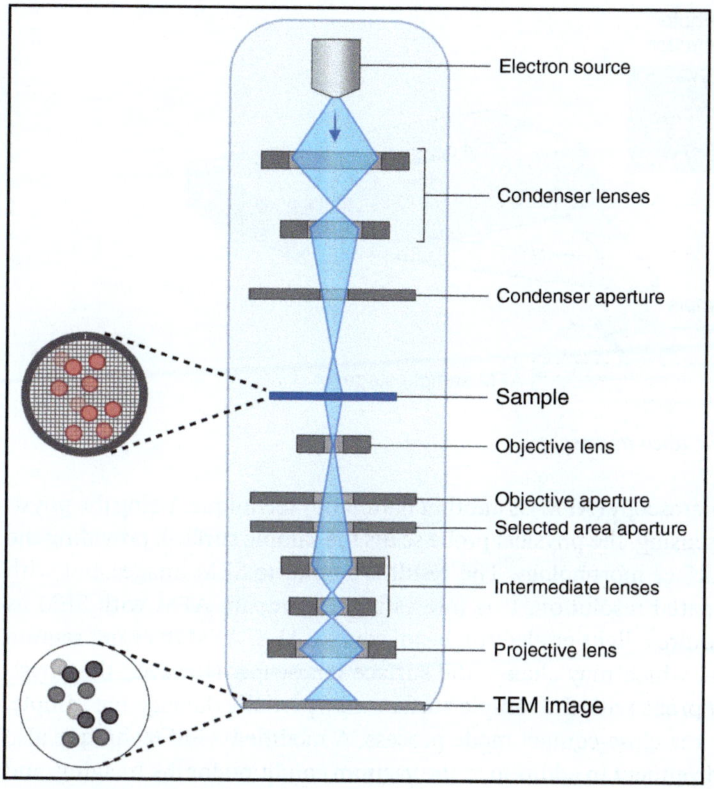

Figure 1.13 Transmission electron microscopy.

are used for the two-dimensional image construction, which is similar to X-ray imaging. In the sample area with a high electron density, the input electron beam would be scattered and may not transmit, generating the dark contrast in TEM image. Therefore, heavy metals are absorbed to the sample to enhance the contrast, and the procedure is called electron staining. The electron staining can also be achieved by an organic dye. After diaminobenzidine (DAB) is stained, through photooxidation, an electron-dense precipitate can be formed to increase the TEM contrast, which is similar to dye staining in the optical imaging.

In SEM, the incident electron beam interacts with the surface atoms of the sample and generates back scattered electrons or secondary electrons. The incident beam is focused on a sample spot and scan the surface, and the detectors are located in the same side of the input beam. As a result, SEM image shows the surface morphology with three-dimensional information especially provided by secondary electrons. The resolution of SEM image is in nanometer range, and usually TEM has higher resolution than SEM. While optical imaging suffers from the diffraction limit in sub-micrometer range, electron microscopy provides much higher resolution. By the imaging resolution scale, electron microscopy could be called as "nanoscopy," rather than microscopy. Both techniques require vacuum condition for the imaging due to the electron beam usage (Figure 1.13).

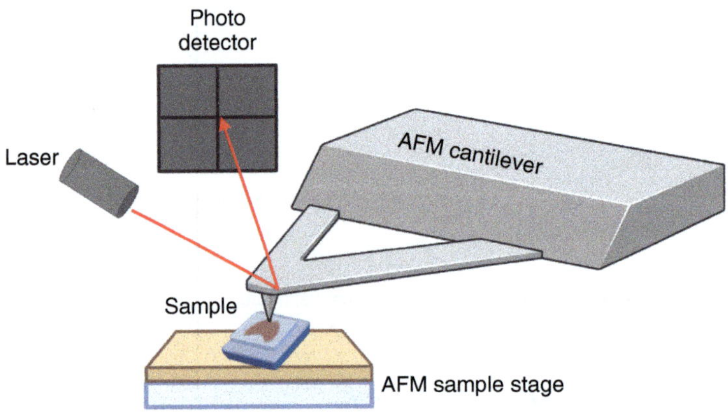

Figure 1.14 Atomic force microscopy.

Atomic force microscopy (AFM) is another nanoscopy technique. Using the physical contact force sensing, the physical probe scans the sample surface, providing the information of surface morphology. The result is similar to SEM images, but with a much higher spatial resolution. It is interesting to compare AFM with SEM as AFM does not require a light or electron beam source. Also, AFM does not require the electron stain, which may change the surface landscape. However, the physical contact of the probe with the sample surface may partially damage the sample, especially during the close-contact mode process. A modified AFM technique also allows liquid environment in addition to the vacuum condition for the imaging, and more biologically relevant samples could be imaged, such as live-cell surface imaging (Figure 1.14).

1.4 Subcellular Imaging

When the object of visualization is too far from our eyes, we use a telescope. If the object is too small, we use a microscope. Superficially, they may seem to use opposite principles, but actually they are similar in a sense that they "magnify" the "too small images" to a sensible size for naked eyes. One is for too small images due to the long distance of the object and the other is for nearby, but physically too small object. If they are similar, can we use telescope instead of microscope for the small object or vice versa? No, we cannot. What is the difference, then? The difference lies in focal distances depending on the position of the object. In a telescope, the focus is on the long distance, and in a microscope, the focus is on the sample slide right under the lens.

Now, let's focus on the microscope for visualizing small objects in biological systems. The basic unit of life is cell. For unicellular organisms, a single cell is an individual or entity. In multicellular organisms, cells gather together to make tissue, and tissues make organ, and organs assemble to make an individual body. The reason why a cell is the basic unit of life is that each cell contains the whole

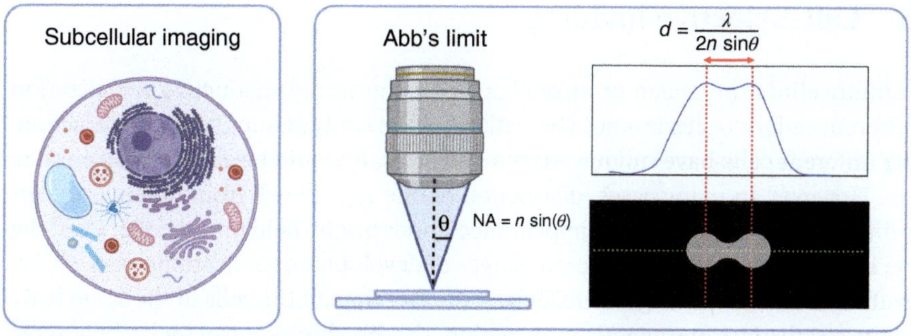

Figure 1.15 Subcellular imaging and Abbe's limit.

genomic information of the individual. In other words, starting from any single cell, in principle, we can reconstruct the whole body.

The usual size of cells in animals or plants is around 10 μm, and unicellular bacteria are about 1 μm in size. While bacteria cell structure is relatively simple, animal or plant cells have complex intracellular structure, called organelles, such as nucleus, mitochondria, lysosomes, Golgi body, and endoplasmic reticulum (ERs). The intracellular organelles are usually about 1 μm or smaller size. When light encounters an object with a similar size to the wavelength, the light path is altered by diffraction. The visible light is in 0.4–0.7 μm (400–700 nm) and if the object is about half micrometer or smaller, the image become blur. This is known as Abbe diffraction limit, named after Ernst Abbe who found it in 1873, and is considered as the physical limit of the optical resolution (Figure 1.15). Therefore, the physical size limit of a microscopic image is about ~ μm range.

To overcome the size limit, several optical and mathematical tricks were developed into "super-resolution" techniques or so-called nanoscopy, which means nanometer-resolution imaging. In addition to the size limit, organelles are usually transparent, so the optical visualization is further challenging, as it is difficult to distinguish different organelles. That is why organelle-selective dyes are widely used for vivid subcellular organelle visualization. In other words, bioimaging is a process of visualizing a biological object, otherwise invisible. Most of the cell images we have in our mind are "stained" images rather than natural cell images. For example, chromosome, as condensed form of DNA, means "color body (chromo-some)" as it is easily stained by dyes. You may have seen the change of the chromosome during the cell division, such as condensation, alignment, and division of DNA. It implies that most of the chromosome images are also obtained from DNA-stained cells rather than intact natural cells. By the same token, if there is a selective dye for each organelle, it would be possible to see specific organelle standing out from a transparent background. These selective dyes are called organelle-selective probes, and if the dyes change their colors upon binding to the target, they can be called as sensors. Therefore, the definition of probes embraces sensors. In other words, sensors are special type of probes in bioimaging, providing the information of change of the target.

1.5 Cell-Selective Imaging

In a multicellular organism or mixed bacteria community, distinctive visualization of different cells or bacteria would be critical for the study of intercellular interaction. If the different cells have unique shapes and sizes, it would be easy to discriminate them. However, in many cases, distinction of one type of cell from others is generally difficult due to their similar appearance under bright-field microscope. Even the same kind of cells may have different stages of development or death process, showing off different morphology. Considering the fact that all the cells in the same body contain exactly the same genetic information, the discrimination of their phenotypic difference is the key for the study.

To overcome the problem, cell-selective probes have been explored for a long time. Antibodies have been the most common probes for the cell distinction and are widely used. Hundreds of antibodies have been developed and validated for cell discrimination and imaging. While useful, due to their high molecular weight of 150 kDa, their access to the intracellular target in live status is intrinsically limited. Even though the binding target of antibodies is on the cell surface, they are usually functionally important enzymes or receptors. As a result, antibodies often induce functional influence in the treated cells, which is not desirable for normal cell study. Alternative solution may be a smart small molecule probe, which may complement the antibodies' weak points, especially for the intracellular target.

Not only for our own cells, we also need to distinguish and visualize foreign life forms, as our body is always interacting with them. For example, our body hosts huge numbers of bacteria as guests in similar or even higher number than our own cells, which is called the microbiome. The bacteria in the microbiome established symbiotic relationships with our body and majority of them are not harmful to us. But, if we get pathogenic bacterial infection, figuring out the identity of the bacteria would be urgent and important for making decision of the proper treatment. The morphological difference may not be informative enough to get a good discriminating information. Media-selective culturing is a standard test for the identification, but the process takes days of time, and also the identification is limited only to the known strains for their culture condition. While polymerase chain reaction (PCR)-based genetic analysis is getting more and more popular for high accuracy and sensitivity, the need for an in-site imaging probe increases for faster analysis and functional monitoring through the visual images. So far, such an efficient and practical cell-selective probe is yet to be developed.

1.6 Tissue and Organ Imaging

When cells gather to make tissues and organs, a tangible physical structure emerges, and macroscopic imaging technique is required. For diagnosis of diseases, often a biopsy (tissue sampling from live body) procedure is required for tissue imaging or biochemical testing. Usually, the tissues are stained with dyes and imaged to determine the disease status. As the test is performed outside of the body, the procedure is called ex vivo imaging. For example, from a surgery for cancer, the excised tissue (suspected as a tumor) is processed through cryo-section or paraffin treatment, and then stained with dyes for visualizing the tumor and healthy tissue. Most likely, the sample is sent to a pathologist who has long-term training and experience to make the call if the tissue is indeed cancer or not. The procedure takes easily an hour or longer, and it is quite difficult to get the results back before the surgery procedure is over. If the sample preparation procedure becomes simpler and faster, and also a user-friendly probe is available, which does not require a pathologist for reading, it would be possible to get the results within the surgery procedure. Not only for tumors, any kind of disease symptoms such as inflammation or infection could benefit by the selective probes.

1.7 Whole-Body Imaging

If the tissue imaging can be performed without removing the tissue from the body, it would be even better. Such an optical imaging in the live body is called intravital microscopy, as a kind of in vivo imaging. The imaging for blood cell flow or extravasation is an example, and unlike the ex vivo imaging, the intravital microscopy allows repeated measurement with minimal invasiveness for long-term monitoring of diseases. Some of the imaging could be achieved from the natural tissue itself, but sometimes it is necessary to use probes to get a clear contrast.

For example, in cancer surgery, it is often difficult to discriminate the exact boundary between the tumor and normal tissue. If there is a selective probe for a tumor to show a clear boundary, it would greatly help the surgeon to decide the excision line for saving maximum healthy tissue for the patient. If the dye was colorless before binding to the tumor, but generate a strong color in the tumor, the probe could also be a sensor for the tumor and carries low background in the normal tissue. The imaging technique used in operation is called intraoperative imaging.

The eventual goal of bioimaging would be a noninvasive (without an open-up surgery) whole-body imaging without a biopsy sampling (for ex vivo imaging). The ideal probe could act as a diagnostic tool to detect disease occurrence with precise position and size information of the target. The probe should not be toxic and also could be used for body response to drug treatment as a prognostic procedure. There is huge room for improvement in the current in vivo imaging with smart probes and improved image process/analysis method.

1.8 Probes in Bioimaging

Probes help to visualize target organelles, cells, tissues, and organs with an outstanding contrast. Sensors are part of probes, and respond to the analyte or environment by changing the color or intensity. Most of the biological images are physically stained images or artificially drawn pictures, which reflect the practical importance of probes in the field. In this book, the history of probe development, their applications in different levels of body, i.e. intracellular organelles, different cells, tissues, and whole body. In later chapters, the probe application in biological environmental changes and diseases, and various imaging techniques both for nonoptical imaging and fluorescence will be described. In perspective, design or discovery of selective probes and the future direction will be suggested.

References

1 Deckman, I., Lechêne, P.B., Pierre, A., and Arias, A.C. (2018). *Org. Electron.* 56: 139–145. https://doi.org/10.1016/j.orgel.2018.02.009.

2 Griffiths, J. (ed.) (1984). *Development in the Chemistry and Technology of Organic Dyes*, Critical Reports on Applied Chemistry, vol. 17. Oxford: Blackwell.

3 He, S., Song, J., Qu, J., and Cheng, Z. (2018). *Chem. Soc. Rev.* 47: 4258–4278. https://doi.org/10.1039/C8CS00234G.

2

Chemical Sensors and Probes for Bioimaging

Sensor is a device or molecule which detects and reports us with the changing physical quantities in the environment. The changing physical quantities could be temperature, pressure, light, sound, chemicals, etc., and the sensing mechanism could be color or electric signal change. Therefore, a sensor works by sensing environmental changes and generating a responsive signal for reporting. In comparison, a probe carries the signal of its presence, but the signal could be either fixed or variable one (as in a sensor). Therefore, the category of probes includes sensors, and sensors are a special type of probes. In engineering field, sensor or probe usually means the device for physical measurement, but in this book, it is mainly focused on molecular sensors and probes.

If the sensor or probe is for bioimaging, most likely the target would be biological materials. In molecular level, the biomolecules could be proteins, carbohydrates, lipids, nucleic acids, or metabolites. In microstructure level, they could be intracellular organelles, extracellular matrix, or the cell itself. In macroscale, they may be tissue as a cluster of cells, functional organs, or eventually entire individuals. Depending on the scale of the target, the choice of probe and sensor, and their operational method would be determined with various strategies.

2.1 History of Dyes in Biological Stains

Historically, what would be the first sensor recorded in the literature? Maybe the silverwares for detecting poison in the food? You may have seen in a movie or read in a book that silver spoons turn color by poison in the food. Silver reacts with sulfur, and the shiny silver surface turns into black (Figure 2.1), so this method can detect only sulfur-containing poisons. Even stronger poisons, if they do not contain sulfur, cannot be detected by silverwares. But, this is an example of sensor for poison outside of body, rather than direct sensor for biological sample itself. How about detecting poison from a dead body? If the test is performed visually in the body, it could be a good example of bioimaging or biosensing, although the test sample is an already-dead body, so called autopsy.

Along with the development of anatomy, the body structure, such as bones, muscles, and the vessel system, were systematically studied. The difference between

Sensors and Probes for Bioimaging, First Edition. Young-Tae Chang and Nam-Young Kang.

$$2\,Ag + S \rightarrow Ag_2S\ \text{(black)}$$

Figure 2.1 Reaction of silver with S.

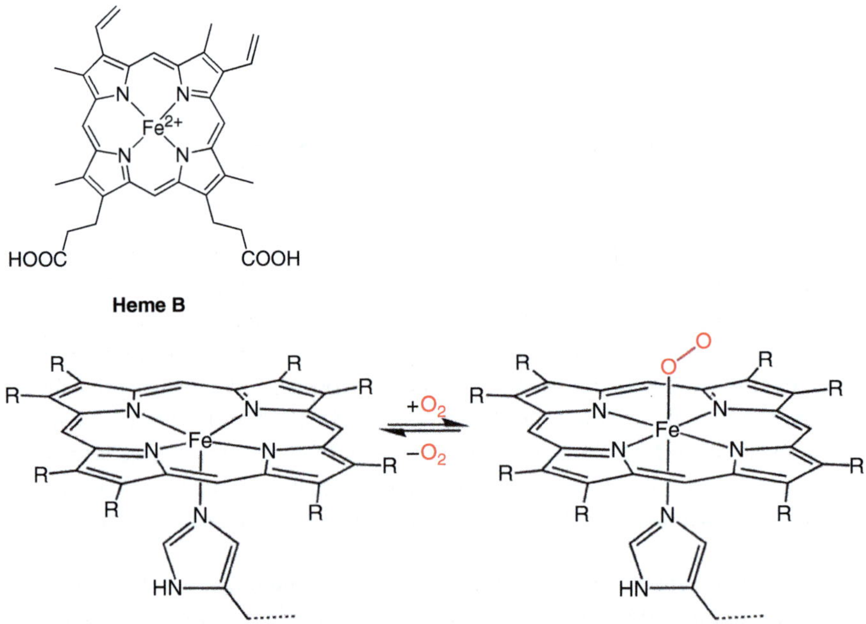

Figure 2.2 Structure of heme B in hemoglobin and binding to oxygen. Source: Adapted from Ref. [1].

arteries and veins was also discovered along with the blood circulation system. The arterial blood has more oxygenated hemoglobin in bright red color, and the venous blood has lower oxygenated hemoglobin in dark red color (Figure 2.2). The color difference is due to the absorption spectra change of hemoglobin depending on the oxygenation status. In that sense, hemoglobin is a naturally occurring optical sensor for oxygen. In many anatomy books, the artery is drawn in red and the vein is in blue, but the used color is rather exaggerated to make a clear contrast between artery and vein. The real color contrast in real artery and vein is not so dramatic. So, the pictures in anatomy books are drawings with enriched human interpretation, not photography.

Among biological sample staining, the iodine–starch reaction has been known for a long time. Iodine (I_2) is not well soluble in water and forms triiodide (I_3^-) by reaction with iodide (I^-) to increase the water solubility. When triiodide contacts with amylose in starch, instantly a blue-black color is formed. Amylose is a linear polymer of glucose with α-linkage, and the linear polymer seems to wrap and stabilize the linear structure of polyiodide (I_n^-, i.e. I_3^-, I_5^-, I_7^- etc.) [1]. The color appearance was explained as the result of charge transfer between iodide/iodine and amylose/polyiodide. Branched polymer amylopectin or a linear polymer cellulose

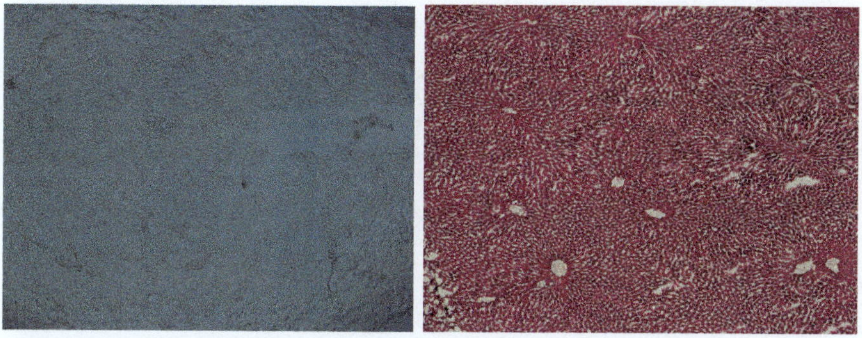

Figure 2.3 Bright-field imaging and stained liver tissue.

with β-linkage does not make such a dramatic color formation. The generated dark color diminishes over time upon amylose digestion with the amylase enzyme. The color changes depend on the amylose length during the digestion, and eventually the color disappears completely. This reminds us the concept of "particle-in-a-box" model for the absorption maximum shift from longer to shorter wavelengths (in Chapter 1). As expected, short amylose, such as maltose (glucose dimer), does not induce the color formation. For bioimaging application, iodine spray has been used for epithelium boundary detection using the glycogen (rich in amylose) contents in gynecology practice [2].

With the invention of microscope, various cell shapes were observed and reported. The general optical microscopy uses thin tissue samples. A bright input light penetrates the sample and make images through the lens. This technique is called transmitted bright-field imaging (Figure 2.3). The image is generated by different amount of light absorption in the tissue, but the contrast for the intracellular structure and cell–cell connection is usually low. To enhance the contrast, various staining methods with light absorbing material (dyes) have been employed.

Cells are wrapped by plasma membrane and many dye molecules cannot penetrate the membrane. Compared to the cell size, the cell membrane is very thin and fragile, and the shape of the cell can be easily distorted. So, before bioimaging, a fixation process is often used. The fixing is usually achieved by formaldehyde or dialdehyde which can make chemical crosslinking between biomolecules such as proteins, making the whole cell structure rigid. Even after crosslinking, the remaining membrane may still block the entry of the dyes. In such a case, a detergent is added to the fixed cell to partially dissolve the membrane, making holes for the free entry of dyes. This procedure is called permeabilization (Figure 2.4).

One of the historic imaging techniques was Golgi staining (also called "black reaction" following the stain's color), developed by Camillo Golgi in the nineteenth century. He is the same scientist who discovered and gave the name of an organelle as Golgi body. Golgi staining reagent was silver nitrate combined with a reducing agent and was used for visualizing the nervous tissue. The reduced silver metal forms black particles and visualize the whole intact nerve cells, including axons and dendrites among the many entangled cells in the nervous tissue.

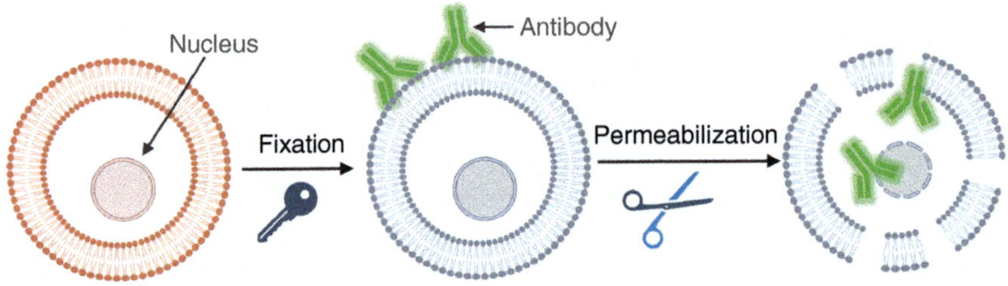

Figure 2.4 Fixation and permeabilization.

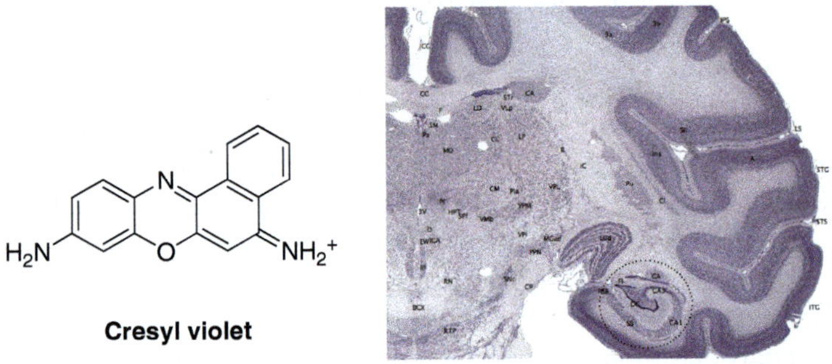

Figure 2.5 Structure of Nissl stains (cresyl violet) and brain section. Source: brainmaps.org / Wikimedia Commons, CC BY 3.0.

For neuron imaging, organic-dye-based Nissl staining is another famous historic method. In Nissl staining (following German neuropathologist, Franz Nissl), basic dye cresyl violet was used to stain a granular structure in neurons, called Nissl body (Figure 2.5). Later, Nissl body turned out to be ribosome, composed of mainly rRNA and abundant in neutrons. Nissl stains visualize mainly the neuron body, rather than the fine structure of axon or dendrite.

Maybe the oldest and also longest surviving biological staining method is H&E staining and it is widely used until today. The selectivity of dyes to intracellular organelles is mainly determined by their hydrophobicity and charge. Inside nucleus, there are high concentrations of DNA molecules with strongly negative charges (due to the phosphate linkage) and thus strongly positive dyes would prefer to stain nucleus. So, if two color dyes are used, usually one of them is a strong basic dye for nuclear staining, and the other one is an acidic dye for cytosol or another organelle staining. In H&E staining, nucleus-staining blue dye "hematoxylin" and cytosol-staining pink dye "eosin" are used as the pair for fixed cells or tissues (Figure 2.6). With the enhanced visualization of the tissue sample, the dye stain methods have been used as the gold standard protocol for clinical sample treatment and pathological readout. However, the process includes chemical fixing of biospecimens, so dynamic biological responses or changes are intrinsically limited to study.

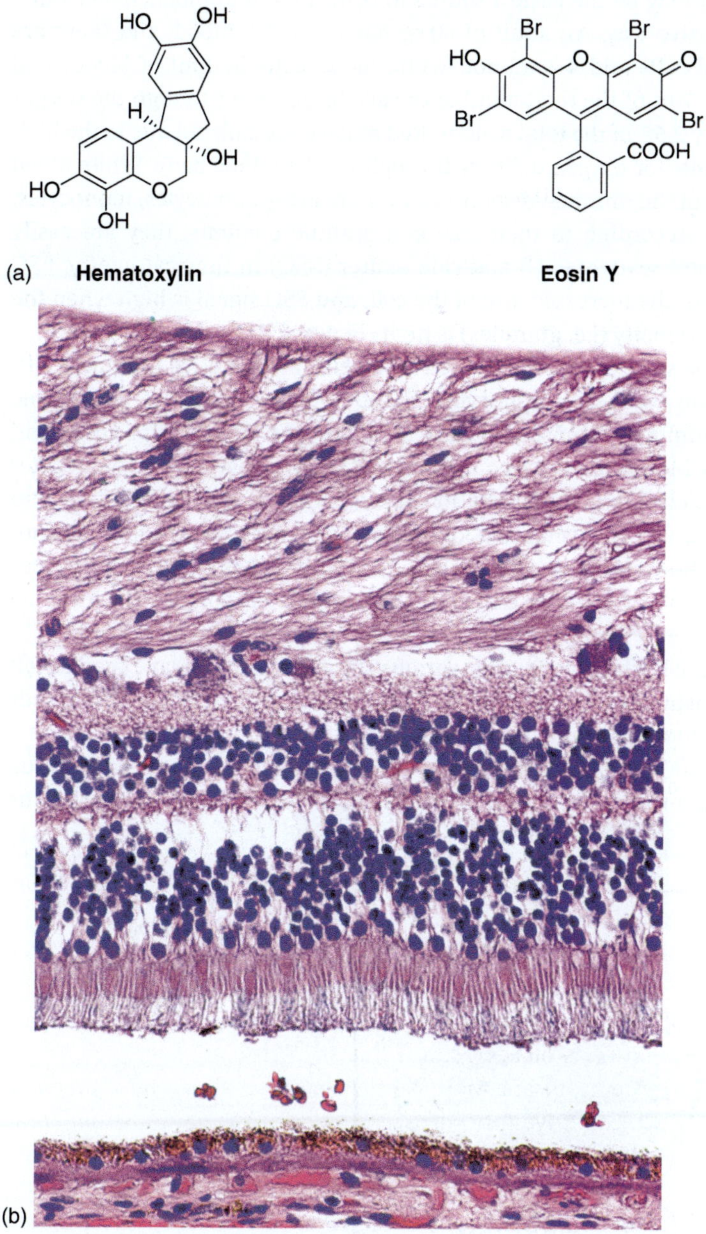

Figure 2.6 (a) Structure of H&E reagents (b) retina image with H&E stain. Source: Librepath/Wikimedia Commons, CC BY-SA 3.0.

2.2 Blood Cell Staining

In our body, blood may be the easiest source to collect large numbers of cells in a relatively less invasive way. An adult of 60 kg has about 5 L blood, and there are 5 billion red blood cells and 4–9 million white blood cells in 1 ml of blood. Red blood cells occupy 70% of the total number of cells in our body, though the weight composition is only 2.5% of the total weight. Red blood cells look red due to the high hemoglobin contents for oxygen delivery throughout the whole body. White blood cells are in charge of the immune response, and there are granulocytes, monocytes, and lymphocytes. According to their size and granule contents, they are easily separated by forward scatter (FSC) and side scatter (SSC) in flow cytometry. FSC signal is higher with the increased size of the cell, and SSC signal is high when the intracellular heterogeneity (i.e. granules) is high (Figure 2.7).

Lymphocytes, the smallest among white blood cells, are adaptive immune cells. There are two main types of lymphocytes: B and T cells. T cells are responsible for the cellular immune response and have subtypes of $CD4^+$ T helper cells and $CD8^+$ T killer cells. Monocytes are precursor cells of macrophages. When monocytes leave the blood vessel near the inflammation or infection site, they differentiate into macrophages with an amebae-shaped body and act as phagocytes. The name monocytes comes from the single nuclear morphology. Granulocytes have multiple lobes in the nucleus and lots of granules that contain chemical weapons for the immune response.

To easily distinguish white blood cells, various staining methods have been developed (Giemsa, Leishman, and Wright methods), but in principle they are minor variations of Romanowsky stain. The basic composition of Romanowsky stain is a basic dye "methylene blue" and an acidic dye "eosin"; DNA in nucleus is stained blue by methylene blue and cytosol is stained pink by eosin (Figure 2.8). The variations

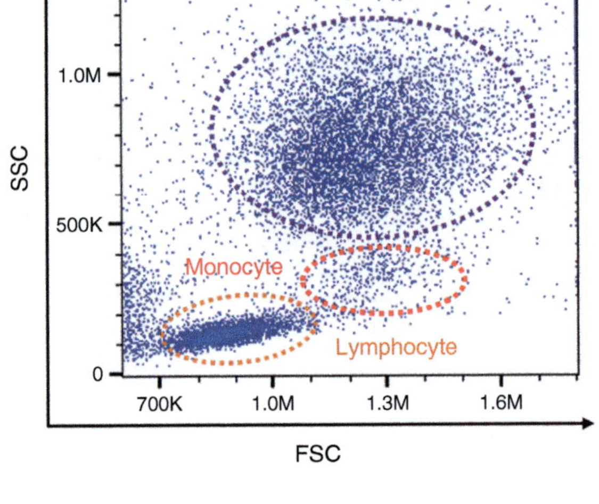

Figure 2.7 Flow cytometry profile of human blood cells based on scattering.

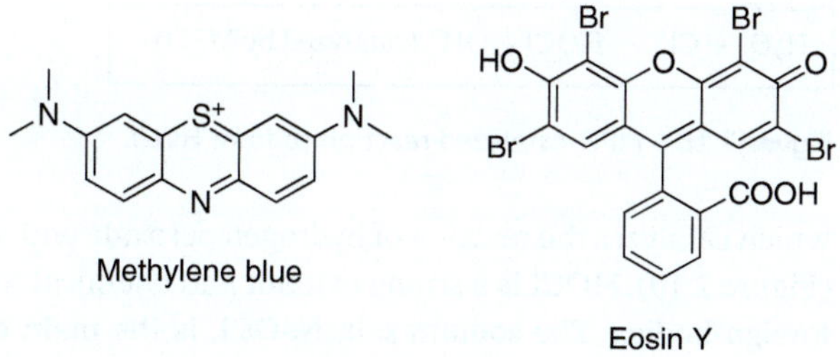

Figure 2.8 Dyes in Romanowsky stain.

are made for optimum cell distinction by mixing single or multiple derivatives of methylene blue and eosin.

In Romanowsky stain, methylene blue and eosin stains also discriminate granules with different charges, in addition to the primary target of nucleus and cytosol. Based on their staining characteristics, granulocytes are divided into three subtypes: neutrophils, eosinophils, and basophils. The most abundant granulocytes are neutrophils, and eosinophils are about 2–3%, followed by basophils at about or lower than 1% of leukocytes. While neutrophils handle general bacterial or fungal infection, the main target of eosinophils is parasites, and basophils secrete histamine for the immune response. Basophils have acidic (negatively charged) granules, and they are stained by methylene blue, showing strong blue granules. Eosinophils have basic (positively charged) granules and are stained by eosin to show off strong red granules. Neutrophils' granules are neither strong acidic nor basic, so they do not show a strong granule staining (Figure 2.9). Interestingly, the granulocytes are first classified by the synthetic dyes and their staining pattern, rather than biological functions or roles, which have been elucidated later.

Romanowsky stain is similar to H&E staining of tissues (hematoxylin stains nucleus and eosin stains cytosol), but especially optimized for blood cell distinction. While the staining is helpful for discrimination of the granulocytes, the subtractive color cannot be used for conventional flow cytometry, which requires fluorescence signal. Myeloperoxidase (MPO) is one of the functional enzymes in granulocytes,

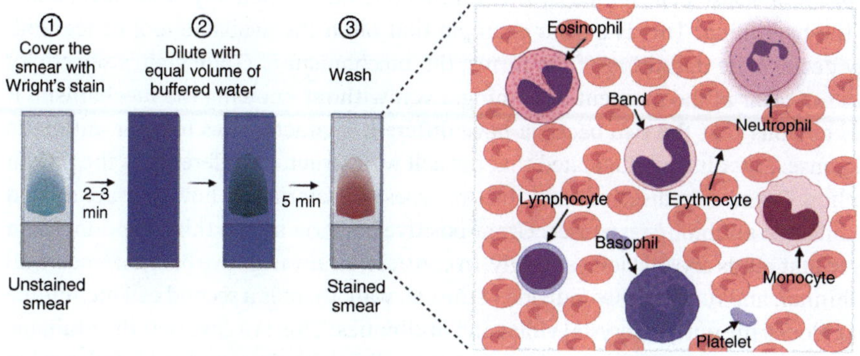

Figure 2.9 Blood cells stained by Romanowsky stain.

$$H_2O_2 + Cl^- \rightarrow HOCl + OH^- \text{ (catalyzed by MPO)}$$

Figure 2.10 MPO-catalyzed reaction to form HClO.

which catalyzes the reaction of hydrogen peroxide with chloride to synthesize HOCl (Figure 2.10). HOCl is a strong oxidant and chemical weapon to attack the invaded foreign bodies. The sodium salt, NaOCl, is the main component of bleach in the kitchen. Each granulocyte contains different amounts of MPO, and the difference of the enzyme activity can be used for granulocyte discrimination. In the automatic flow cytometry in hospital, a fluorescence substrate of MPO [3] or its product HOCl [4] is used for the discrimination of granulocytes, replacing the historic stains.

2.3 Bacteria Staining Using Gram Method

Bacteria are unicellular life forms that are about 1 μm in size and lack a nucleus, a compartment to encapsulate genomic DNA. Compared to mammalian cells (10 μm size), bacteria are much smaller, but fast in proliferation. Bacterial cells divide in every 20 minutes under optimum conditions, while mammalian cells divide in a day or so. As some bacteria infect humans and livestock animals, inducing diseases, detection and identification of the type of bacteria is a very important issue. The most standard method is Gram staining, developed by Hans Gram in the nineteenth century. Gram stain discriminates the bacterial cell wall structure, and its composition is crystal violet and safranin (Figure 2.11). By crystal violet staining, followed by an alcohol washing, Gram-positive bacteria still maintain a strong blue color, while Gram-negative bacteria lose the blue color. As a counterstain agent to stain all types of cells, light pink safranin is added at the end. As a result, Gram-negative cells show a pinky safranin color, but the strong crystal violet color in Gram-positive bacteria surpasses the weak safranin color.

Similar to granulocyte case, it is very interesting to observe that the classification of bacteria is also determined by the staining dye rather than any biological factors in the beginning. This is a clear example that often the available tool or technology decides the definition of life forms; the mechanism of Gram stain's selectivity was not clear at the moment of naming. Even without knowing the mechanism, it was obvious that the two bacteria have different characteristics in their antibiotic response. Later it was elucidated that the cell wall structure difference is the reason for how Gram stain discriminates the two types of bacteria and how they are related to different antibiotic responses. Gram-positive bacteria have a thick peptidoglycan layer that binds crystal violet strongly. In contrast, Gram-negative bacteria's cell wall is thinner, and furthermore, outside of the cell wall, there is a second cell membrane that blocks the entry of crystal violet and antibiotics. This is a case that the available tool opened up the application first, and the full mechanism study was followed at a later time.

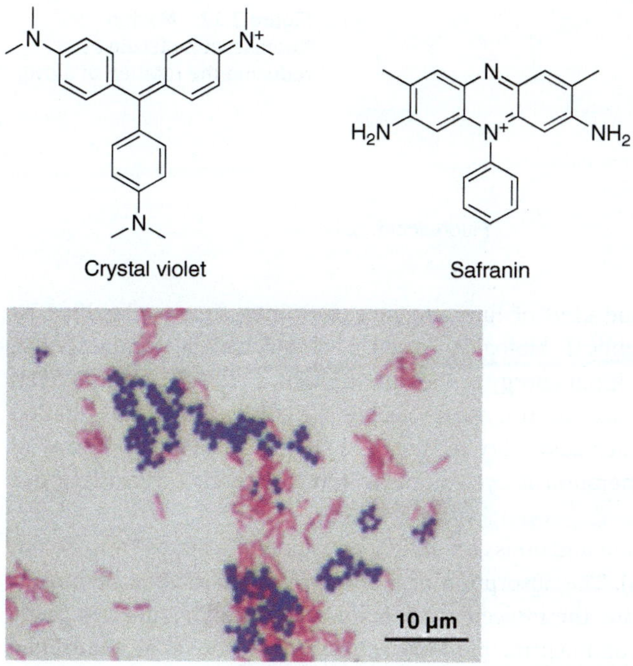

Figure 2.11 Gram stain agents and bacteria. Source: Y tambe/wikimedia Commons, CC BY-SA 3.0.

2.4 Fluorescent Sensors and Probes

Gram stain or H&E stain binds to and visualizes the target area of the cell by providing a strong color contrast for bright-field microscope. Through dyeing, the structure of cells become much more vivid with higher color contrast than the unstained sample. The dyes work based on acid–base property or charge status of the probe molecules and determine the organelle staining pattern. With the advancement of modern biology, the molecular-level imaging has also become possible using antibodies as the probe for specific proteins or carbohydrates as the target.

By the way, the color we see from the stained samples are subtractive colors through absorption of certain wavelength from the while light input. As the color contrast is determined by the ratio of absorbed color to input light, if the ratio is less than 1% (or higher than 99%), it would be difficult to get a clear contrast. Similarly, if the dye's absorption is not intense for a certain wavelength or the input light is too weak as in the case of nighttime, the color contrast fades away. To overcome the limits, probes with higher sensitivity and live-cell compatibility have emerged in the form of fluorescent probes.

The material that absorbs certain wavelength of visible light is defined as a dye. Many dyes lose their absorbed energy through molecular motion such as vibration or rotation, releasing heat. But, some dyes avoid the waste of energy by limited motion, and release the saved energy in the form of light. This phenomenon

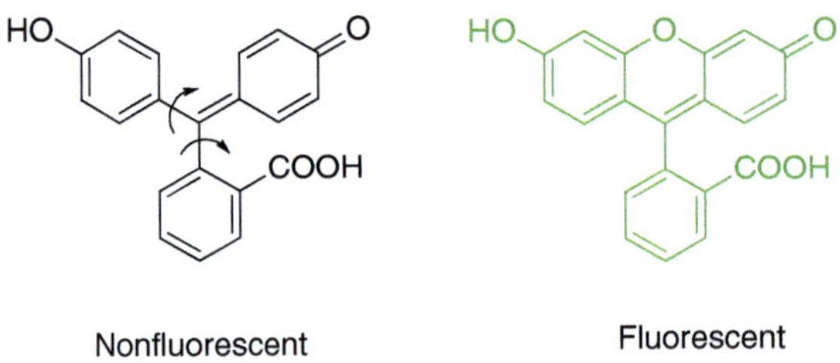

Figure 2.12 Making fluorescent molecule by reducing the rotation of a dye.

is called fluorescence, one kind of luminescence. Luminescence is classified by its input energy, i.e. chemical, biological, electric, or mechanical luminescence. Fluorescence obtains the input energy in the form of light, so fluorescence also can be called as photoluminescence. If a dye molecule is rigid with minimal rotation or vibration, it has a higher possibility to fluoresce. Therefore, a common way of fluorescent molecule generation is cyclization of rotatable bonds of existing dye molecule (Figure 2.12).

The process of absorption and emission of light is usually illustrated by a Jablonski diagram (Figure 2.13). The absorption of light by a dye takes about femtoseconds, and during this time, the molecular shape of the dye is assumed not being changed by the Franck–Condon principle. For the next picosecond range, the excited molecule relaxes a little by releasing some portion of the absorbed energy, and this process is called internal conversion. If the input light wavelength is broad, the first excited states are varied, but they may relax to the same final excited state. Usually, the fluorescence light wavelength is longer than the absorbed light due to the energy loss during rearrangement of the molecule at the excited state. The resulting wavelength difference between absorption and emission is called Stokes' shift. The lifetime of fluorescent molecules is usually nanoseconds. The light-emitting process itself takes about femtosecond similar to the absorption process, and nanosecond lifetime is the waiting time until all the excited molecules emit the fluorescence. During the absorption and internal conversion, the electronic spin remains singlet, but sometimes the excited electron moves to a triplet state with a lower energy level through an intersystem crossing. When the waiting time for the emission gets longer

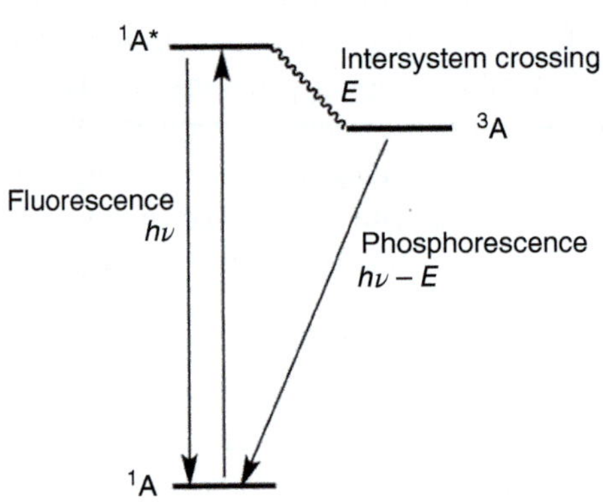

Figure 2.13 Jablonski diagram. Source: Ref. [5]/Smokefoot/CC BY SA 3.0.

to microseconds (or even longer to hours to days) due to the intersystem crossing, the process is called phosphorescence.

The configuration of absorption measurement is the linear arrangement of light source–sample–detector, and the absorption is recorded by the subtraction of the detected light from the input light. In comparison, for the fluorescence measurement, the detector should not be in the light path to avoid the interference by the input light. Typically, the detector is located in right angle from the light path (Figure 2.14). Another possibility is time-resolved measurement using pulse laser as an input and the internal conversion time as the time gap. In this format, the input light is given for femtosecond, and after waiting picosecond for the internal conversion, the measurement of emission starts and lasts for nanoseconds. Usually the fluorescence spectrum is shown as a pair of absorption and emission. Interestingly, absorption unit is universal regardless of machine, but fluorescence emission unit is arbitrary (relative fluorescence unit [RFU]). Therefore, different RFU values are obtained from different machines or conditions.

Strictly speaking, the pair of emission spectrum should be "excitation spectrum," but absorption spectrum is commonly used instead of excitation spectrum. Excitation spectrum is an emission measurement by scanning the excitation wavelength through a fixed emission wavelength channel. If all the absorbed light contributes to the emission, absorption spectrum and excitation spectrum should be in identical shape. If some of the absorbed light does not contribute to the emission, it is

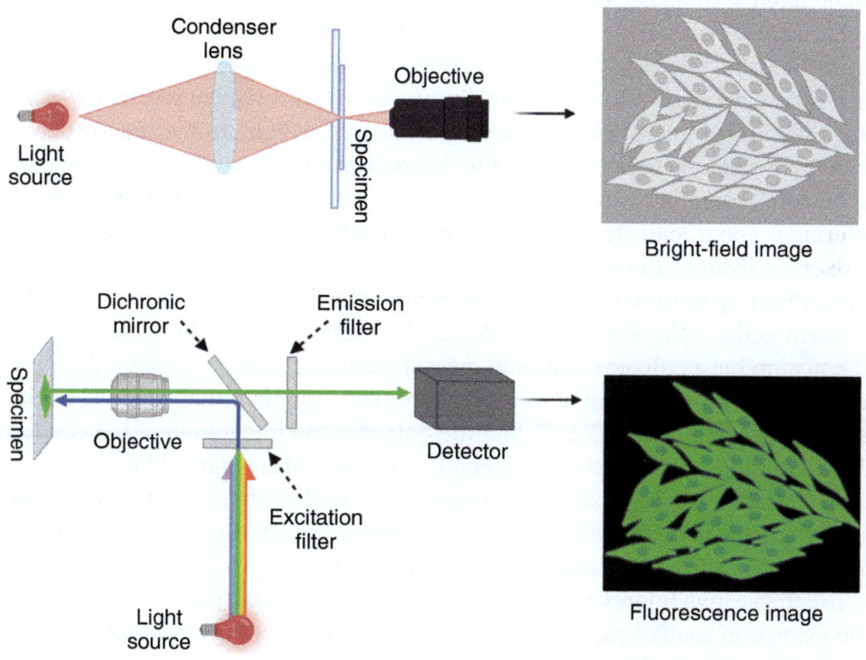

Figure 2.14 The instrument configuration of absorption and fluorescence measurement.

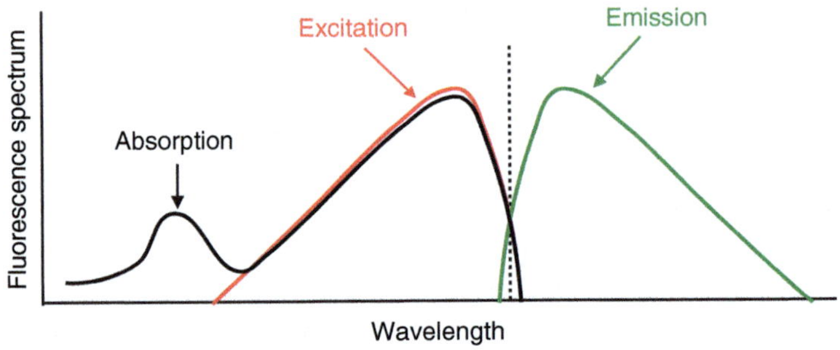

Figure 2.15 The spectra of absorption, excitation, and emission.

necessary to measure both absorption and excitation spectra and compare to dissect them (Figure 2.15).

There is an interesting observation in fluorescence spectra, called mirror image rule. When the molecular structure is rigid, i.e. the molecular shapes at ground state and at excited state are similar, the excitation and emission spectra are often symmetric mirror images, reflecting the preference of transition to the same vibrational mode [5]. When the electron is excited from S0 to S1 states, the preference of transition among vibrational modes ($0 \rightarrow 0^*$, $0 \rightarrow 1^*$, $0 \rightarrow 2^*$, $0 \rightarrow 3^*$, etc.) is different, and the most favored transition determines the λ_{ex}. The excited electrons are relaxed to 0^* in S1, then the emission transition ($0^* \rightarrow 0$, $0^* \rightarrow 1$, $0^* \rightarrow 2$, $0^* \rightarrow 3$ etc.) may have a similar trend with the excitation, and the relative energy gap with preference generates the mirror images in the spectrum. And the two spectra have usually steeper slope near the mirror plain compared to the remote area (Figure 2.15).

In terms of instrument configuration, fluorescence measurement is similar to scattering measurement as of the right-angle position of the light source and the detector. In scattering, photons interact with electrons in the dye molecule and change the direction, sometimes accompanied by energy loss or gain. The required time of scattering is about femtosecond, so a time-resolved measurement may distinguish fluorescence from scattering.

The efficiency of fluorescence of a molecule is defined by QY (quantum yield, Φ), which is the ratio of emitted photon and absorbed photon. As the absorption and emission have different units and measurement methods, a standard dye with a known QY is commonly used as a reference. The input is usually defined by the absorption at a single wavelength and the output is the integration of the emitted light across the whole emission wavelength. It is noteworthy that the lost energy by internal conversion (Stokes shift) is not considered as a loss in QY. That means lower energy photon (longer wavelength photon in emission) and higher energy photon (shorter wavelength in excitation) are considered the same in terms of photon number. To achieve high fluorescence intensity, the molecular absorption should be high (high extinction coefficient, ε) and so should be the QY. So, the brightness of a fluorescent dye is defined by "$\varepsilon \times$ QY." The brightness of representative fluorescent probes is summarized in Figure 2.16.

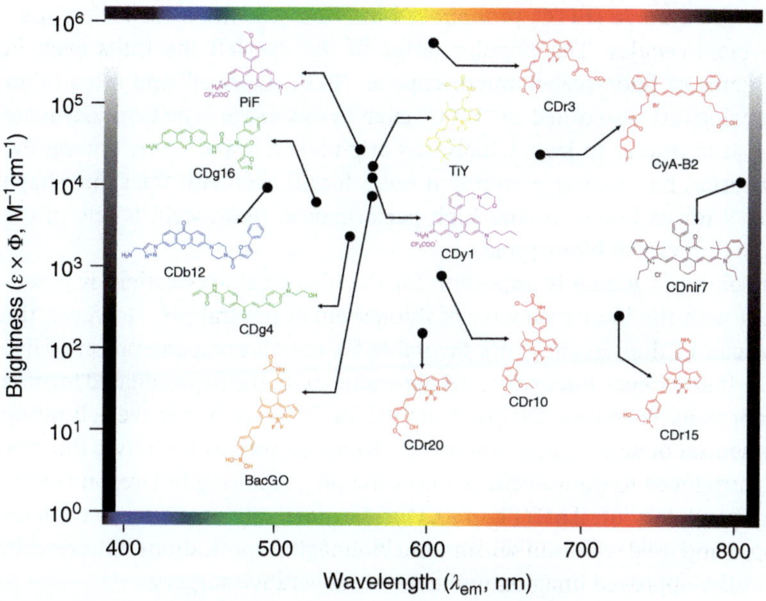

Figure 2.16 Brightness of representative fluorescent probes.

Optical microscope requires a light source to penetrate the sample and lenses to focus and collect the light. With white light as the input source, the image has a bright background. Fluorescence microscope needs to detect relatively small amounts of emitted light, separated from the input light, so a darkroom environment is necessary. So, fluorescence image has a bright target signal with the dark background. As the fluorescence does not rely on the ratio of input and output lights, but directly counts the output photons against an absolute black background, the sensitivity limit of fluorescence, in principle, can be a single photon or a single molecule. Also, fluorescence is in general more sensitive to environmental changes in comparison to absorption, and many fluorescent probes have a high possibility of working as a sensor. Due to these advantages, many sensors and probes are developed for fluorescent version of bioimaging, especially for the live cell and tissue, and even for the whole body. Combined with the development of fluorescence microscopy hardware, a new era of fluorescent probes has fully blossomed.

2.5 Representative Fluorescent Compounds for Bioimaging

Fluorescein ($\lambda_{ex}/\lambda_{em}$ = 491/515 nm) is one of the most common fluorescent dyes in biological application with its quite high molar extinction coefficient (ε) of 88 000 cm^{-1} M^{-1} and high quantum yield of 0.95 in 0.01 M NaOH [6]. Fluorescein is also used in highlight pens for daily life usage. Its amine-reactive derivative, fluorescein isothiocyanate (FITC, pronounced as "fit-c"), is the most widely used dye to label biomolecules. The popular usage of the dye left the traits even in the channel name of fluorescence microscope as "FITC channel" and often other fluorescein analogs are also called as "FITC" even in case there is no isothiocyanate (ITC) moiety. It is similar to jeep, which was originally a brand name among the Chrysler car series, but became a common noun for all cars with the same shape and function. With its low price and high performance, fluorescein is one of the first choices in fluorescent bioimaging.

The water solubility, which is important for the biological application, is practically excellent with the ionic character of fluorescein at neutral pH. However, the anionic character of fluorescein is not favorable for membrane penetration of live cells. Also, the fluorescence intensity is pH dependent with the deprotonated form as the fluorescent state. Therefore, the optimum pH for fluorescein is above 9, limiting the usage to neutral or acidic environment. To overcome the pH limit, two fluorine atoms were introduced to fluorescein to lower the pK_a, resulting in Oregon Green, which is bright even at neutral pH (Figure 2.17). Many derivatives of fluorescein have been developed and used widely in sensors and bioimaging applications. Fluorescein itself is a US-FDA-approved imaging probe for intraoperative surgery.

Rhodamine is a cousin of fluorescein, with N instead of O in the dye scaffold. Both of them belong to the xanthene class of dyes, and rhodamine is among the oldest synthetic dyes for fabric dyeing. They have similar high absorption and rhodamine's fluorescence intensity is practically not sensitive to pH, unlike fluorescein. The replacement of O to N makes rhodamine a more positive dye than fluorescein, and the general cell permeability is also much better than fluorescein. The wavelength of rhodamine can be easily adjusted by alkylation of N or by introduction of a ring structure. Tetramethyl rhodamine (TMR) is the most popular rhodamine derivative in biological applications [7], and its ITC derivative is commonly called TRITC (pronounced as "trit-c"). Texas Red is a representative longer wavelength rhodamine with a fused ring structure [8] (Figure 2.18).

The essential fluorophore of fluorescein and rhodamine is xanthene, and the attached phenyl group is actually not involved in the fluorescent property of the dyes, but acts as a protecting group from chemical attacks to the joint position. Usually, the phenyl group carries COOH, which may form a nonfluorescent spiral

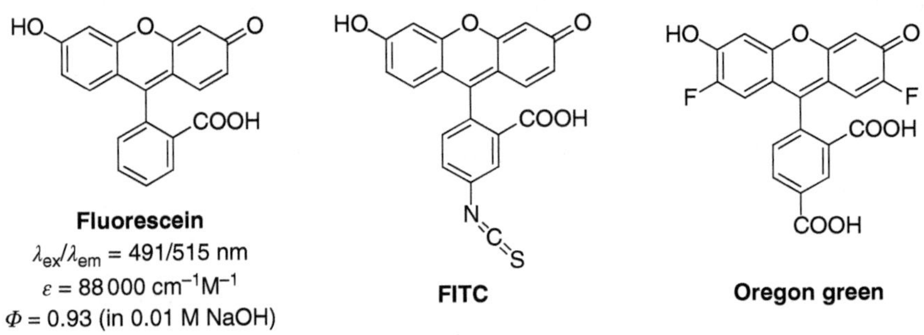

Figure 2.17 Structure and optical information of fluorescein, FITC, and Oregon Green.

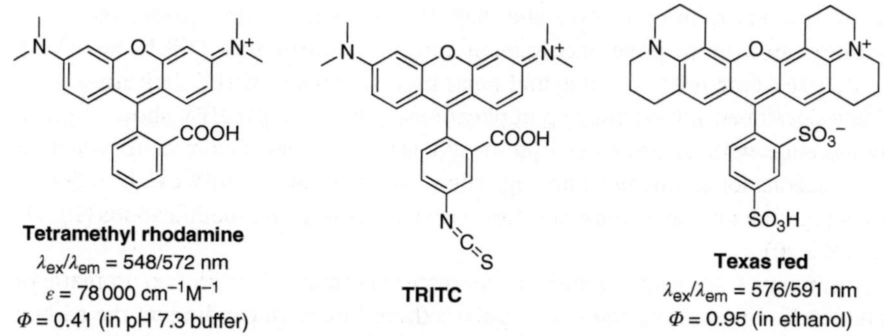

Tetramethyl rhodamine
$\lambda_{ex}/\lambda_{em}$ = 548/572 nm
ε = 78 000 $cm^{-1}M^{-1}$
Φ = 0.41 (in pH 7.3 buffer)

TRITC

Texas red
$\lambda_{ex}/\lambda_{em}$ = 576/591 nm
Φ = 0.95 (in ethanol)

Figure 2.18 Structure and optical information of rhodamine.

lactone structure in equilibrium with the fluorescent open form. If COOH is removed from rhodamine, the resulting dye is called rosamine. Interestingly, the quantum yield of rosamine is much lower (1–10%) than rhodamine due to the free rotation of the phenyl ring. It means COOH also acts as a physical anchor for mechanical rotation of the phenyl ring. As evidence shows, if COOH is replaced with a simple methyl group, the fluorescence is still as high as that of rhodamine or fluorescein. While rhodamines generally localize in the cytosol, rosamines tend to localize in mitochondria in live cells. Without negatively charged carboxylate, rosamine carries only delocalized positive charge on the xanthene structure, which may be favorable to mitochondria (reasonable hydrophobicity and positive charge: Section 3.6). By the same token, the ester of rhodamine (i.e. rhodamine 123), of which the negative charge is removed, also tends to localize in mitochondria, and behave similar to rosamine. Even though the quantum yield of rosamine is lower than that of rhodamine in free solution, if rhodamine binds to macromolecules in the cell, and thus restrict the rotation of the phenyl ring, the fluorescence could be increased, providing the possibility of an in vivo sensor (Figure 2.19).

4,4-Difluoro-4-bora-3a,4a-diaza-s-indacene, or difluoroboron dipyrromethene (BODIPY) is relatively young introduction into fluorescent tool boxes and has become popular due to its excellent photostability and high molar absorption and fluorescence quantum yield [9]. With difluoroboron structure, BODIPY seems chemically unstable, but actually quite stable in various conditions. Compared to negatively charged fluorescein and positively charged rhodamine, BODPIY is neutral in charge and thus nonpolar and hydrophobic. Therefore, the water solubility

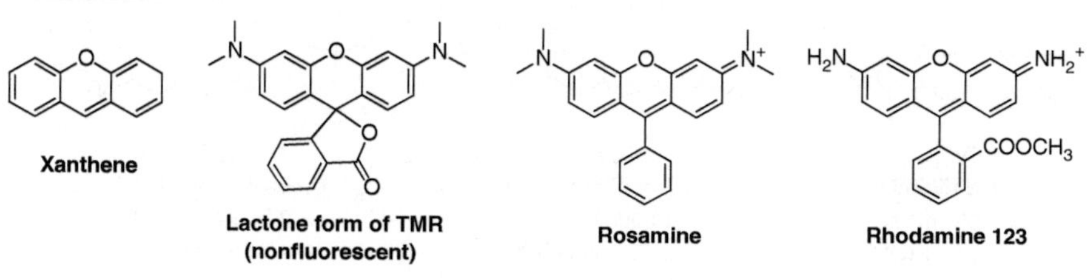

Xanthene

Lactone form of TMR (nonfluorescent)

Rosamine

Rhodamine 123

Figure 2.19 Xanthene, lactone form of TMR, rosamine, and rhodamine 123.

would be lower than other dyes and may not be friendly with aqueous biological environment with the preference to membrane or lipid structure. Still, by modifying the BODIPY core with targeting and polar motifs, many BODIPY derivatives were widely developed for bioimaging applications. The basic BODIPY shows a green fluorescence with sharper absorption and emission than fluorescein, which is advantageous for multicolor imaging. The wavelength of BODIPY can be adjusted by conjugation or electron-donating and -withdrawing group modifications [10, 11] (Figure 2.20).

Cyanine dyes are a large family of fluorescent compounds with two aromatic or heterocyclic rings connected with a polymethine linker. Depending on the carbon number in the polymethine structure, they are classified as Cy3 [12], Cy5 [13], and Cy7 (Figure 2.21). The wavelength of cyanine dyes increases stepwise as the polymethine size increases by about 100 nm. To fill up the gap in the wavelength, ring-expanded derivatives were systematically introduced with the names Cy3.5, Cy5.5, and Cy7.5, with about 30 nm longer wavelengths compared to the original structure (Figures 3.14 and 3.47) The first US-FDA-approved dye, indocyanine green (ICG), belongs to the Cy7 class. With the amazingly high molar absorption coefficient, Cy3 and Cy5 have been extensively used for genomic and proteomic labeling and analysis. Cy3 and Cy5 have moderate quantum yields of 4% and 27% in PBS [14]. Cy7 also has a higher molar absorption coefficient of ~200 000 cm^{-1} M^{-1} and 28% of quantum yield [15].

Among shorter wavelength dyes than that of green fluorescein, coumarin is a representative blue dye. The core structure of coumarin is oxobenzopyran, and the carbonyl acts as an electron-withdrawing group. Hydroxyl or amino group attached on the opposite site serves as an electron-donating group. With stronger electron

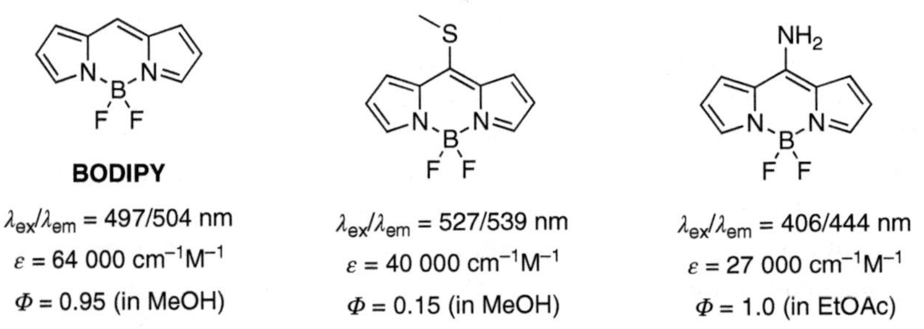

Figure 2.20 BODIPY and color.

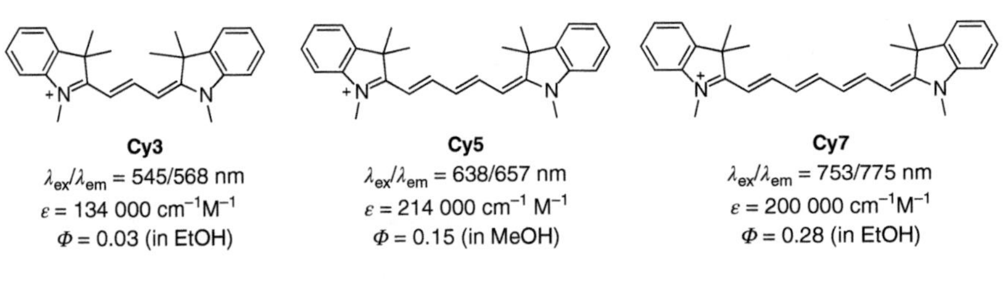

Figure 2.21 Structure of Cy3, Cy5, and Cy7.

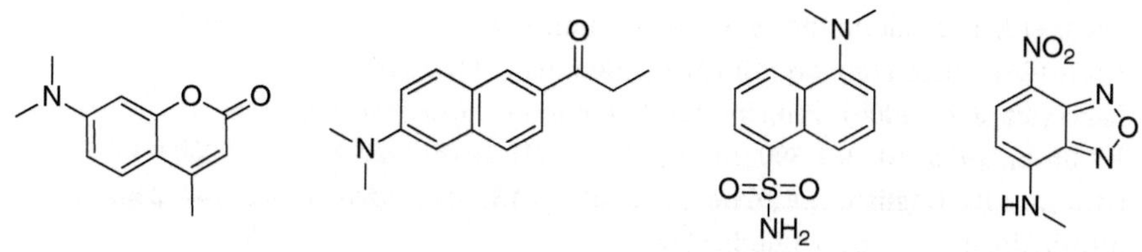

Figure 2.22 Structure of coumarin, prodan, dansyl amide, and NBD.

donating effect, the amine version has longer wavelength than the hydroxyl version. With a reasonable extinction coefficient and quantum yield, coumarin is the first choice of blue fluorescent dye for labeling [7].

Naphthalene-based dyes are another class of blue dyes. With electron-donating and -withdrawing groups in opposite positions, the naphthalene dyes have a strong dipole moment, resulting in general red shift of emission wavelength in polar solvents. The bathochromic phenomenon is explained by the better stabilization of polar excited state in a polar solvent. The further stabilized, the bigger Stokes shift, emitting longer red shifted light. This solvatochromic property of naphthalene dyes has been applied to polarity sensor development. Prodan is the representative example of the polarity sensor with the longer emission wavelength in a polar environment [16].

For labeling purpose, dansyl chloride has been widely used for amine-containing biomolecules. Dansyl chloride is based on a naphthalene scaffold with low fluorescence, but after reaction with amine, the product becomes fluorescent. The resulting dansyl amide when excited with ultraviolet (UV) light, emits a green light with a long Stokes shift. Due to the small size and turn-on effect upon reaction with amine, dansyl chloride has been used for amino acid labeling and high-performance liquid chromatography (HPLC) analysis. The large Stokes shift is convenient for excitation and emission light separation, but the quantum yield is moderate [17].

NBD chloride is another popular amine-labeling dye with a green color emission of the reaction product. The excitation wavelength of NBD-amine is similar to fluorescein, and the quantum yield is much higher than that of dansyl amide. So, NBD is considered as a smaller but similar property dye to fluorescein without a strong pH dependency [18] (Figure 2.22).

References

1 Madhu, S., Evans, H.A., Doan-Nguyen, V.V.T. et al. (2016). *Angew. Chem. Int. Ed.* 55: 8032–8035. https://doi.org/10.1002/anie.201601585.

2 Reich, O. and Pickel, H. (2021). *Eur. J. Obstet. Gynecol. Reprod. Biol.* 261: 34–40. https://doi.org/10.1016/j.ejogrb.2021.04.011.

3 Suzuki, K., Ota, H., Sasagawa, S. et al. (1983). *Anal. Biochem.* 132: 345–352. https://doi.org/10.1016/0003-2697(83)90019-2.

4 Shepherd, J., Hilderbrand, S.A., Waterman, P. et al. (2007). *Chem. Biol.* 14: 1221–1231. https://doi.org/10.1016/j.chembiol.2007.10.005.
5 Lakowicz, J.R. (2006). *Principles of Fluorescence Spectroscopy*, 3e, 5.
6 J. Lumin. 1975, 10, 381-390. https://doi.org/10.1016/0022-2313(75)90003-4.
7 Grimm, J.B., English, B.P., Chen, J. et al. (2015). *Nat. Methods* 12: 244–250. https://doi.org/10.1038/nmeth.3256.
8 Brouwer, A.M. (2011). *Pure Appl. Chem.* 83: 2213–2228. http://doi.org/10.1351/PAC-REP-10-09-31.
9 Tram, K., Yan, H., Jenkins, H.A. et al. (2009). *Dyes Pigm.* 82: 392–395. https://doi.org/10.1016/j.dyepig.2009.03.001.
10 Goud, T.V., Tutar, A., and Biellmann, J.F. (2006). *Tetrahedron* 62: 5084–5091. https://doi.org/10.1016/j.tet.2006.03.036.
11 Bañuelos, J., Martín, V., Gómez-Durán, C.F.A. et al. (2011). *Chem. Eur. J.* 17: 7261–7270. https://doi.org/10.1002/chem.201003689.
12 Klochko, O.P., Fedyunyayeva, I.A., Khabuseva, S.U. et al. (2010). *Dyes Pigm.* 85: 7–15. https://doi.org/10.1016/j.dyepig.2009.09.007.
13 Michie, M.S., Götz, R., Franke, C. et al. (2017). *J. Am. Chem. Soc.* 139: 12406–12409. https://doi.org/10.1021/jacs.7b07272.
14 Mujumdar, R.B., Ernst, L.A., Mujumdar, S.R. et al. (1993). *Bioconjugate Chem.* 4: 105–111. https://doi.org/10.1021/bc00020a001.
15 Levitz, A., Marmarchi, F., and Henary, M. (2018). *Molecules* 23: 226. https://doi.org/10.3390/molecules23020226.
16 Vequi-Suplicy, C.C., Coutinho, K., and Lamy, M.T. (2014). *Biophys. Rev.* 6: 63–74. https://doi.org/10.1007/s12551-013-0129-8.
17 Guy, J., Caron, K., Dufresne, S. et al. (2007). *J. Am. Chem. Soc.* 129: 11969–11977. https://doi.org/10.1021/ja0738125.
18 Uchiyama, S., Santa, T., and Imai, K. (2000). *Analyst* 125: 1839–1845. https://doi.org/10.1039/B005217P.

3

Organelle-Selective Probes

3.1 Introduction

Cell is the basic unit of life, and itself is a universe. The size of a common human cell is about 10 μm. The biggest cell in humans is egg cell with 150 μm diameter. Sperm cell has about a 50 μm long tail and a 3–5 μm long elliptical shaped head. The longest cell is neuron cells and sometimes the length of an axon is even longer than 1 m. To understand the cell structure more realistically, let us imagine enlarging the cell size a billion times. In this scale transformation, 1 nm becomes 1 m. Then, the cell diameter is about 10 km (similar to a medium-sized city) and the cell membrane thickness is about 4 m. The cell membrane is composed of a phospholipid bilayer, keeping the hydrophobic tail inside and the hydrophilic head group outside. This amphiphilic property of phospholipids is almost same with the components of detergents or soaps. Therefore, cell membranes and soap bubbles may share a common feature. The usual soap bubble boundary thickness is about 1 μm and estimated to reach 0.1 μm just before rupture. If we enlarge a 1 cm sized soap bubble into 10 km diameter (a million times enlargement), the membrane becomes about 1 m thick, which is similar dimension of cell membrane thickness. So, soap bubbles are a reasonably good model of cells in terms of size and thickness of membrane, but the membrane orientation of soap bubbles is opposite to that of cells. Unlike cells, soap bubble membrane keeps the hydrophilic part inside and sticks out the hydrophobic part to the outside of the bilayer, with a thin layer of water in between (Figure 3.1). With the progressive evaporation of water, the bubble boundary becomes thinner and thinner, and eventually the soap bubble ruptures. Despite the opposite boundary orientation, considering the fragile soap bubble, you may feel how thin the cell membrane is and how flexible the cell shape might be.

In this cell city model of 10 km size, a water molecule may look like a handheld boomerang with two wings of 10 cm length each. Even though water is one of the smallest molecules in the biological system, due to its hydrophilic property, water cannot easily penetrate the 4 m thick hydrophobic barrier of the cell membrane. For the smooth penetration of water through the membrane, a special water channel protein (aquaporin) is required. Without aquaporin, cells may feel thirsty even in the midst of a water environment. Not only water, any hydrophilic material

Sensors and Probes for Bioimaging, First Edition. Young-Tae Chang and Nam-Young Kang.

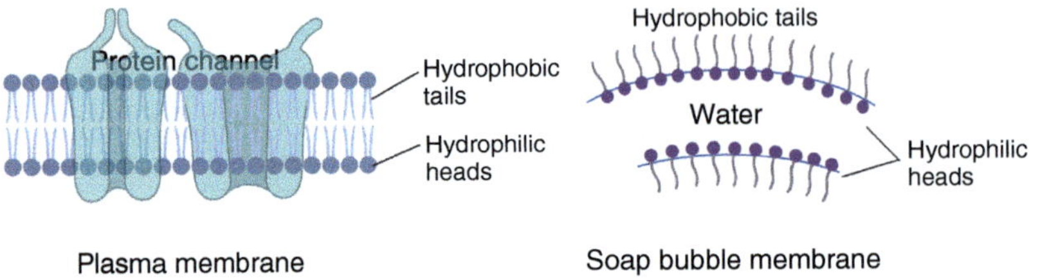

Figure 3.1 Structure of cell membrane and soap bubble.

needs a special transporting system to cross the cell membrane. Metal ions also need channels to move in and out of the cell. Through channels, the concentration gradient may decide the net direction of the materials' movement. For the uphill movement of ions, against the concentration gradient, specific pump proteins and operating energy are required. Na-pump utilizing ATP (adenosine tri-phosphate, representative biological battery molecule) is such an example. For helping the movement of bigger molecules, there are specialized transporter proteins, either in energy-dependent or gradient-dependent manners. For the harmonized metabolism of the cell, many molecules need to be moved in and out, and there are hundreds of channels, pumps, and transporters on the cell membrane. In general, useful nutrients need to be imported, and waste or toxic metabolites should be removed from the cells.

Some molecules, however, have the ability to penetrate the membrane by themselves. For example, ethanol has both a hydrophobic ethyl part and a hydrophilic OH group. So, it can penetrate the cell membrane even without a channel or transporter's help. As the process can happen passively, ethanol can reach almost every part of our body without a physical limitation. Some of them may reside in the membrane itself making the boundary structure loose. The dumb and loose feeling from drinking alcohol may be induced by the loosened membrane structure in our body. For design of efficient drug molecules, the efficient uptake of drugs by the target cells is important. Through a survey of thousands of marketed drug molecules, Christopher Lipinski, a medicinal chemist, proposed a set of rules for efficient cell uptake of drug molecules [1]. According to the rule, for preferred cell membrane penetration, the molecule should be smaller than 500 molecular weight (by size, it may be smaller than 1 m in the city model) and the hydrophobicity should be in an optimum range ($2 < \log P < 5$, where P is a partition coefficient of the molecule between *n*-octanol and water). If the molecule is too hydrophilic ($\log P < 2$), the molecule cannot enter the cell membrane's inner space; hence, there is no chance of crossing it. If the molecule is too hydrophobic ($\log P > 5$), the molecule enters any membrane inner space it meets for the first time, and sticks there, not crossing it. An example is bitter powder drugs. The hydrophobic drugs bind to the cell membrane on the tongue and stimulate the bitter taste sensory buds. The taste does not easily go away for a long time even after washings with water, due to the hydrophobic nature. Therefore, unless the target of the drug is on the tongue, the too high hydrophobic property of the drug candidate may not be desirable in terms of proper drug delivery (Figure 3.2).

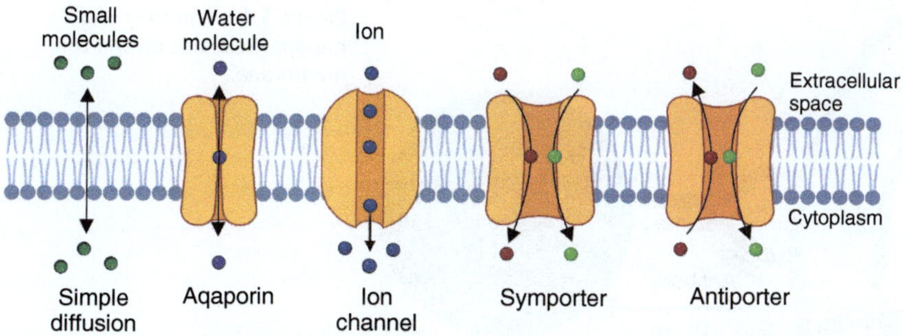

Figure 3.2 Penetration of molecules through membranes.

For maintaining the harmony of cell city, sensing environmental changes through signaling molecules and proper response are important. Some signaling molecules enter the cells and deliver the message directly to the target site, and hydrophobic steroid hormones are such an example. However, many signaling molecules also cannot enter the cells easily due to their large size or hydrophilic nature. To deliver the signal into the cell, there are special proteins called receptors on the surface of cell membrane. They are membrane-embedded proteins with hydrophobic waist (which may be about 4 m in size matching the cell membrane thickness) and hydrophilic head and legs. If the head is on the outside of the membrane, the signaling molecule (a hormone or a neurotransmitter) binds to the head side, the structure of the receptor is changed to deliver the signal into inside the cell. This process is called as "signal transduction." The transferred signal could be receptor's leg structure change (conformational change or chemical modification) or secretion of a second messenger into the intracellular space. Some of the receptors have enzyme function in the leg part and catalyze the chemical reaction in the cytosol. The reactions can be removal or attachment of phosphate groups, or protein cleavage or modification, etc. (Figure 3.3).

Antibodies are commonly used for cell surface membrane protein recognition, so it would be good to compare the size with the cell membrane. Antibody is a Y-shaped protein of approximately 8 m long and 4 m wide. If an antibody binds to a receptor on the cell membrane with head and leg, the two partners may have a comparable size. Sometimes, antibodies are used to deliver nanoparticles to the target cells. Usual size of nanoparticles for biological studies is 10–100 m. While the small nanoparticles

Figure 3.3 Membrane structure with embedded protein and signal transduction.

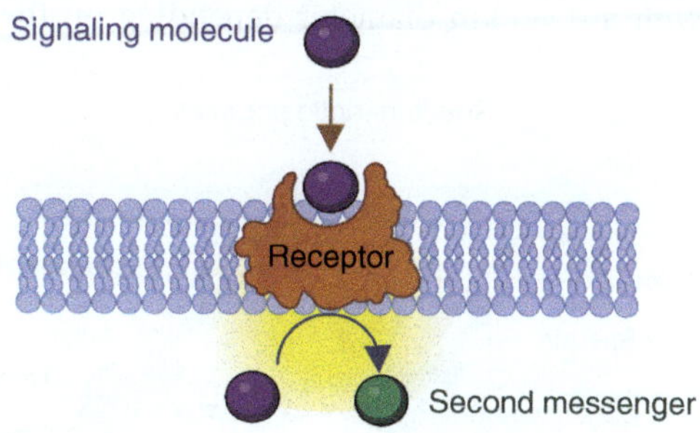

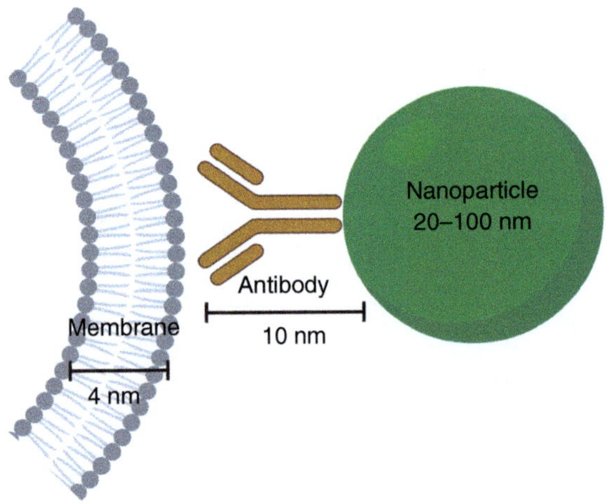

Figure 3.4 Antibody and nanoparticle size on the cell membrane.

have a comparable size to antibodies, for big nanoparticles, one may need to think about an optimum antibody-to-nanoparticle ratio for efficient application. Also, if a 100 m size nanoparticle is labeled with a single antibody, you may also wonder which part may drive the movement or targeting of the conjugate (Figure 3.4).

Beneath the cell membrane, there is a protein network to strengthen and maintain the fragile membrane structure. This network comprises actin and spectrin, covering the whole membrane area by putting an inward pressure. With a counter outward pressure, tubulin provides a microtubule structure to sustain the membrane. Actin is a globular protein of 6–7 m in size, which is connected together into double-stranded fibers. Microtubules have an empty column structure with a 25 m outer and 13 m inner diameter. These proteins are not only fixed structural proteins, but are also dynamically changing proteins responsible for the abrupt cell morphology changes in cell movement and cell division processes (Figure 3.5).

Inside cells, there are further compartmented structures by the intracellular membrane, so called organelles. First of all, nucleus is the information center where all the genetic information of the cell is stored in the form of DNA. To keep the high contents of genetic information, the nucleus is several kilometers in diameter with double layer of membrane structure. The extended membrane from the nucleus forms a layered network structure of endoplasmic reticulum (ER) or Golgi body. Mitochondria are the powerplants in the cell, and have a cylindrical shape with double layer of membranes. The size of mitochondria varies from 1 to 10 km with a 0.5–1 km diameter depending on the cell condition. There are also many

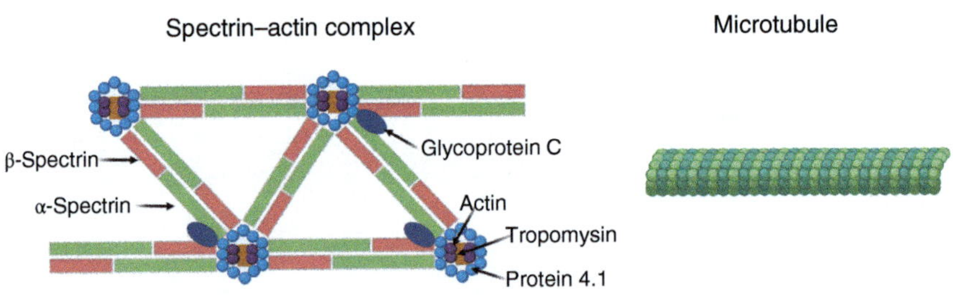

Figure 3.5 Structure of spectrin–actin network and microtubule.

vesicles such as endosomes, lysosomes, and peroxisomes with various sizes and functions. These vesicles usually have a single-layered membrane, unlike nucleus or mitochondria. Each organelle has its own specific role and contains different chemical materials and biological machineries. Most of the enzyme-catalyzed biological reactions occur in the defined organelles, with ensured high concentration of the substrates and optimum reaction conditions such as pH and specific ion concentration. Due to the active regulation of the ion gradients, most of the organelle membranes maintain an electrical potential through the membrane.

Except organelles, the free space inside the cell membrane is cytosol. Cytosol is filled with various sizes of protein machineries, with several meters as the average size. The biggest machine in the cell is 30 m long ribosomes. Most of the machines are made of protein, but interestingly the main component of ribosome is RNA with accessory proteins assembled. As protein synthesis is the very basic function to build biological body, the fact that RNA is the key catalytic machine (i.e. enzyme) for the process (not by protein enzyme) implies the importance of RNA in the early stage of emerging life, so called origin of life.

Usually, cytosol is imagined as seawater with swimming fish (proteins), but actually the environment is more similar to a thick soup. The solute amount inside the cell is 300–400 $g\,l^{-1}$. If we assume the density of cytosol is about 1 (similar to water), it is literally half water and half fish situation. The crowded intracellular situation was impressively well described by David Goodsell (https://ccsb.scripps.edu/goodsell), who is a scientist painter. Therefore, the diffusion rate for molecules to collide with each other is significantly slow, compared to dilute solution. While slow diffusion is a negative factor for biological reaction rate, due to the high concentration of large molecules, the macromolecular crowding effect may have the opposite effect. The high contents of macromolecules may reduce the volume of free solvent, and thus effectively increase the concentration of reactants, inducing accelerated biological reactions [2] (Figure 3.6).

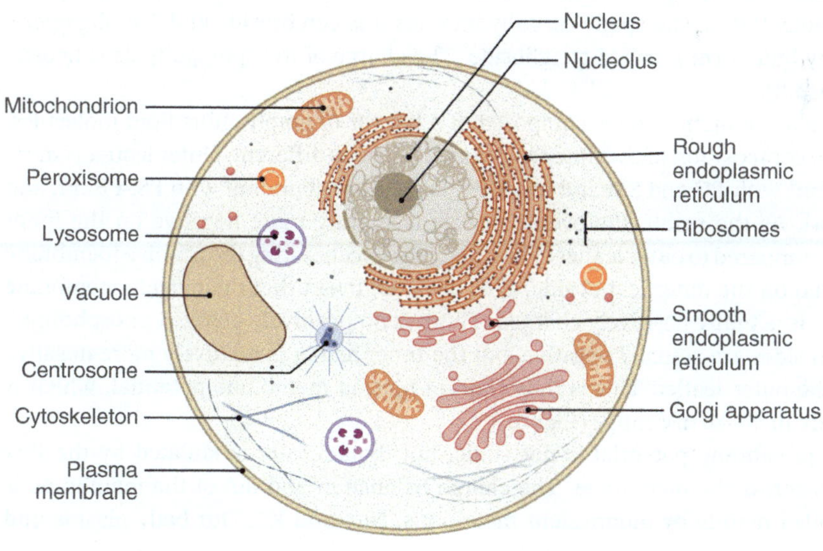

Figure 3.6 Cell structure in summary.

3.2 Cell Plasma Membrane

Cell plasma membrane defines the boundary of the cell. Inside the membrane is the real cell territory, and outside is the environment. In 1944, the famous physicist Schrodinger defined life as that which resists decaying into disorder and equilibrium [3]. In other words, the closed system of life maintains order and low entropy inside by consuming energy and increasing environmental entropy. Following the definition, the plasma membrane could be the boundary of a life in terms of energy, entropy, and material exchange. This activity is called "metabolism," and is considered the central function of the live cell.

Plasma membrane is composed of a phospholipid bilayer, keeping inside hydrophobic and the surface hydrophilic. The amphiphilic property of the membrane is similar to that of surfactants such as soaps or detergents. This may be due to the fact that life started in the sea (water-based aqueous media) and chose the hydrophobic material as the separating material between the cell interior and the environment. The most abundant contents of plasma membrane material are glycerol phospholipids, followed by sphingolipids. Glycerol has three hydroxyl groups and usually two of them are connected to fatty acid tails through an ester bond, and one of them is connected to various polar phosphate head groups. The fatty acid in hydrophobic tails is commonly a 16–20 carbon chain with various unsaturation levels. The head group is either negatively charged or neutral depending on the structure. The phospholipids with a neutral head group are phosphatidylcholine (PC) or phosphatidylethanolamine (PE), and PC is usually the most abundant phospholipid in the plasma membrane. The negatively charged phospholipids are phosphoric acid (PA), phosphatidyl serine (PS), phosphatidylinositol (PI), and phosphatidylinositol phosphates (PIPs) (Figure 3.7).

The second largest group of membrane lipids is sphingolipids. The backbone of a sphingolipid is sphingosine and its ester form with various sizes of fatty acids is ceramide. The head group of ceramides are commonly choline (SM: sphingomyelin) or carbohydrates. The single carbohydrate form is cerebroside and the oligomeric carbohydrate form is called ganglioside. The charge of the sphingolipids is neutral (Figure 3.8).

The plasma membrane is composed of a bilayer of amphiphilic lipid molecules, and the composition of the inner and outer leaflets is different. Outer leaflet is more abundant with PC and SM and inner leaflet is more abundant with PS, PE, PI, and PIPs [4]. As the result, the overall charge is slightly more negative on the inner leaflet compared to outer leaflet of the membrane, generating the positive membrane potential on the outer leaflet side. This does not mean that the outside membrane surface is actually positively charged. There is no positively charged phospholipid, so both sides are actually negative, but the inner leaflet is relatively more negative than the outer leaflet. This is the origin of plasma membrane potential, which is typically in 40–80 mV range (Figure 3.9).

The membrane potential is not static, but dynamically modulated by the flux of ions across the membrane. The charge balance in and out of the membrane is controlled mainly by monovalent metal ions, Na^+ and K^+. Our body plasma and

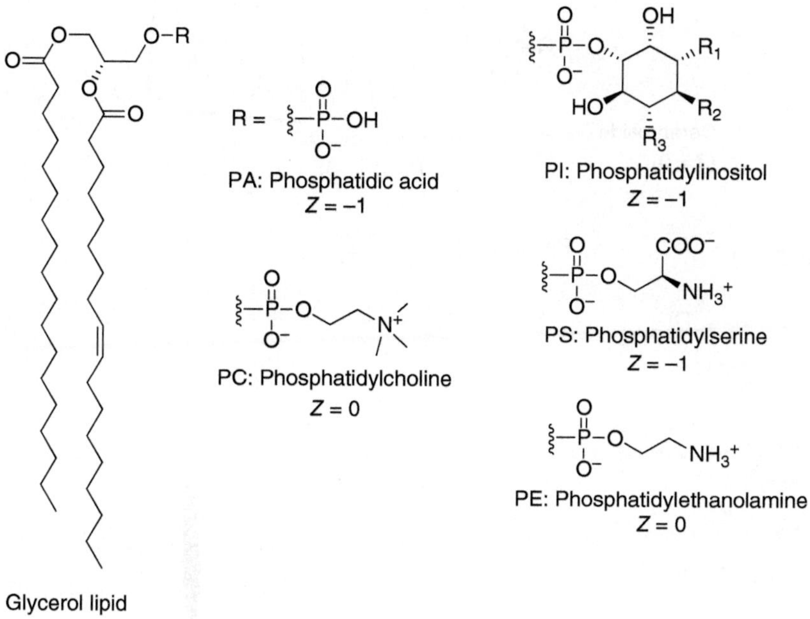

Figure 3.7 Structure of glycerol phospholipids.

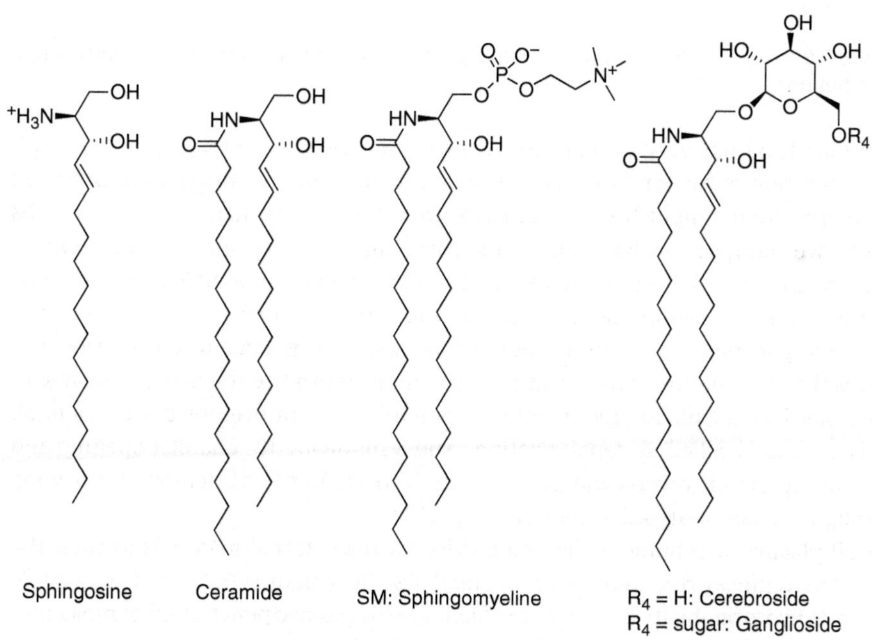

Figure 3.8 Structure of sphingolipids.

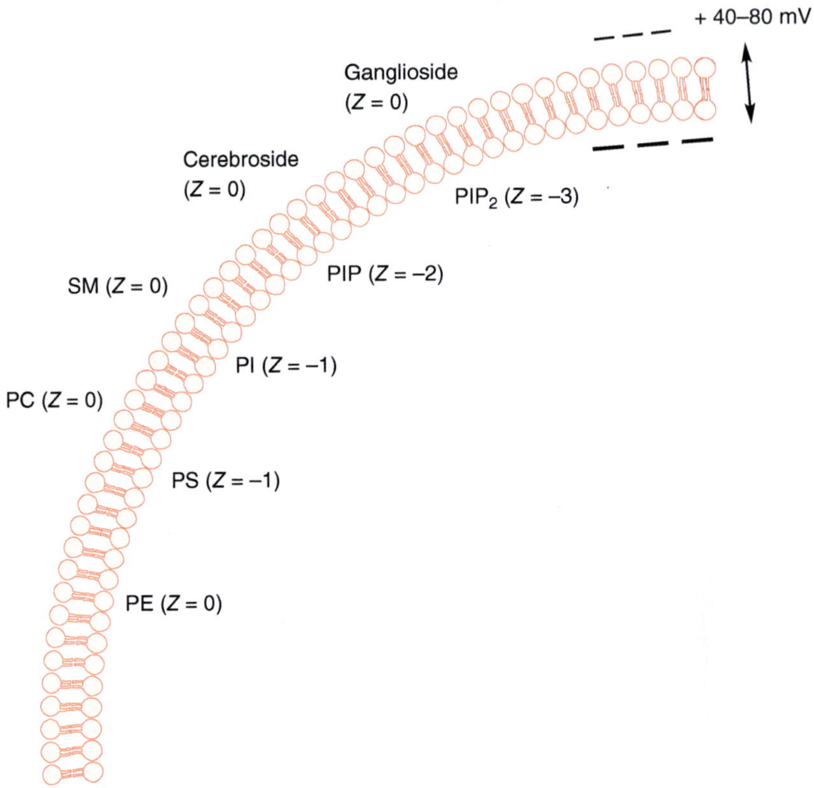

Figure 3.9 Asymmetric distribution of phospholipids in the plasma membrane and charge distribution.

amniotic fluid has very similar ionic composition with that of sea water, i.e. high concentration of Na^+ and Cl^-. This may be evidence that the origin of cells started in the sea. To distinguish the inner space from the environment, the ancient cells may have pumped out Na^+. But at the same time, to keep the osmotic pressure balance, they may have imported K^+ to a similar concentration of Na^+ outside. The resting status of membrane potential is determined by these ion concentrations bound to phosphate head groups of phospholipids. Once a cell receives a signal, it opens the Na^+ channel, and the imported cations neutralize the negative charge of the inner leaflet, making the membrane potential zero or even reversed potential. This process is called as depolarization. And a following K^+ channel opening and Na^+ pump action reverts the potential back to resting level, temporally passing through a hyperpolarization state (Figure 3.10).

Cell plasma membrane is the first barrier for the external materials to enter the cell. The Lipinski rule indicates that small size (less than 500 Da) and reasonable range of hydrophobicity ($2 < \log P < 5$) facilitate the passive penetration of molecules through the cell membrane. As these are simple physical parameters, the rule may also apply to artificial liposome system for judging the passage of small molecules through the membrane. In cells, due to the overall negative charge of the cell

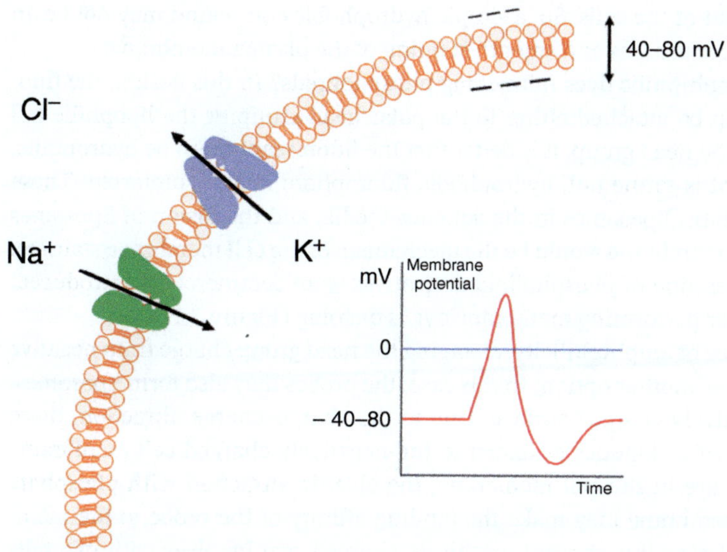

Figure 3.10 Membrane potential dynamics and monoionic cation change.

membrane surface, negatively charged molecules may experience electric repulsion with the head group of phospholipids on the membrane. Even when they reach the outer leaflet of the plasma membrane, the relative negative cross-membrane potential may further hamper the entry of the molecules. In contrast, the positively charged molecules have advantages for the access to the membrane surface and also the penetration through the membrane due to the membrane potential. Therefore, in general, positively charged molecules tend to have better entry into the cell than negatively charged molecules. If the positive charge is localized and strong, the polar group effect may dominate as the hydrophilic character and the cell permeability may drop. So, a weakly positive or delocalized positive charge is preferred for better membrane penetration. Neutral molecules are free from this consideration. So, the preferred property for cell uptake is either neutral or weakly positive charge with reasonable hydrophobicity.

For visualizing the plasma membrane, how could we design the probe? How about super hydrophobic fluorescent compound with $\log P > 5$? Well, this kind of molecule may have solubility problem in aqueous media, so it may form oil drops or precipitates most likely in nanometer size, i.e. nanoparticles or nano aggregates. When the nanoparticles meet the plasma membrane, the contact site may be strongly stained, often accompanied by an irregular aggregation, even in micrometer size interfering the smooth staining of the membrane. Due to the uneven delivery, some cells may not have the chance of contact with the nanoparticles, especially if they are located in the inner side of cell clusters. Even though they reach the plasma membrane, they can be internalized into the cell and may compete with the staining of intracellular membranes (ER and Golgi body) or lipid droplets (LDs). With these factors, the resulting images may be quite heterogeneous and ugly with chunk of aggregates all over in and out of the cells. So, a simple hydrophobic compound may not be an ideal option for continuous and smooth staining of the plasma membrane.

How about amphiphilic dyes mimicking phospholipids? In this design, the fluorescent motif can be attached either to the polar head group or the lipophilic tail group. If it is to the head group, it is better that the fluorescent motif be hydrophilic. If the attachment is to the tail, hydrophobic fluorophore may be preferred. These molecules may form liposomes in the aqueous media, and the fusion of liposomes with the plasma membrane would be the mechanism of the cell membrane staining. While early generation of phospholipid derivatives were commercially introduced, search for a better performing membrane dye is ongoing (Figure 3.11).

In a similar idea of amphiphilicity, changing the head group charge from negative to positive may be another option. In this case, the probes may also form liposomes, but with a positively charged surface. Due to the charge–charge attraction, these probes may have fast kinetics of fusion to the negatively charged cell membrane. Also, once they are in the cell membrane, the electric attraction with phosphate groups in the membrane may make the binding affinity of the probe to the membrane stronger. Using this concept, positively charged cyanine dyes with two tails of C18 have been widely used for representative plasma membrane probes with a broad range of color options (Figure 3.12).

Neuronal cells have a small cell body (cell soma) with long dendrites and axons. In the brain tissue, the complicated circuit of neurons is composed of crosslinked dendrites and axons. Due to the high number and density of the connection, visualization of the neuronal structure has been challenging. The cyanine series of membrane dyes have been used for neuronal cell imaging and tracking of the long axon or dendrites by the slow diffusion through the plasma membrane. In

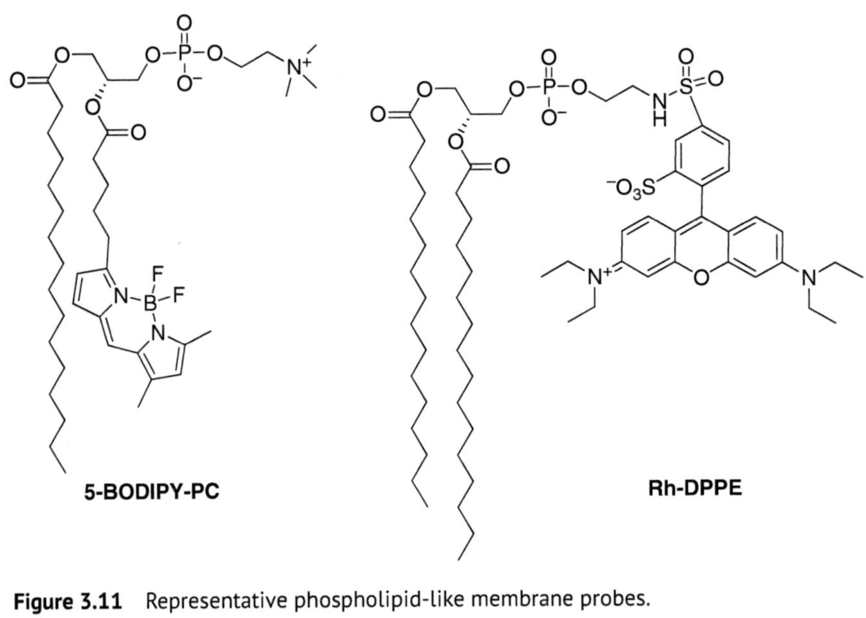

Figure 3.11 Representative phospholipid-like membrane probes.

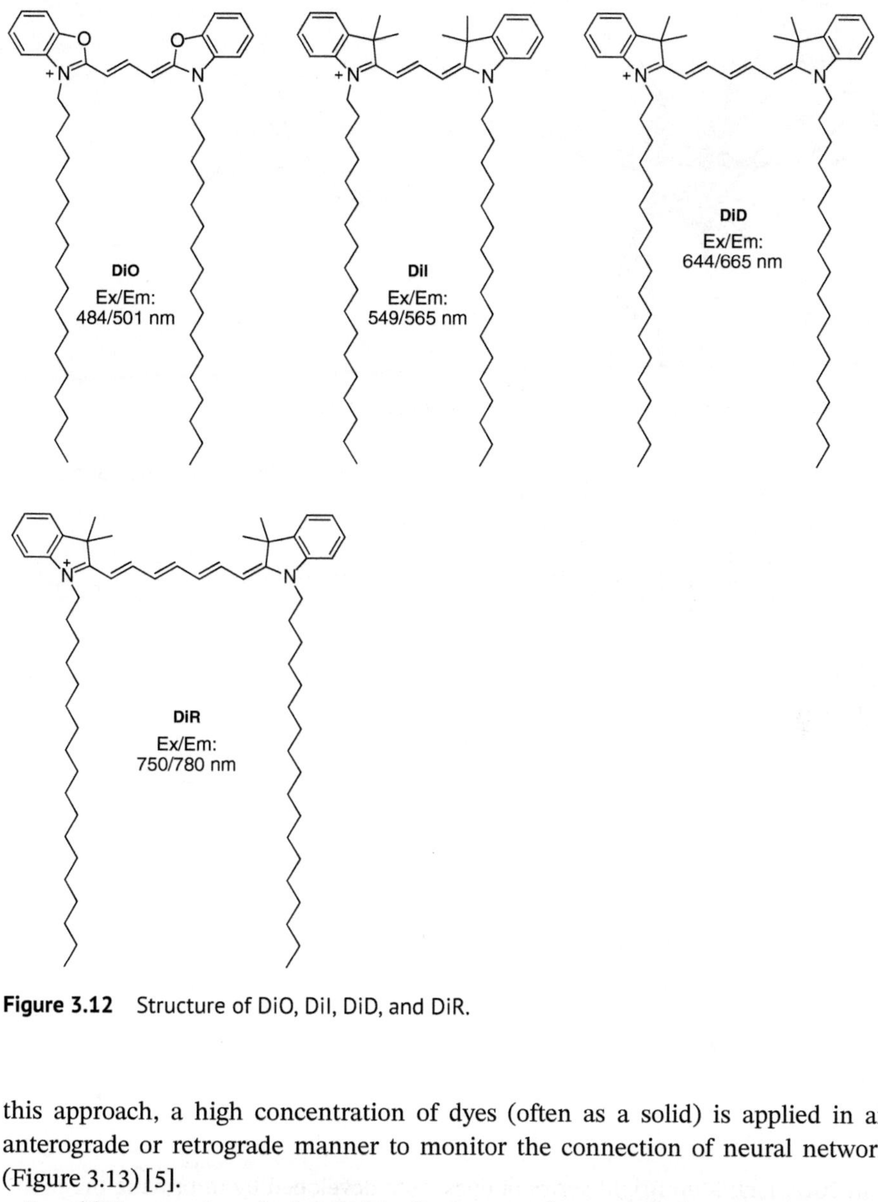

Figure 3.12 Structure of DiO, DiI, DiD, and DiR.

this approach, a high concentration of dyes (often as a solid) is applied in an anterograde or retrograde manner to monitor the connection of neural network (Figure 3.13) [5].

In addition, various hydrophobic fluorescent dyes coupled to hydrophilic or charged groups have been developed for plasma membrane probes [6]. But many of them showed poor membrane staining efficiency with low signal-to-noise ratio [7]. Also, due to their low water solubility, they tend to easily form aggregates on the membrane and inside cells [8]. Good membrane dyes may require high brightness and even distribution throughout the membrane without disturbing the natural structure of plasma membrane using a small quantity. Also, long-term stability on the membrane is also important without undergoing endocytosis or other uptake into the inner space of the cell.

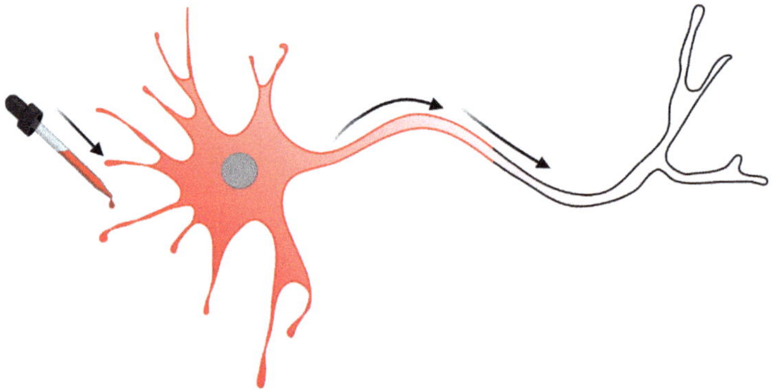

Figure 3.13 Imaging of neural structure with DiX series dye.

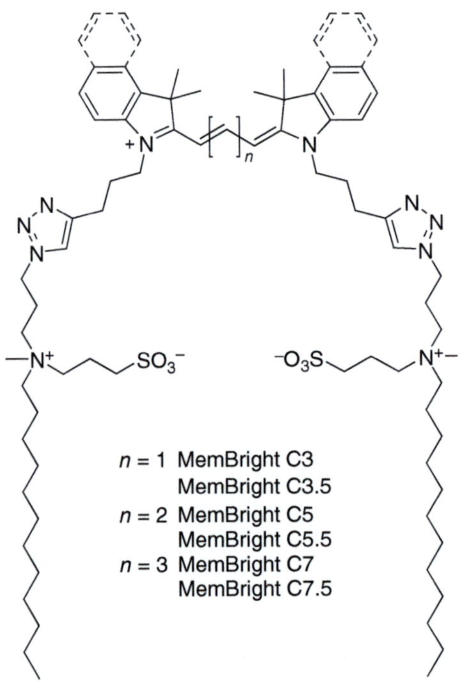

Figure 3.14 Structure of MemBright dyes.

In 2019, new MemBright series of dyes were developed by improving the problems of previous membrane probes, by incorporating zwitterionic head group for better solubility and less aggregation (Figure 3.14). Conventionally, membrane dyes required several micromolar range of application, but MemBright dyes could yield bright membrane images in 2D and 3D format only at 20 nM concentration [9].

Cell plasma membrane is not only composed of phospholipids, but also embedded with many receptor proteins as shown in Figure 3.3. In addition to the embedded proteins, there are peripheral membrane proteins temporarily bound to the lipid bilayer. Some of the sphingolipids act as anchors for the peripheral proteins or carbohydrates. The receptor proteins often assemble together to form a domain, called

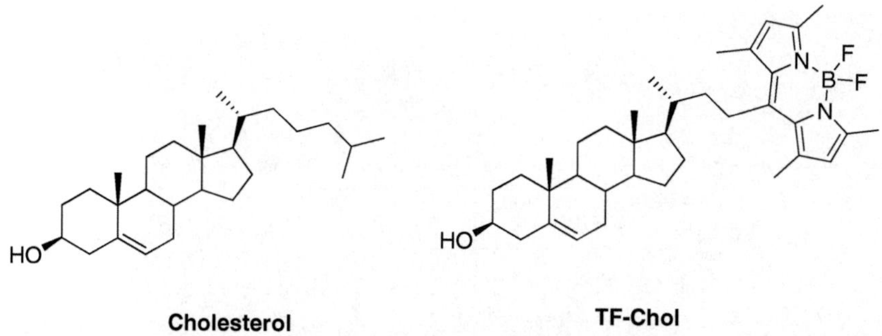

Figure 3.15 Cholesterol structure and cholesterol-derived probe for lipid raft.

"lipid raft" area of 10–40 nm diameter. In addition to the protein clusters, the lipid raft area is relatively abundant with sphingolipids and cholesterol. Cholesterol insertion makes the membrane more rigid, which is similar to the role of stones in a mud wall. Another important aspect for membrane rigidity is the unsaturation of lipid tail. Usually, higher unsaturation (common in glycerol lipids) tends to induce more mobile character of the membrane. Overall, the lipid raft area is more rigid and thicker than the surrounding glycerol lipid area. The landscape may look like a high-density raft is floating on the water surface.

With the possible role of cholesterol in the plasma membrane, cholesterol-conjugated fluorophores have been developed for the study of lipid raft domain. The systematic study of lipid distribution using artificial giant unilamellar vesicle (GUV) and cell membrane-derived GUV demonstrated lipid raft (called ordered lipid domain) area was preferably stained by cholesterol-based probes (Figure 3.15). In comparison, glycerol lipid (Figure 3.7) and DiX series probes (Figure 3.12) preferably stained free floating (called disordered lipid domain) membrane area [10]. While lipid rafts domain is enriched with sphingolipids, the localization of sphingolipid derivatives is environment and lipid composition dependent. Alternatively, sphingolipid-targeting protein can be used for selective lipid raft probe. For example, cholera toxin is known to bind to GM1 ganglioside, a glycosylated sphingosine structure, so fluorescently labeled cholera toxin has been widely used for lipid raft imaging [11]. Recently, the positively charged carbohydrate polymer chitosan has been labeled with fluorescein and demonstrated to noninvasively visualize the lipid raft domain (Figure 3.16) [12].

3.3 Endosome and Lysosome

The plasma membrane is not a static structure, but is always undergoing dynamic changes. For example, part of the plasma membrane buds off inside the cell to form a vesicle, called endosome. This process is called "endocytosis." Endosomes are surrounded by a lipid bilayer, which was originally part of the cell membrane. However, the ingested materials inside endosomes are originally from the environment. If we consider cytosol as "inside" of life, the internal space of endosomes is filled

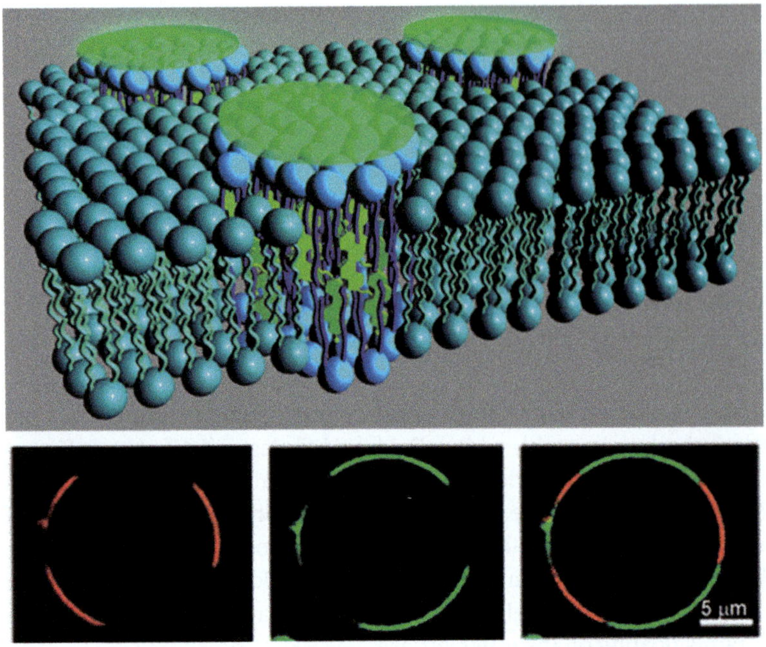

Figure 3.16 Confocal image of lipid raft in GUV model. Red-glycerol lipid and Green chitosan. Source: Jiang et al. [12], from American Chemical Society.

with "outside" or environmental materials. Cell plasma membrane has different compositions on the inner and outer leaflets, and endosome's membrane is also reversed. So, endosome is an inside-out structure!

The early endosomes (pH 6–7) recruit proton pump proteins, and using ATP, protons in the pump change the inner space more and more acidic. Some of the early endosomes return to and fuse with the plasma membrane for recycling through exocytosis, which is a reversible process of endocytosis. The acidified late endosomes have about pH 5–6, and they further develop into lysosomes of pH 4–5. Lysosomes contain various digestive enzymes and take part in the active lysis of waste or toxic materials in the cell. Sometimes, the late endosomes fuse with already-formed lysosomes to finish the digestion of the exogeneous materials (Figure 3.17). Usually, the size of lysosomes is smaller than 1 μm, but active macrophages often form several micrometer size of lysosomes [13]. In addition to the digestive functions, lysosomes also take care of recycling resources and signaling process for cell growth, division, differentiation, and cell death.

The acidity of the internal lumen is important for the lysosomal enzyme function, representing the active status of lysosome. Any mis-regulation and malfunction of the hydrolytic enzymes could be the cause of lysosomal storage disorder, inducing various metabolic diseases. Utilizing the acidic lumen characters, most of the lysosome probes usually carry amine groups, such as morpholine or dialkylamine. The probes may passively enter the cells and penetrate lysosomal membrane in an equilibrium of partitioning. Once they are in the lysosome, the amine group of the

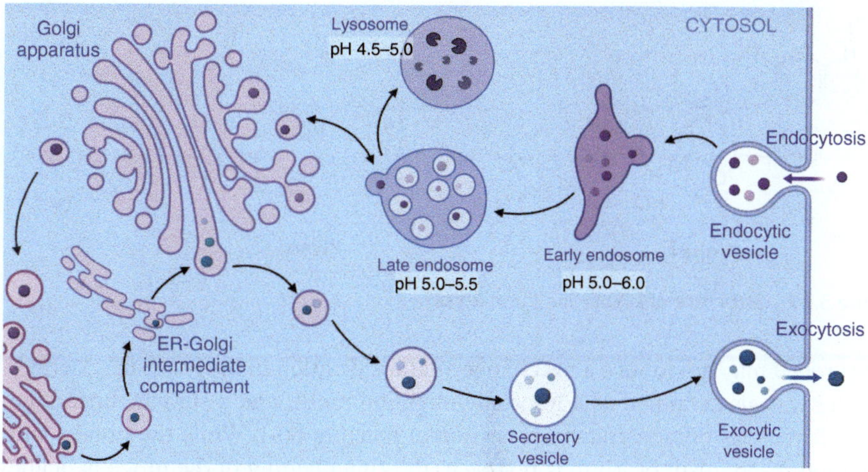

Figure 3.17 Formation and maturation of endosome and pH changes.

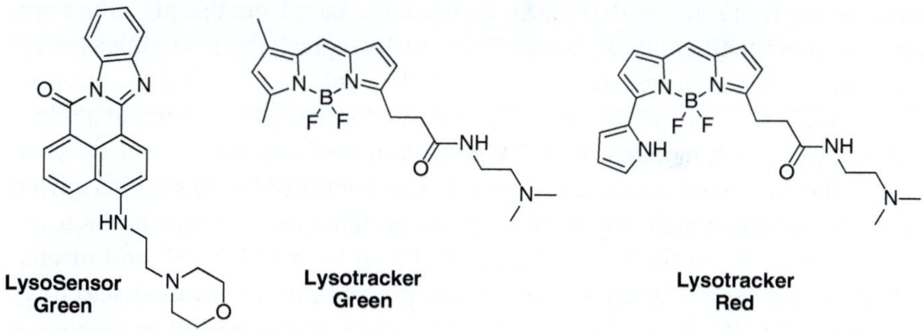

Figure 3.18 Structure of Lysotrackers.

probe is protonated and the charged state probe is captured inside the lysosome with low membrane permeability. Most of the neutral fluorophores can be modified with the amine group to target lysosomes using this strategy. Also, the protonation of the amine group can block the photoinduced electron transfer (PET) effect of the probe, often turning on the fluorescence signal of the probe. The accumulated probes show off the fluorescence signal reflecting the active lysosome status [14]. The representative lysosome probes are described in Figure 3.18 [15]. One issue with using amine-targeting motifs for lysosome probe is their basification of the lysosomal environment, known as lysosomotropism [16]. High concentrations of probes may increase the pH of lysosomes and disturb their original function, eventually inducing cell death. Therefore, such probes may not be a good option for long-term monitoring of the cells.

Lysosomal proteins are mainly categorized either into soluble hydrolases and membrane proteins, and most proteins are targeted to the lysosome by glycosylation [17]. Utilizing glycosylation as the strategy, N-linked glycans were developed for lysosome probes with LysoProbe I as an example [18]. Without the amine group,

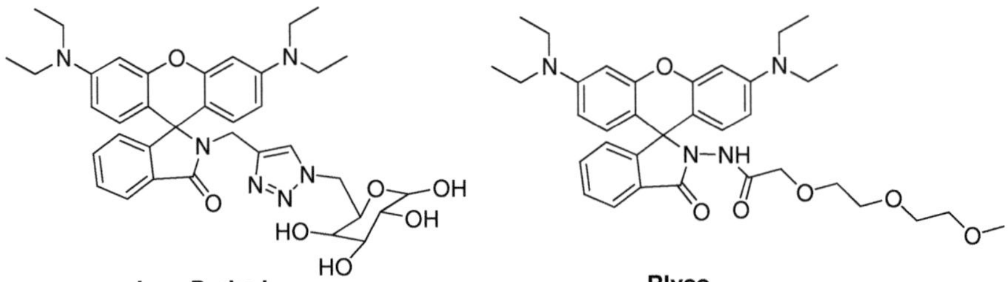

Figure 3.19 Structure of LysoProbe I and Rlyso.

LysoProbe I may not induce a pH change after localization in the lysosome. Methylcarbitol was also claimed as a lysosome-targeting motif, and a similar rhodamine probe Rlyso was demonstrated for lysosomal imaging [19]. While the uptake and retention mechanism of Rlyso is not clear, the ring opening of the probe in acidic condition is an important fluorescence turn-on process (Figure 3.19).

The critical difference between lysosomes and early endosomes is pH, where lysosomes are more acidic than early endosomes. Based on the pH difference, amine groups with different pK_a were utilized to distinguish them. Modified benzyl amine (pK_a = 9.8) is more basic than morpholine (pK_a = 7.4), and the stronger basic probe can be captured by less acidic early endosome [20]. In another study, a piperazine (pK_a ~ 10) modified BODIPY probe **6** showed a late endosome selective staining. The fluorescence turn-on of probe **6** was monitored by pH titration, giving a pK_a value of 5.08, which may be through the protonation of aniline position. So, capturing the probe in the late endosome is driven by the high pK_a and turning on fluorescence seems to happen at a lower pH condition with dual selectivity mechanism [21]. However, the clear discrimination among early/late endosome and lysosome is always challenging (Figure 3.20).

3.4 Nucleus and DNA

Deep inside cells, there is the information center, nucleus. Nucleus is an isolated organelle by membrane and interestingly the membrane is double layered with nuclear pores for material exchange with cytosol. So, topologically, nuclear interior is connected to the cytosol. In other words, the intranuclear space is inside of life, and it makes sense considering the importance of the organelle as the central information storage space of life. Interestingly, our genetic information is composed of digital format written in DNA using quaternary system, in comparison to a silicon-based computer's binary system. The functional unit of genetic information is a gene, and conventionally one gene contains the information for one protein. Proteins are the basic building blocks of our body and also catalytic machines for all biological reactions in our body. The genetic information we receive from our parents is one set of information for making all the proteins we need. So, our life is built on a dual layer system of information (DNA) and machinery (protein). When

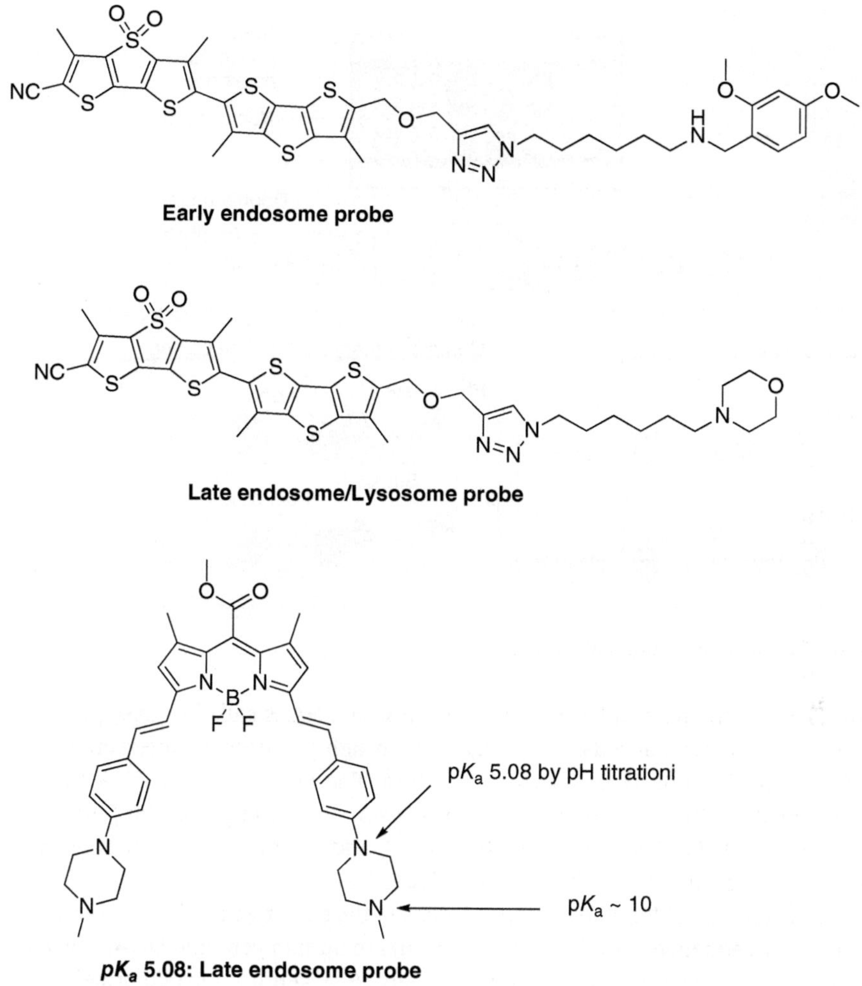

Figure 3.20 Structure of early endosome and late endosome/lysosome probe.

cells divide, the genetic information is copied and condensed into chromosomes. In human, we have 46 chromosomes and during the cell division, 46 pairs of duplicated chromosomes are formed and evenly distributed to the two daughter cells. So, all the cells keep the exactly same information in the nucleus and, in principle, we can get all the necessary information to build our whole body from just a single cell.

The collection of all the genetic information of a cell or an individual is called "genome," and the sequence analysis of the total genome is called "genome project." DNA is composed of deoxyribose sugar backbone and the monomers are connected through a phosphate bond. Each monomer has a base (A, T, C, G) which carries the genetic information. The bases make pair of A with T and C with G, and the linear polymer of DNA has complementary strands bound together as a double stranded

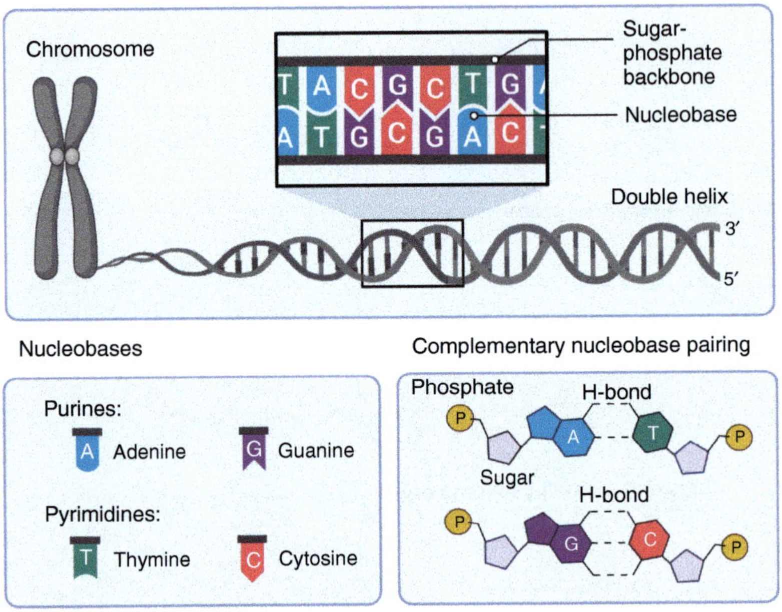

Figure 3.21 DNA structure and double helix.

form. That is why the number of monomer unit in DNA is called as base pair. The base pairing makes a stacked planar structure through π–π aromatic interaction, perpendicular to the long DNA polymer direction. The two strands of DNA make a right-handed double helix as proposed by Watson and Crick [22]. The phosphate groups are located outside of the double strand, and has major and minor grooves continuously running through the helix (Figure 3.21).

With the successful Human Genome Project in the early twenty-first century, now we know the sequence of three billion base pairs in human genomic DNA. Through the genome sequence analysis, surprisingly only 2% of genome encode proteins, i.e. only 2% of genome is allocated for genes by the conventional definition. Also, total number of genes turned out to be only 20 000–30 000. That means our body is composed of only 20 000–30 000 of different proteins, which is much smaller than what we expected before the Human Genome Project. Even a yeast cell has 6000 genes and human has only four to five times more genes? The total number of human genes is similar to that of fruit fly (*Drosophila*) or worm (*Caenorhabditis elegans*)! Does that mean the complicated human biology can be built on simpler rules than we believed? Maybe it is the opposite. We have thought one protein may have one major function. But, the genome project result implies that each protein's function may be much more complex than we thought, and they may work through interconnected mutual regulation and partnerships. Also, 98% of the noncoding DNA in the genome may have other functions in our body, and the conventional definition of gene should be expanded to cover the new discovery. The total length of three billion DNA in the human genome is about 1 m long if connected into a single strand. As we get each set from father and mother, totally we have about 2 m long DNA

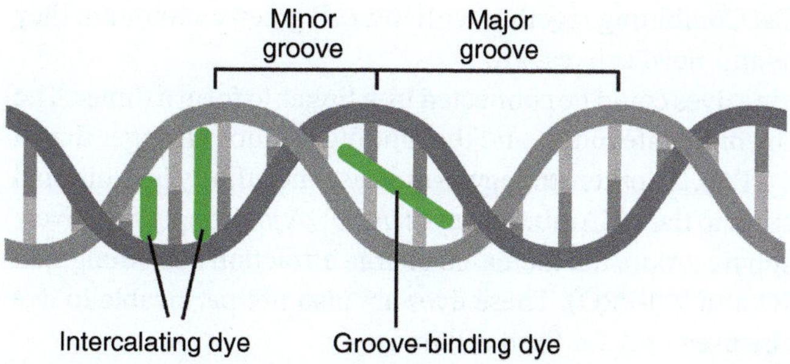

Figure 3.22 Binding mode of DNA dyes.

per cell. Compared to the size of cell and nucleus, the DNA length is amazingly long. It is necessary to effectively fold DNA molecules to fill into the refined space of nucleus. The highly information-rich nucleus could be considered as a library filled with millions of books.

DNA targeting is an obvious option for nuclear imaging due to its very high concentration in the nucleus. Many DNA sensors have been developed and used for nuclear imaging. Depending on the binding mode to the DNA molecules, they can be classified into several groups. Among these, the main two categories are intercalating probes and minor-groove-binding probes (Figure 3.22).

The first main class of DNA sensors is intercalating dyes between the base pair layers. These dyes are usually flat aromatic fluorophores, and can form π–π interaction with the base pairs up and down. The intercalation can happen only to double-stranded DNA (dsDNA), not single-stranded DNA (ssDNA), so these probes are usually dsDNA selective. While useful, the intercalation may interfere with normal DNA repair and replication, so most of the DNA intercalating dyes are mutagenic (tend to induce mutation). Some anticancer drugs are developed based on this DNA intercalating property, as cancer cells divide quickly and have higher chance to get damaged by the intercalating agent. Propidium iodide and ethidium bromide are representative DNA intercalating probes (Figure 3.23). Upon binding to dsDNA, they emit strong red fluorescence. These dyes are not permeable through intact cell membrane, so their nuclear imaging is only limited for the dead cells or

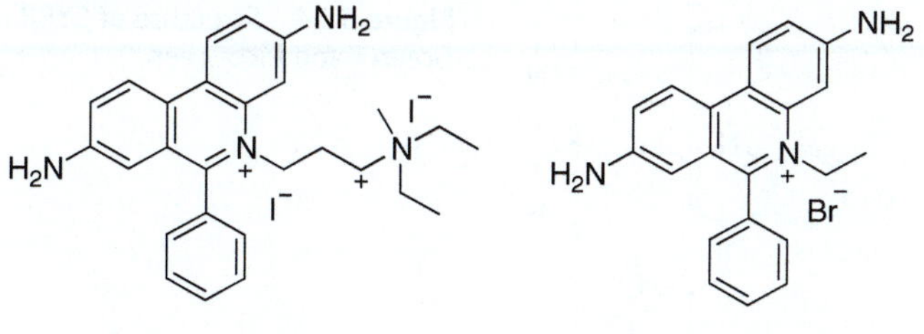

Figure 3.23 Structure of propidium iodide and ethidium bromide.

membrane damaged cells. Combining together with live cell green compound, they composite a popular live-and-dead cell test kit.

This kind of intercalating dyes could be connected by a linker to form a dimer. The dimer can bind to DNA by bidentate mode, and the binding is much stronger than a monomer (e.g. PO into POPO). Another strategy to increase the affinity is additional positive charge introduction to the DNA-binding dyes. As DNA is strongly negatively charged due to the phosphate group, the increased charge attraction can strengthen the binding (e.g. PO-PRO and YO-PRO). These dyes are also not permeable to live cell membrane and can be used only for fixed cell images.

Among the DNA intercalating dyes, SYBR Green I is optimized for the fluorescence increase upon DNA binding, and popularly used for DNA gel chromatography or real-time quantification in polymerase chain reaction (PCR) [23]. While widely used, some of the commercial dyes have limited information open to the public. For example, an improved derivative of SYBR Green, PicoGreen is also a dsDNA probe whose structure is known (Figure 3.24). However, the structures of the ssDNA-selective probe OliGreen and the RNA-selective probe RiboGreen are not known yet. In a similar way, the SYTO series of live-cell compatible nucleic acid probes and the SYTOX series of fixed-cell version probes have limited information on their structure and selectivity between DNA and RNA, making it difficult to choose for the application.

The second main class of DNA probes is minor groove binders. Compared to open shallow structure of major groove, minor groove is narrow and relatively deep, so it may be easier to make a specific binding with a probe. Positive charge would also help to increase the affinity to the negatively charged DNA. The most popular live-cell nucleus probe is a minor groove binder, Hoechst dyes. In water, the fluorescence of these dyes is low, but when they bind to DNA, a bright blue color is emitted. DAPI (4,6-diamidino-2-phenylindole) is another popular blue fluorescent DNA minor groove binder. The cell permeability of DAPI is lower than Hoechst and is used for detecting the membrane-damaged cells. Due to these two popular blue DNA dyes, in most of the cell images, nucleus images are almost always blue (Figure 3.25).

For multicolor imaging with other organelle probes, the need for a non-blue nucleus probe has emerged. One of the approaches is combinatorial fluorescence library combined with high-throughput screening to find a non-blue nuclear probe. Through cell-imaging-based screening, nucleus staining candidates are preselected,

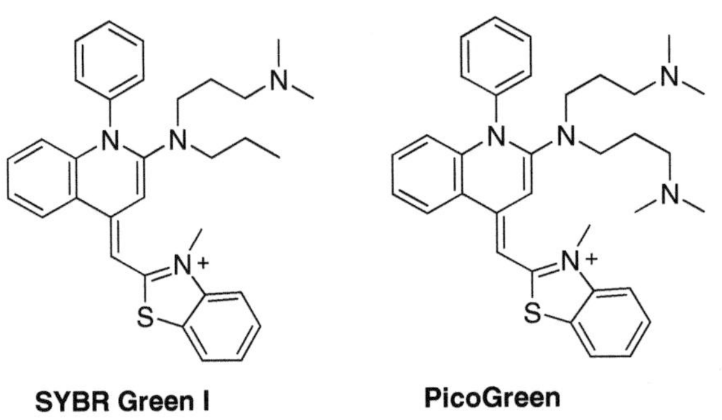

Figure 3.24 Structure of SYBR Green I and PicoGreen.

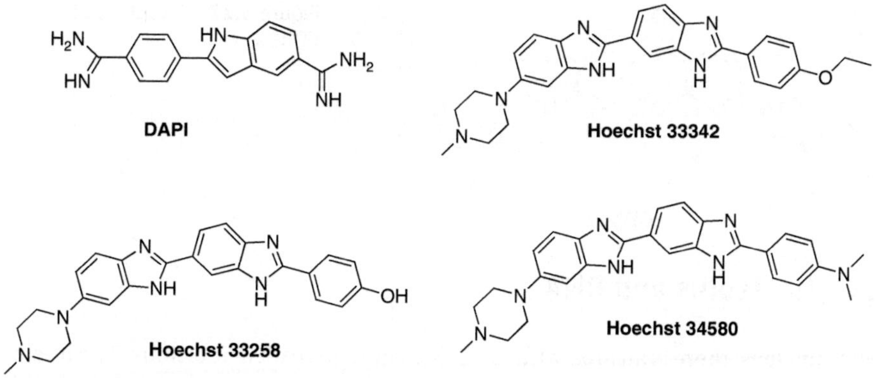

Figure 3.25 Structure of Hoechst dyes and DAPI.

and the selectivity to DNA was confirmed for the first green nucleus imaging probe (**A41B2**) development [24]. A red nucleus probe DRAQ5 was also introduced and widely used for cell-based screening [25], but had limitations due to its higher toxicity than Hoechst. Later improved green probe **C61** [26] and red probe DEAB-TO-3 [27] were introduced for versatile multicolor imaging with other organelle probes. Nevertheless, the utility of the longer-wavelength probes has still been far exceeded by the original blue Hoechst dyes. In 2014, a series of Hoechst conjugates with longer-wavelength dyes were introduced for various color options [28]. In 2015, even longer silicon-rhodamine was introduced to Hoechst, and stimulated emission depletion (STED)-based super-resolution imaging was demonstrated broadening the application scope [29] (Figure 3.26).

The popular Hoechst dyes also have problems with biological functions. They hamper the normal cell division during mitosis probably because of their strong DNA binding. Once the dividing cells finish the DNA synthesis, chromosomes condense and align in the center of the cells. Then, the chromosomes need to separate and move to each daughter cell, but Hoechst-treated cells showed an arrest at the chromosome separating stage. This means the real-time monitoring of cell division is not possible using Hoechst. To solve the problem, a new blue dye CDb12 was developed by M-phase selective dye screening. CDb12 has stronger blue fluorescence in condensed form of DNA chromosome over chromatin, and does not interfere with multiple cell divisions [30] (Figure 3.27).

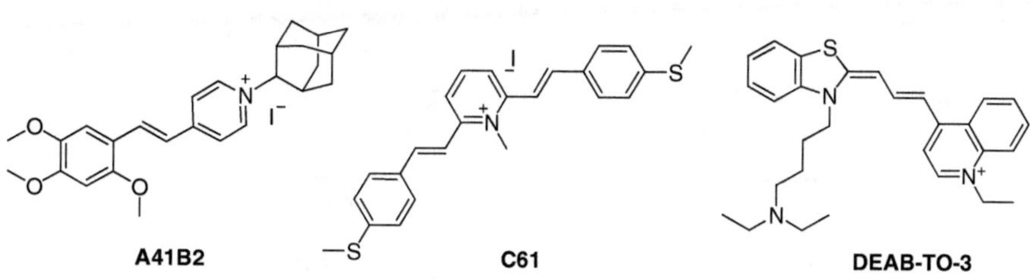

Figure 3.26 Green and red nucleus probes.

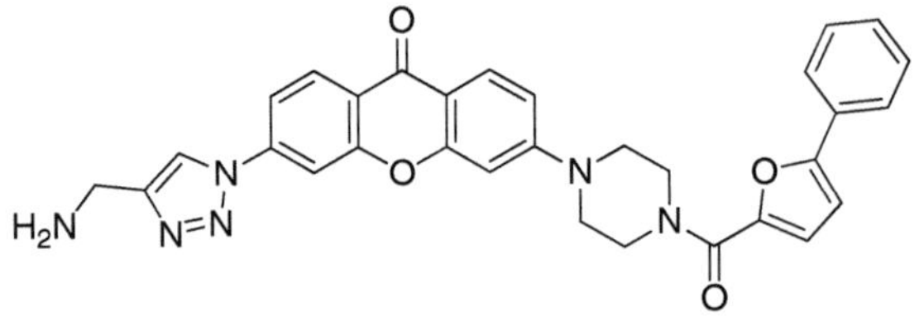

CDb12

Figure 3.27 Structure of CDb12.

3.5 Nucleolus and RNA

Inside nucleus, there is an area with low DNA contents, called nucleolus. The nucleolus is not isolated by membrane structure, but it has high RNA content instead of DNA. When a gene is used to make a protein, the first step is transferring DNA information to mRNA (messenger RNA). This step is similar to DNA replication, but the product is RNA not DNA, and the process is called "transcription." RNA and DNA are very similar to each other, and share same base structure for A, C, and G. Base T in DNA is slightly modified into U with one methyl omitted in RNA. But the Watson and Crick base paring is not affected by this difference. So, as in dsDNA, DNA and RNA with complementary sequences can also form double strand and the binding is even stronger than dsDNA. By the same token, two complementary RNAs also can form dsRNA, and the binding is even stronger than dsDNA or DNA–RNA complex. In animal cells, dsRNA is not commonly found. Therefore, once such a structure is detected, it would be interpreted as an invasion by a foreign body such as a virus, and the host defense system will be activated. Another main difference between DNA and RNA is the sugar backbone. DNA uses deoxyribose and RNA uses ribose with an extra hydroxyl group at 3′ position. Because of this extra OH group, the phosphate stability in RNA is million times lower than that in DNA (Figure 3.28).

The synthesized mRNA is transported to cytosol, and the information is decoded by ribosomes to synthesize proteins. This protein synthesis process is called "translation" and ribosome is the actual production machine for proteins. After

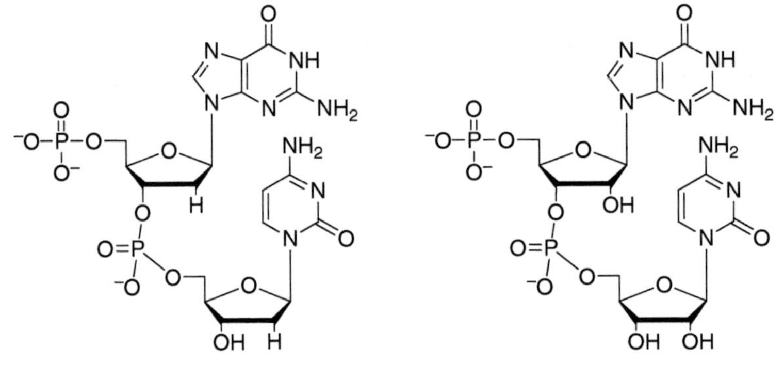

Figure 3.28 Structure of DNA and RNA.

protein synthesis is finished, mRNA is degraded, so mRNA is a temporary information molecule for a gene encoded in DNA. Interestingly, the backbone of ribosome is composed of RNA (so called rRNA) and decorated by proteins. Protein synthesis is one of the key functions of life, and the fact that the catalyst for protein synthesis is not protein enzyme, but RNA, casts the clues for the origin of life: RNA world hypothesis where RNA played the key role of the ancient life.

Nucleolus is the site of rRNA synthesis; therefore, nucleolus is an RNA-rich area in the nucleus. So, for the nucleolus-specific imaging probe, RNA will be an ideal target for probe design. But, due to the structural similarity between DNA and RNA, selective probe design for one of them is not easy. Even RNA-selective dyes in test tube do not always show nucleolar staining over nucleus in cell imaging. The commercial SYTO dyes are labeled as DNA- or RNA-selective probes, but due to the unclear selectivity guide, the choice of proper probe has been difficult.

The practical nucleolus probes were first developed by an unbiased cell-imaging-based screening of styryl dye library (Figure 3.29). The selected probes showed several fold higher RNA selectivity over DNA in in vitro test [31], but the origin of high selectivity in the live mammalian cells for nucleolar staining over nucleus is not fully understood [32]. The apparent visualization of nucleolus by the RNA probes may be due to relative high concentration of rRNA compared to relatively low concentration of other RNAs such as mRNA or tRNA. The developed RNA dyes have been applied to various important biological studies. Green RNA probe **E36** was used for monitoring large RNA–protein particle export during nuclear envelope budding [33]. Red RNA probe **F22** has been used for RNA imaging in the phase separation of prion-like RNA-binding proteins [34] and monitoring RNA entanglement during the protein condensate formation induced by RNA [35]. The details of this study are described in Section 3.12. By combining DNA binder **TO** and **C61**, the hybrid probe Styryl-TO showed a clean nucleolar stain, demonstrating the difficulties of the design of RNA-selective probe over DNA, or vice versa [36].

A certain dye binds both to DNA and RNA, but with different response. For example, acridine orange emits 525 nm fluorescence upon binding to DNA and

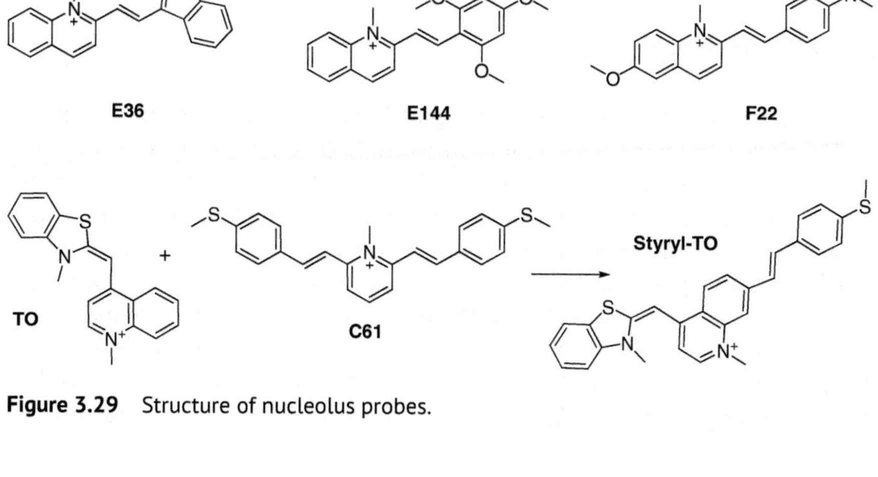

Figure 3.29 Structure of nucleolus probes.

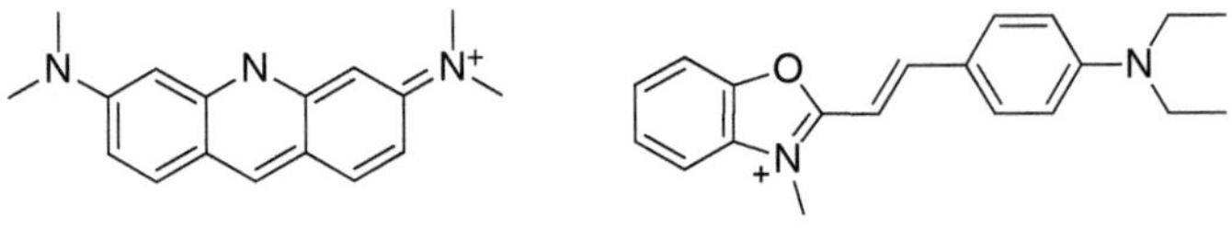

Acridine orange **Probe 1**

Figure 3.30 Structure of acridine orange and probe **1**.

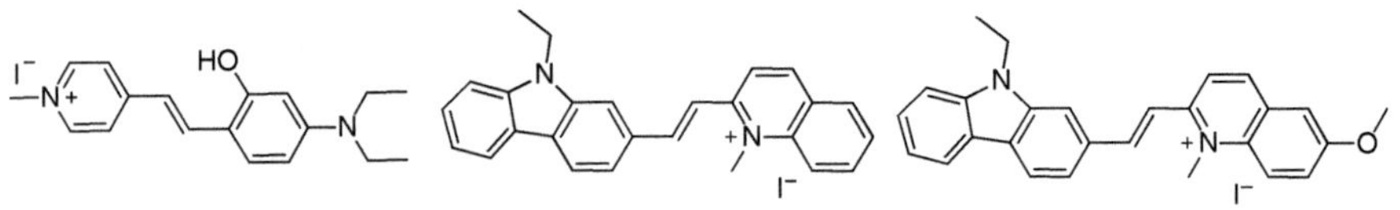

132A **107E** **107F**

Figure 3.31 Structure of malaria probes.

650 nm upon binding to RNA [37]. An interesting red fluorescent probe **1** was introduced for real-time cell cycle monitoring. Probe **1** responds both to DNA and RNA with similar fluorescence enhancement in vitro. In live cells, probe **1** stains nucleolar structure during the interphase, but stains DNA during the mitosis phase. By nucleic acid hydrolyzing enzyme test, it was confirmed that the probe binds to RNA during the interphase and binds to DNA during the mitosis phase, showing the structural discrimination of the nucleic acids [38] (Figure 3.30).

Malaria is a parasite, which uses red blood cells as the host. As red blood cells lack nucleus, the detection and imaging of infected malaria has been tried using various DNA probes such as DAPI [39]. As a complementary staining, RNA-selective probes were elucidated to enhance the detection of malaria in various stages of life cycle [40]. Among them, **132A** diffuses throughout the cytoplasm and provides detailed live-cell staining of malaria cells without any need of lysis buffer or detergent, and has been successfully used for high-throughput screening of antimalarial natural product screening [41] (Figure 3.31).

3.6 ER and Golgi Body

The nucleus is not a topologically isolate organelle with a bilayer of membrane structure. The nuclear membrane is extended to cytosolic direction and forms folded and stacked sac structure, called endoplasmic reticulum (ER). The ER has single bilayer membrane structure and the lumen of ER is topologically "outside" of cell, like the inner space of endosome or lysosome. The ER is the largest membrane-based organelle, occupying almost half of the total membrane of a cell. The ER membranes are closely contacted with mitochondria, Golgi, and cell membrane. If the ER membrane buds off into cytosol, it may form an intracellular vesicle which is difficult to distinguish from endosome. ER is also the storage space for Ca^{2+}, which is maintained very low in cytoplasm. The secretion of Ca^{2+} from ER to cytosol upon responding to signaling molecules constitutes an important signal transduction mechanism.

Depending on the surface appearance, ER is classified as rough endoplasmic reticulum (rER) and smooth endoplasmic reticulum (sER). The rER has many ribosomes on the surface and the protein synthesis actively occurs in rER. sER is usually located in outer space of rER, and synthesized proteins transport from rER to sER carrying protein folding. Some proteins are synthesized and secreted into the ER lumen, and the budded off vesicles act as carriers of the protein moving to the plasma membrane direction. Once they reach the cell surface, the membrane of vesicle fuses to the plasma membrane, pouring the proteins and contents outside of cell, i.e. secretion. This process is called as exocytosis and the process is reversible to endocytosis.

Outside of ER, there is another stacked sac structure, called Golgi body (following the discoverer's name). Golgi body contains various metabolic enzymes and modifies the synthesized proteins, and the process is called post-translational modification (PTM). PTM includes methylation, hydroxylation, acetylation, lipidation, SUMOylation, ubiquitination, glycosylation, phosphorylation, or sulfation. The traveling vesicles from rER to the plasma membrane usually pass through sER and Golgi body through repeated membrane fusion and budding process. Interestingly, the membrane thickness is different along the travel path; the thickness order is ER (3.8 nm) < Golgi body (4.0 nm) < plasma membrane (4.3 nm) [42]. Thus, the directional driving force of vesicle traveling may rely on the thickness gradient. As a whole, the ER membrane is thinner and flexible than the plasma membrane (Figure 3.32).

As ER is the most extensive network of folded membrane in the cell, membrane targeting could be a reasonable approach. For selective ER probe design, membrane thickness or flexibility could be a potential target. Following the idea, medium size of hydrocarbon tail structure may prefer ER over the plasma membrane. While the plasma membrane probe DiO has an 18-carbon chain (so, the detailed structure is $DiOC_{18}$), $DiOC_6$ is one of the examples for traditional ER imaging probes. But $DiOC_6$ also responds to mitochondrial membrane potential, so once $DiOC_6$ enters cell, the first staining occurs in mitochondria by the electrostatic interactions and slowly shifts to ER by physical partition equilibrium [43]. So, the overall dynamic character made the imaging complex and the quality of ER image was rather blurry [44]. Also,

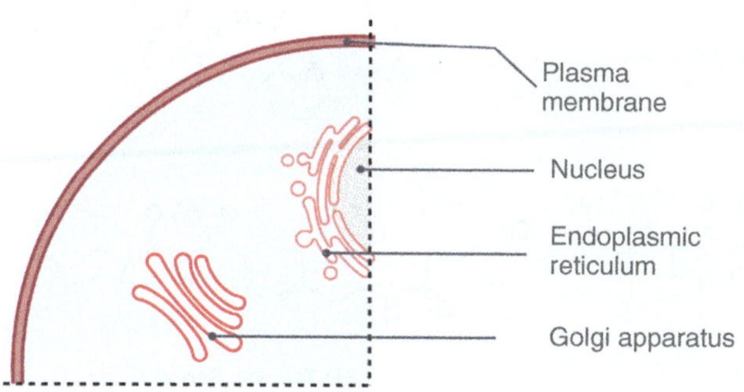

Figure 3.32 Structure of ER, Golgi body, and plasma membrane.

selectivity of the ER probe against other membrane structures such as Golgi body or plasma membrane is always challenging with partial overlap (Figure 3.33).

The breakthrough for ER-specific probe is from a diabetes drug, glibenclamide, which is known to bind to ER through SUR (sulfonylurea receptor) [45]. Utilizing the structure of glibenclamide, BODIPY-attached first-generation ER probes, ER-Tracker Green and Red were introduced [46] (Figure 3.34).

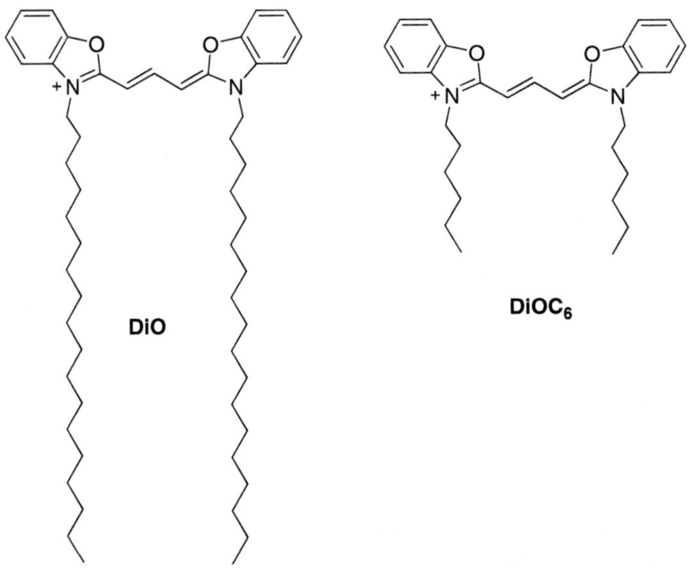

Figure 3.33 Structure of DiO and $DiOC_6$.

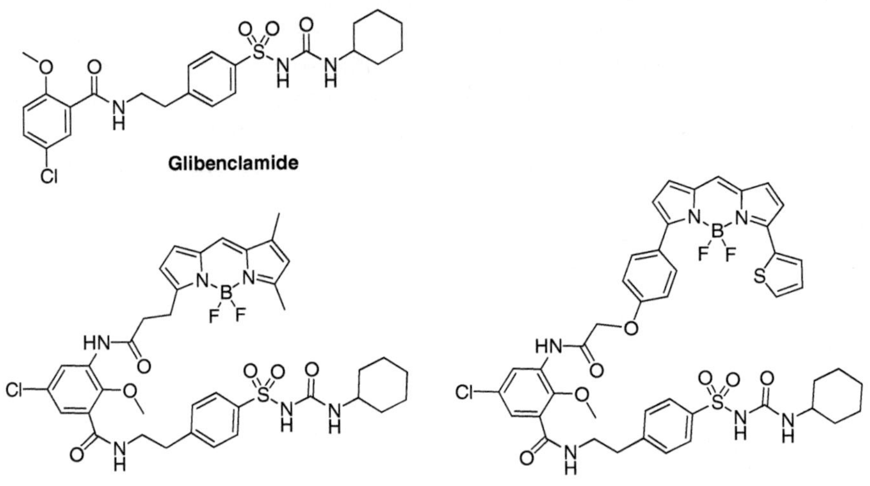

Figure 3.34 Structure of glibenclamide and ER-Trackers.

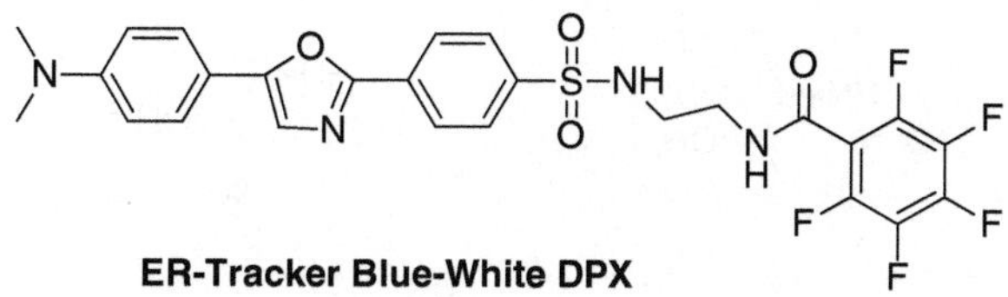

Figure 3.35 Structure of ER-Tracker Blue-White DPX.

By further structure–activity relationships (SAR) studies of glibenclamide, sulfonamide motif is known to be essential for ER binding, through the overexpressed ATP-sensitive potassium channels on ER membrane. Based on this knowledge, a simpler but more efficient probe ER-Tracker Blue-White DPX was developed and widely utilized. Pentfluorophenyl group in ER-Tracker Blue-White DPX can react with thiol and form covalent bonding with ER proteins [15] (Figure 3.35).

As an alternative ER-targeting motif, propyl chloride was suggested, while the mechanism is not yet very clear. TPFL-ER is a fluorene-based ER probe and through two-photon imaging, the ER change during cancer cell apoptosis was monitored [47]. $NRER_{Cl}$ is a Nile-red-based ER membrane probe and has been used for lipid order imaging under oxidative and mechanical stress [48]. Both these probes contain propyl chloride as the ER-targeting motif, and chloride pump in ER is assumed as the molecular target of the probes. As a relatively new motif, an extensive study is yet to be carried out (Figure 3.36).

For Golgi body targeting, the medium thickness of the membrane between the ER and plasma membrane has been utilized. A medium-sized ceramide (an amide of sphingolipid) was known to be preferred to localize in the Golgi body, and fluorescently labeled ceramide has been used as a Golgi probe. If the carbon chain of the tail in ceramide gets longer, the probe may prefer the plasma membrane than Golgi body membrane (Figure 3.37).

For the post-translational modification of proteins in Golgi, many proteins use cysteine residue for anchoring. Hijacking the anchoring system, a simple cysteine-labeled fluorescent molecule was tested and proved to localize to Golgi body [49]. While the mechanism is not fully proven, the thiol requirement for targeting suggests binding to sulfhydryl receptor site through disulfide bond formation. COX-2 (Cyclooxygenase-2) is overexpressed in Golgi body of cancer cells. A COX-2 inhibitor, IMC (indomethacin) was used for Golgi-body-targeting

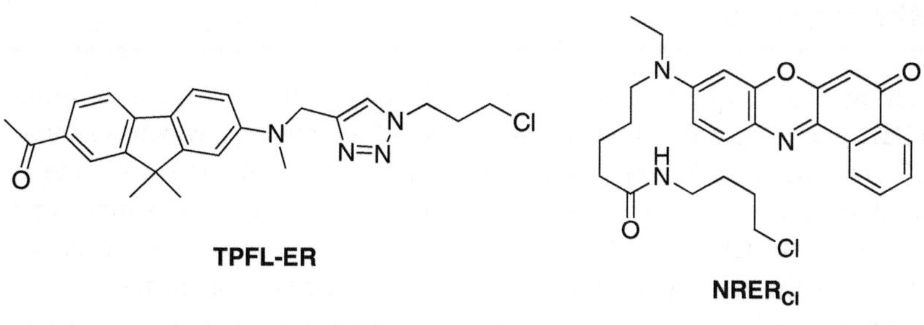

Figure 3.36 Structure of TPFL-ER and $NRER_{Cl}$.

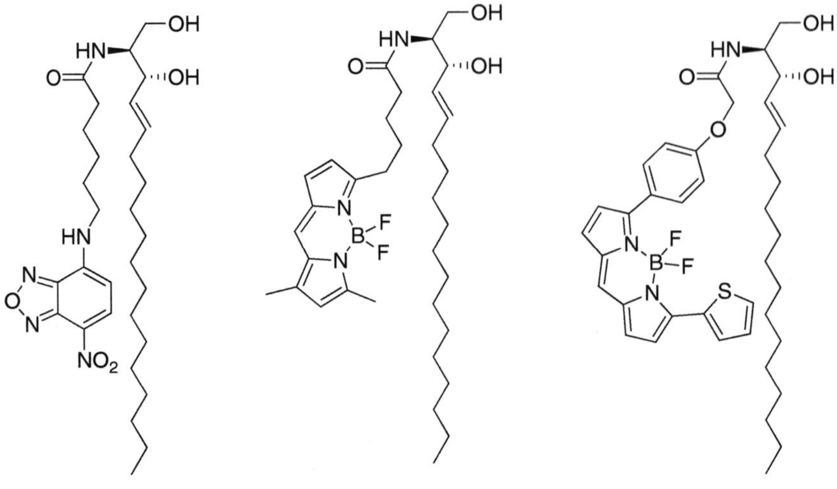

NBD C6-Ceramide BODIPY FL C5-Ceramide BODIPY TR C5-Ceramide

Figure 3.37 Structure of Golgi body probes.

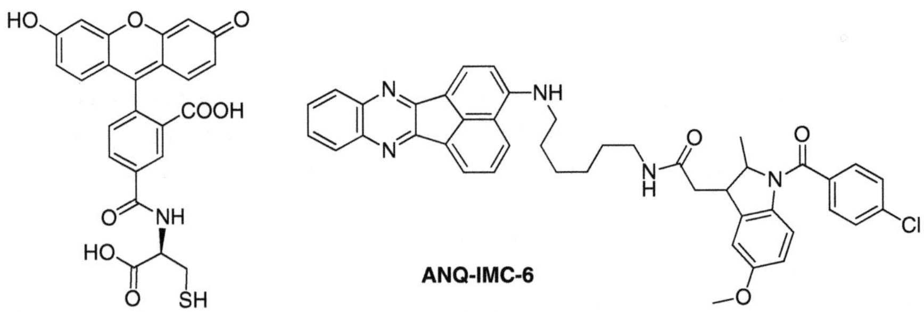

ANQ-IMC-6

Fluorescein-cysteine-1

Figure 3.38 Structure of cysteine-fluorescein and ANQ-IMC-6.

motif and coupled to ANQ dye. The conjugated probe ANQ-IMC-6 showed good Golgi body localization in the cancer cells [50] (Figure 3.38).

3.7 Mitochondria

Mitochondrion is a unique organelle in the cell. Unlike other organelles, mitochondria have a topologically isolated double layer of membranes, which is similar to that of Gram-negative bacteria. The inside space of the inner membrane is called the matrix, and there is an intermembrane space between the inner and outer membranes. In Gram-negative bacteria, the intermembrane space is called periplasm. When the cell breaks down glucose through glycolysis in the cytosol, the product pyruvate is imported inside the matrix and used as a fuel in citric acid cycle (also known as Krebs cycle or tricarboxylic acid [TCA] cycle) to generate the high-energy material, NADH (nicotinamide adenine dinucleotide + hydrogen), by reducing NAD^+. NADH enters electron transport chain (ETC) in the inner membrane of mitochondria, pumping protons into the intermembrane space. The accumulated proton gradient generates a strong mitochondrial membrane potential of 150–180 mV across the inner membrane with the negative charge in the matrix side [51]. Due to the pH gradient of ~0.9 pH units, the matrix becomes more alkaline than the intermembrane space [52]. The accumulated protons pass through ATP synthase on the inner membrane and generate ATP in the matrix. The ATP synthase can also act as a proton pump using ATP as the energy source, which is the reverse reaction of ATP synthesis. The whole process of ATP synthesis is called oxidative phosphorylation (Figure 3.39).

Compared to glycolysis, oxidative phosphorylation uses glucose energy much more efficiently. Chemically speaking, the process is a full oxidation of the carbohydrates with oxygen, i.e. a controlled burning of carbon sources to obtain energy. The oxidative process may generate reactive intermediates, called reactive oxygen species (ROS), such as superoxide, hydrogen peroxide, hydroxyl radical, and single oxygen. The ROS materials can damage biological molecules and be toxic to the cell. So, it is an efficient process for energy utility, but a potentially dangerous process for the biological system.

Despite the potential toxicity of oxygen, some bacteria may have achieved the ability to use oxygen for efficient fuel burning, i.e. oxidative phosphorylation. This super high efficiency of energy generation may have provided the bacteria with a great evolutionary advantage. For other competing cells, there may have been one of the two options: acquiring the ability in the cell or adopting the bacteria as a symbiotic partner. As we see the current ecosystem, the symbiosis has become dominant at least for the multicellular organisms. An interesting hypothesis, but is there evidence for the symbiosis?

Several circumstantial evidences exist. First of all, the ATP synthase location and function are very similar between Gram-negative bacteria and mitochondria, supporting the origin of mitochondria as bacteria. The proton gradient is formed in the intermembrane space in mitochondria and in the periplasm in the Gram-negative bacteria. In Gram-positive bacteria, the role of the outer membrane is played by

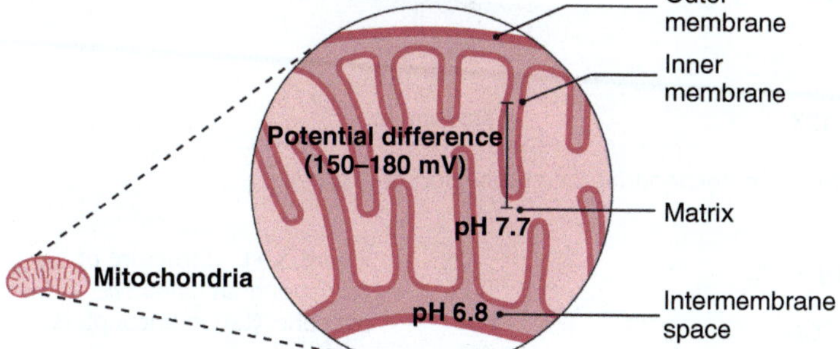

Figure 3.39 Structure of mitochondria and membrane potential.

the cell wall. Second, mitochondria carry their own genes independent of the host, although many others were transferred to host genome. So, the symbiosis provided the host with power plant outsourcing, segregated by double layers of membrane, and the mitochondria with specialized function and reduced burden for genomic replication.

As a result, the mitochondrial inner membrane keeps a high membrane potential when the cell is energetically functional with a strong negative charge on the inner leaflet of the inner membrane. This membrane potential is the main target for the mitochondrial probe design. Therefore, probes with reasonable hydrophobicity for biological membrane permeability and delocalized positive change tend to localize in the inner leaflet of the mitochondrial inner membrane. One example is rhodamine 123, where the carboxylic acid in the rhodamine is neutralized through esterification and the total charge becomes positive. The mother structure of rhodamine has an overall neutral charge (one positive and one negative charge) and tends to spread out through the cytosol. By the same token, removing the carboxylate group from rhodamine makes rosamines, which are in general mitochondrial staining dyes (Figure 3.40).

The mitochondrial staining of these dyes is sensitive to the mitochondrial membrane potential, and removal of the potential induces the release of the dyes from mitochondria. Mitochondrial oxidative phosphorylation uncoupler such as carbonylcyanide-3-chlorophenylhydra-zone (CCCP) or carbonylcyanide-4-trifluorometh-oxyphenylhydrazone (FCCP) renders the inner membrane permeable to protons and abolish the mitochondrial membrane potential (Figure 3.41).

Carbocyanine dyes without extra charged motifs also tend to bind to mitochondria due to their delocalized positive charge. Depending on the hydrophobicity, their localization is under competition among different organelles. For example, $DiOC_6$ binds to mitochondria first, but the equilibrium is slowly moved to ER. Due to the dynamic characteristics of mitochondria, comparison between different probes

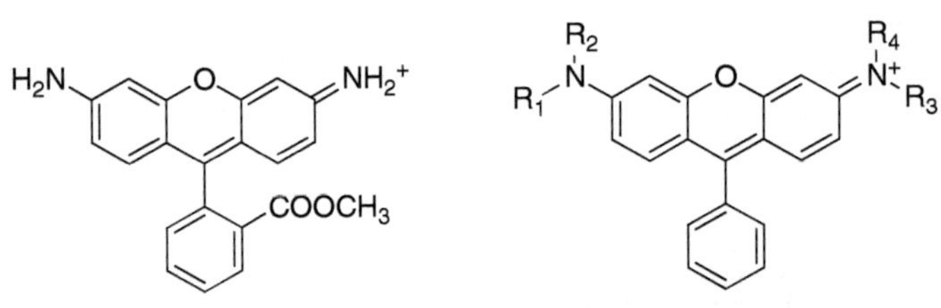

Rhodamine 123 **Rosamine**

Figure 3.40 Structure of mitochondrial membrane potential probes.

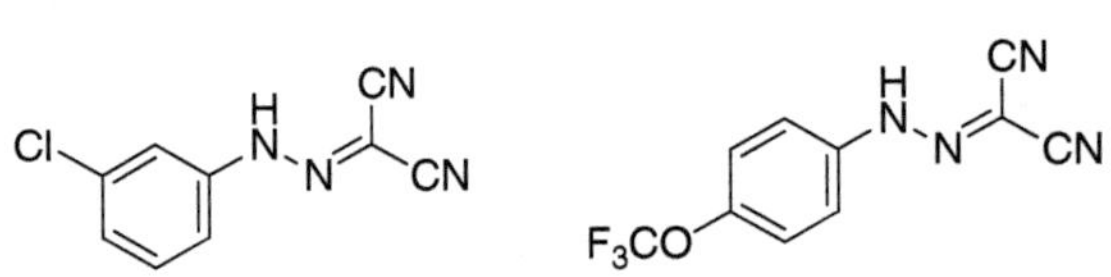

CCCP **FCCP**

Figure 3.41 Structure of mitochondrial oxidative phosphorylation uncouplers.

especially in quantitative measurement is quite difficult. The most commonly used standard mitochondrial probes are MitoTracker dyes available from ThermoFischer [15]. Among them, MitoTracker Green is less sensitive to membrane potential and thus is used for measuring the absolute number of mitochondria (Figure 3.42).

In contrast, JC-1 is a sensitive sensor for mitochondrial membrane potential. At a low concentration, JC-1 shows green fluorescence, and when the mitochondrial membrane potential is high, the accumulated JC-1 molecules form J-aggregate to give out red fluorescence. As a result, the red/green ratio is positively correlated with the mitochondrial membrane potential, and JC-1 is widely used for mitochondrial membrane potential image probe (Figure 3.43).

Not only dye probes, conjugated drugs or complex molecules targeting mitochondria are other emerging functional molecules. For the delivery of the various cargo molecules to the mitochondria, hydrophobic and positively charged motif, triphenyl phosphonium (TPP), is most popularly used [53] (Figure 3.44).

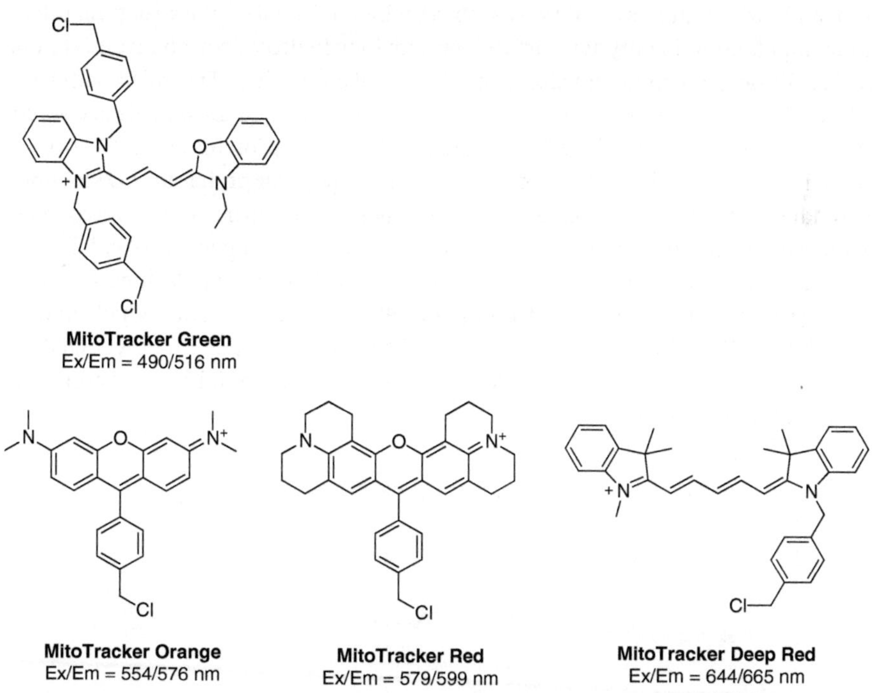

Figure 3.42 Structure of MitoTrackers.

Figure 3.43 Structure of JC-1.

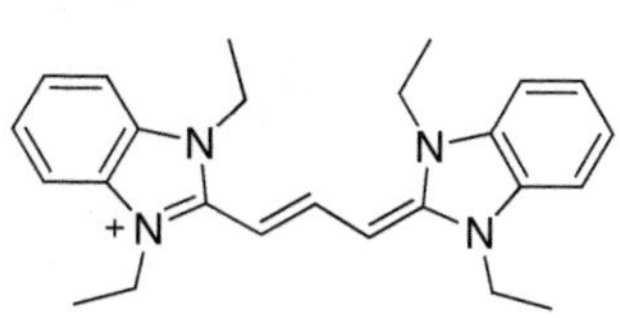

JC-1

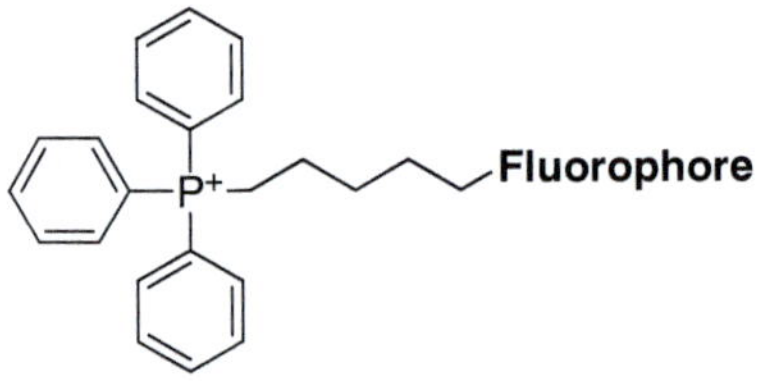

Figure 3.44 Mitochondria-targeting motif TPP.

3.8 Lipid Droplet

Lipid droplets (LDs) is a monolayer phospholipid-enclosed structure enriched with neutral lipids and considered as a cellular energy storage organelle. Lipids are one of the three main components of the living things together with amino acids (building blocks of peptides and proteins) and carbohydrates. The definition of lipids is any biological molecules which are not soluble in water, but soluble in organic solvents. It is interesting to compare the nomenclature with amino acids (acids with an amine group) and carbohydrates (hydrated carbons), which is based on the structural information. Lipids contain fatty acid derivatives with long hydrocarbon chain and sterols with fused rings, which do not share any structural similarity. LD is filled with triacylglycerols (glycerol ester of fatty acids) and sterols inside and the outside is layered with PC [54]. The surface of LD is decorated with various proteins (Figure 3.45).

Oil red O [55] and Nile red [56] have long been two popular probes for LD staining. Nile red has very low fluorescence in aqueous media, and turns on its fluorescence in hydrophobic environment, and thus considered as a fluorogenic probe. Nile red also has strong solvatochromic properties in solvents of different polarities. One disadvantage of Nile red is its broad excitation and emission spectra, which makes multicolor labeling difficult. A more recent addition is a green pentamethyl BODIPY dye, BODIPY 493/503, with small Stokes shift and narrower excitation and emission spectra [57] (Figure 3.46).

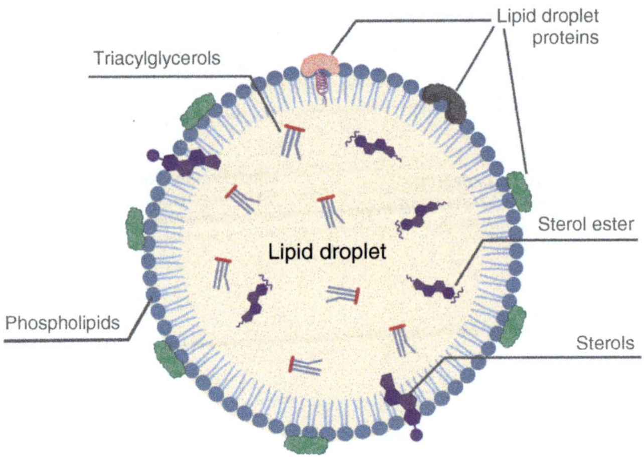

Figure 3.45 Structure of LD: diacyl glycerol and sterols.

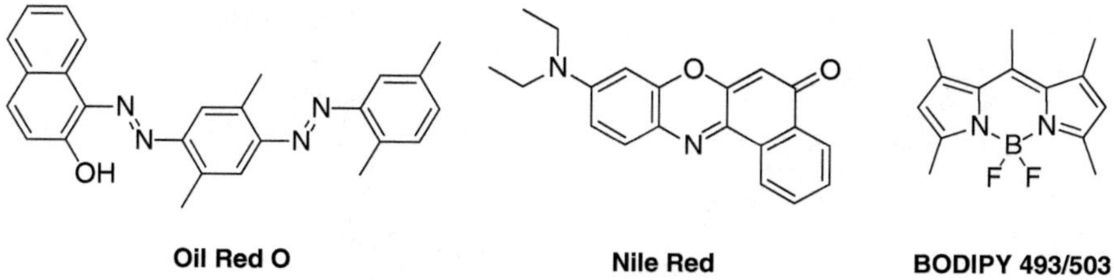

Figure 3.46 Structure of Oil red O, Nile red, and BODIPY 493/503.

Figure 3.47 Structure of StatoMerocyanines.

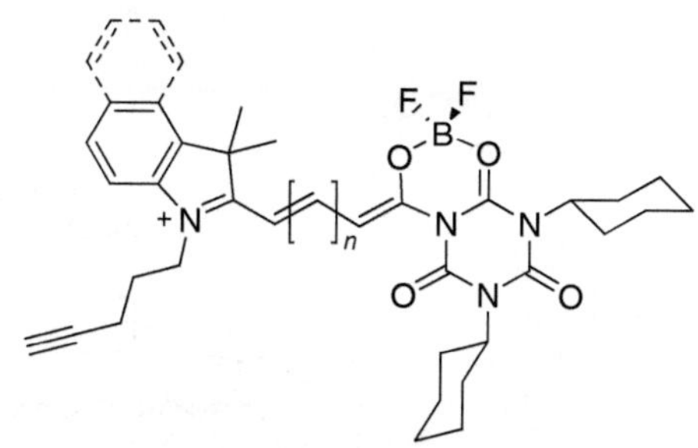

In general, the LD probes have low solubility in water, often making the live-cell imaging difficult. A better water solubility and wavelength control for multicolor imaging was in need. In 2018, a series of LD probes with improved specificity were developed in the name of StatoMerocyanines [58]. Two cyclohexyl groups were attached to merocyanine dyes to increase the lipophilicity and at the same time the bulkiness to prevent quenching by ππ-stacking. Through the conjugation length control of the merocyanines from Cy3 to Cy7, a broad range of multicolor labeling has become possible (Figure 3.47).

Although design is a possible approach to develop a LD probe, SAR study is another approach for digging out the general trend for the selectivity. Through a systematic quantitative structure–activity relationships (QSAR) study of BODIPY library, 52 new LD dyes were elucidated with 92% of success rate of prediction for the organelle selectivity (Section 3.14).

3.9 Peroxisome

Peroxisome is the vesicle where many biological oxidations occur. Especially, a long-chain fatty oxidation and breakdown by beta oxidation is a characteristic event in peroxisomes, while medium-sized fatty acid oxidation mainly occurs in

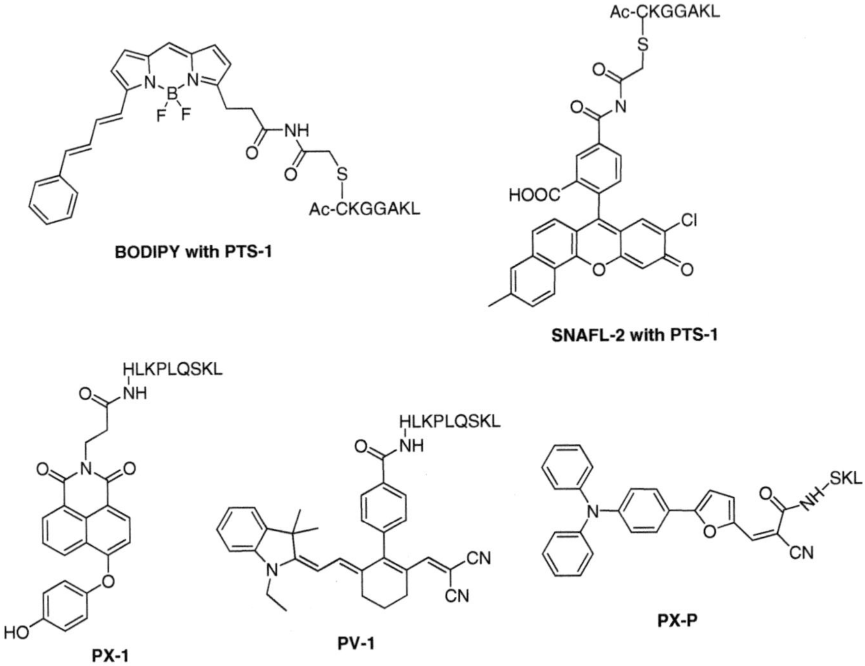

Figure 3.48 Structure of peroxisome probes.

mitochondria [59]. Peroxisomes also contain large quantities of catalase which decomposes H_2O_2 generated by mitochondria during the cellular respiration [60].

Peroxisome-targeting signal (PTS) is the endogenous signal peptide to localize proteins to peroxisome. Most peroxisomal proteins including catalase have a c-terminal tripeptide PTS-1 (consensus of S/C/A-K/R/H-L), and are imported into peroxisomes after translation and folding in the cytoplasm [61]. Fluorescently labeled peptides of CKGGAKL with fluorescein, BODOPY, and SNAFL-2 were demonstrated to localize to peroxisomes [62].

A similar peptide sequence HLKPLQSKL was used to provide peroxisome-localizing sensors. PX-1 carries peroxynitrite-detecting motif for reactive nitrogen species (RNS) [63], and PV-1 with molecular rotor was used for viscosity visualization in peroxisomes [64]. In 2021, even shorter signal peptide SKL (Ser-Lys-Leu) was used to design peroxisome-targeting polarity sensor, PX-P, and the correlation of nonalcoholic fatty liver and polarity changes in peroxisomes was studied [65] (Figure 3.48).

While the unique biological reactions in peroxisomes could be a clue to design a metabolism-derived probe, such a functionally active probe is yet to be developed.

3.10 Cytosol

Right inside the plasma membrane is the cytosol. There are many compartmentalized organelles in the cytosol; the topologically connected cytosol is one space of real "inside" of the life. Due to the very high solute contents of 30–40% in cytosol, the cytosol is not like a free ocean where you can swim, but more like a sticky mud environment. Nevertheless, there is no physical barrier in the cytosol, and molecules can reach each other by diffusion (much slower than in pure water or aqueous culture media, though).

Ideal cytosolic probes may be hydrophilic and have low binding affinity to any organelles in the cell. Considering the negatively charged surface of any membrane of organelles, highly negatively charged molecules are ideal to keep the electric repulsion from the organelle surface. The problem is how to deliver such a charged molecule into the cell across the cell membrane. Protection of carboxyl acid group with AM (acetoxymethyl) was suggested as the solution. For example, calcein is a fluorescein derivative with four carboxylic acid groups. By protecting the four acids with AM and two hydroxyl groups with acetyl, calcein-AM is a fully protected hydrophobic precursor for calcein. Calcein-AM can penetrate the cell membrane in and out easily, and the protecting groups are hydrolyzed by esterase activity in the cell. The resulting calcein can be trapped inside the cell and also does not bind to other organelles due to the highly negative charges, realizing an ideal cytosol staining property. The hydrolysis of the protecting group is catalyzed by esterase enzyme, and the staining of calcein occurs only in healthy cells. It was systematically studied that the cell retention increases proportionally to the number of carboxylate groups in the probe [66].

Carboxyfluorescein diacetate succinimidyl ester (CFDA-SE) is another popular cytosolic probe. CFDA is an acetyl-protected fluorescein and SE is an amine reactive function group. As a protected hydrophobic dye, CFDA-SE penetrates the cell membrane smoothly and the acetyl group is hydrolyzed by esterase similar to calcein-AM case. Even though the released fluorescein molecule may not be hydrophilic enough to reside inside the cell, SE can react with amine groups of macromolecules inside the cell such as the lysine residue of proteins. As a result, the fluorescein molecule can be anchored on the macromolecule and reside in the cell for a long period. The labeling could be either soluble proteins in the cytosol or membrane proteins on the organelle surface, but due to the broad reactivity of SE group to various amines, the staining pattern is usually spread out cytosolic distribution. As the turn-on effect of the cytosolic probes is dependent on the active esterase, these dyes could be used for indicating healthy cells as the component of live-dead staining kit (Figure 3.49).

3.11 Extracellular Vesicle

Extracellular vesicles (EVs) are emerging in the spotlight organelle-like structure found not in the cells, but outside of cells. EVs are lipid-bilayer-based liposome-like particles mainly released from almost all cell types from mammalian cells to bacteria. Liposomes are believed to be the origin of cells, which may have acquired genetic material and replication ability, and eventually evolved into cells. With the similar topology of EVs, they could be considered as mini cells, but they cannot replicate

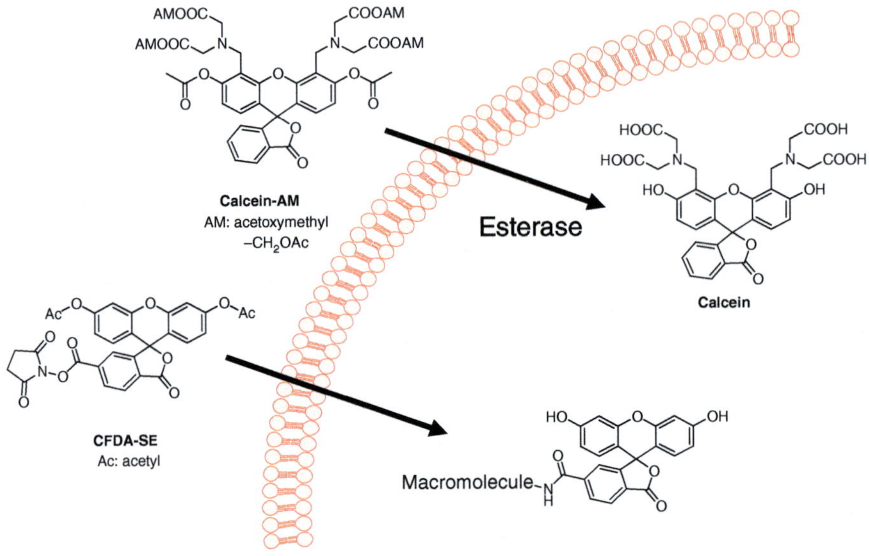

Figure 3.49 Calcein-AM and CFDA-SE in cell.

due to the lack of genomic information. In multicellular organisms, EVs are found in biological fluids, including blood, urine, and cerebrospinal fluid. They are also released in vitro cell culture into the media.

EV size ranges from the physically smallest possible unilamellar liposome (20–30 nm, wiki) to a regular cell size (10 μm) or bigger, although the majority of EVs are smaller than 200 nm. EVs are classified by their size and origin into exosomes, microvesicles, and apoptotic bodies [67]. Exosomes originate from the endosomal space inside cells. The multivesicular body is an endosome with smaller vesicles inside (multilayered intraluminal vesicles). When the multivesicular body moves to the cell surface and fuses with the plasma membrane through exocytosis, the inner vesicles are released into the extracellular space as exosomes [68]. Since the size of exosomes is limited to that of the parent multivesicular body, exosomes are usually smaller than other EVs (30–150 nm) [69]. Microvesicles are formed by ectocytosis and the lipid bilayers are of plasma membrane origin. Considering the biogenesis route, they are also called as ectosomes. The size of microvesicles ranges from 30 to 1000 nm [70]. Technically speaking, platelets and even red blood cells (which do not have genomic DNA!) can be defined as microvesicles. Apoptotic bodies are released by dying cells during the apoptosis, and the process is more like fragmenting rather than budding. During the apoptosis, it is known that the phosphatidylserine in the inner leaflet is exposed to the outer leaflet, and the apoptotic body shows the same trend. An apoptotic body is usually bigger than EVs, often reaching to micrometer size (50–5000 nm) [71]. Cancer cells also release micron-sized EVs and they are termed "oncosomes" [72]. Compared to endosomes with inside out structure, the topology of EVs is the opposite, same as cells. The inside composition is same as the cytosol of mother cells. In that sense, EVs can be considered as partial or mini cells with incomplete compositions, especially the genomic DNA (Figure 3.50).

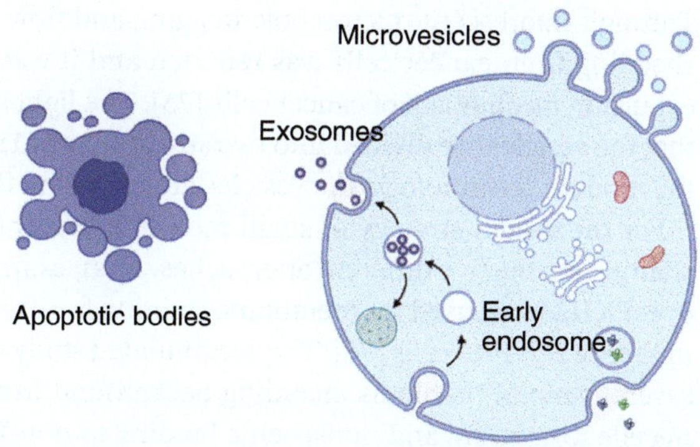

Figure 3.50 Origin of EVs.

EVs are believed to play various functional roles such as eliminating unwanted materials from the cells or cell-to-cell communications. For example, EVs from human and mouse mast cells carry mRNAs and miRNAs (as exosomal shuttle RNAs) and deliver to recipient cells for translation or translational regulation [73]. EVs contain proteins, nucleic acids, and other metabolites, and their composition is different depending on their origin and the given situation. Considering the highly enriched contents inside, EVs are emerging as an important source of novel biomarkers. For EV analysis, various separation methods have been tried so far, including ultracentrifugation, size-exclusion chromatography, size-dependent filtration, capillary electrophoresis, flow fractionation, and affinity matrix methods. However, different methods generate uneven results in yield and purity of EVs, and a good qualified separation method has yet to be established.

The characterization of EVs is also challenging without a good standard established yet. The analysis can be either a pooled sample or a single particle analysis. In a pooled sample analysis, the contents of EVs are extracted and analyzed, e.g. proteins by antibody and DNA by PCR. Total particle count can be obtained from light-scattering techniques as the reference for the quantification. Still, small EVs (<100 nm) are difficult to detect or quantify. The pooled sample approach requires reference samples to compare, and the ratio between the population or samples is the practically obtainable result. The distribution of each particle's information is lost in the pooled sample approach.

Single particle analysis is potentially more informative reflecting the character of each EV particle, such as size, surface protein, and genetic material. The possible techniques are microscopic imaging or flow cytometry (for large EVs, usually >200 nm) with fluorescent labeling. For smaller EVs, electron microscopy is a possible option and new cutting-edge techniques are under development. The suggested guidelines for EV characterization are (i) at least one membrane-associated marker as evidence of the lipid bilayer, (ii) at least one cytoplasmic marker to show that the particle is not merely a membrane fragment such as a micelle, and (iii) at least one "negative" or "depleted" marker: a marker of a non-EV particle or a soluble molecule not thought to be enriched in EVs [74].

For optical probe applications, even before the terms in EVs are established, there was an observation of detecting of EV with various fluorescent styryl labeling.

Through fluorescent microscopic imaging and flow cytometry analysis, the vesicle shedding from cancer cells was reported and it was suggested as an efficient drug expulsion mechanism of cancer cells [75]. The light scattering measurement reveals that the vesicles are divided into two size groups of 150 and 800 nm range. Following the modern terminology, the vesicles are called small and large oncosomes.

For the design strategy of small molecule labeling of EV, membrane and lumen staining are the two different approaches. As an example of membrane dye, Cy3 class dye, PKH26 was used for membrane staining for EVs, demonstrating the phagocytic uptake of EVs into cells [76]. The accumulated study with membrane dyes uncovered several general problems including background from unbound dye, aggregate and micelle formation, and nonspecific binding to non-EV [77]. Lipid dyes can be used for labeling parent cell membrane and the EVs can be isolated afterwards. However, it is not clear if the membrane dyes stain all the EVs evenly or affect the formation of EVs. The long half-life of membrane dyes compared to that of EVs makes long-term monitoring of EVs more complicated, as the lipid signals may be recycled and redistributed through membrane fusions, limiting their application to short-term study [78].

Cytosolic staining probes stain EV lumen according to their topological similarity. As an example, calcein-AM, a cytosolic probe, was demonstrated for general labeling of EVs [79]. Another cytosolic probe, CFDA-SE, was used in combination with PKH26 for EV labeling and nanoscale flow cytometry [80]. The probes require the lumen esterase action to convert to fluorescent product and retention in EVs. Therefore, the possibility of false-positive EV staining is low, but may limit the application only to subpopulation of esterase active EVs [81].

So far, EV probes for different characters or components are not available yet. As a relatively young field, even robust or standard probes for EVs are yet to be developed.

3.12 Non-membrane-Bound Condensate

Most of the organelles are compartmentalized by a phospholipid bilayer similar to the plasma membrane. Interestingly, compartments without a membrane structure were discovered as phase-separated biomolecular condensates. The membrane-free condensates exhibit distinct biochemical properties that can absorb and accumulate specific proteins and nucleic acids.

Localized protein synthesis in neuron often requires assembly and transport of translationally silenced ribonucleoprotein particles (RNPs). Some of them are exceptionally large and are an example of the condensate. The large RNP harbors synaptic protein transcripts (mRNA template for protein synthesis) and travels from the nucleus to the synapse at the end of the axon. While the RNP initiates the translation at the synapse area, where the complex is assembled was not clear. RNA probe **E36** (Section 3.5) was employed for the tracking of the large RNP to identify the original location. Through the fluorescence imaging, the granules were observed to exit the nucleus by budding through the inner and the outer nuclear membranes in a

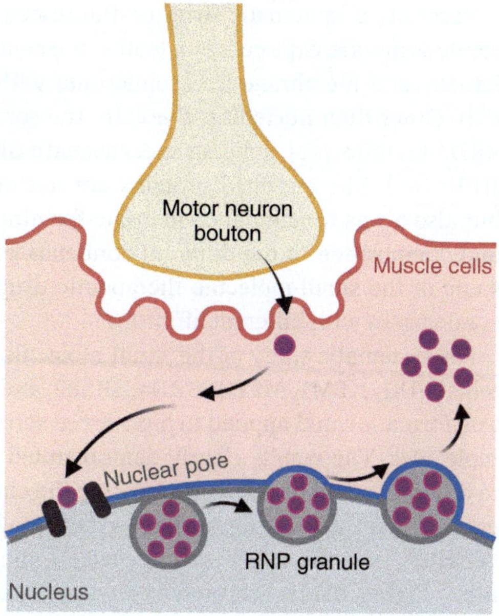

Figure 3.51 Nuclear envelope budding.

nuclear egress mechanism. Therefore, nuclear envelope budding is an endogenous nuclear export pathway for large RNP [33] (Figure 3.51).

Prion-like RNA-binding proteins (RBPs) are largely soluble in the nucleus, but form pathological condensates when mis-localized to the cytoplasm. By employing a red RNA probe **F22** (Section 3.5), the factor to keep the proteins soluble in the nucleus was elucidated as nuclear RNA. In a previous study, it was found that RNA critically regulates the phase behavior of prion-like RBPs. Low RNA/protein ratios promote phase separation into liquid droplets, whereas high ratios prevent droplet formation in vitro. Reduction of nuclear RNA levels or genetic ablation of RNA binding causes excessive phase separation and the formation of cytotoxic solid-like condensates. In summary, the nuclear RNA buffers the phase separation behavior of prion-like RBPs [34].

Stressed cells shut down translation and form stress granules via a network of interactions involving G3BP protein. Under non-stress conditions, G3BP forms a compact state stabilized by electrostatic interactions between the intrinsically disordered acidic tracts and the positively charged arginine-rich region. RNA dye **F22** elucidated the G3BP and RNA interaction. Under the stress condition, the released unfolded mRNAs induce a conformational transition of G3BP through protein–RNA interactions. Subsequent crosslinking of G3BP clusters drives RNA molecules into networked RNA/protein condensates. It was found that G3BP condensates impede RNA entanglement and recruit additional client proteins that promote SG maturation or induce a liquid-to-solid transition that may underlie diseases. Condensation coupled to conformational rearrangements and heterotypic multivalent interactions may be a general principle underlying RNP granule assembly and monitoring RNA entanglement during the protein condensate formation induced by RNA [35].

Recently, a systematic drug or fluorescent probe mapping to various nuclear condensates are explored. Nucleolus, the center for rRNA synthesis, has long been known as a membrane-free condensate with enriched FIB1 and NPM1 proteins [82]. Other than nucleolus, there are transcriptional condensates with MED1 and BRD4 proteins [83], splicing speckles with SRSF2 [84], and heterochromatin with HP1α [85]. The enriched proteins are not only biomarkers for each condensate, but also act as functional condensate-forming scaffold. Although the classification and composition of the different condensates are yet to be systematically studied, some of the small-molecule therapeutic drugs are known to accumulate in these condensates with different affinities.

For systematic study of the small molecule preference, the isolated scaffold proteins (FIB1, NPM1, MED1, BRD4, SRSF2, and HP1α) were used for in vitro condensate formation and applied to high-speed screening with various fluorescent library molecules. The results clearly demonstrated that each condensate has preference to different small molecules, providing the clue for selective probe development for each condensate. If drug molecules have preferred accumulation in a specific condensate where the target protein coexists, the effective active concentration will be much higher, conferring preferred result. To the contrary, if the drug localized condensate is different from that of the target protein, due to the sequestered effect, the drug molecules may not be effectively used for the target, but have high chance of off-target effect [86]. Compared to the canonical high-affinity protein-drug induced by molecular interaction, the concentrates seem to harbor distinct chemical environment to attract the preferred small molecules. Nevertheless, if the molecular fingerprint of the preferred small molecule for each condensate is elucidated, design of the selective probe or drug molecules will be possible.

Currently, the studies are focused on the nuclear condensates, and it is expected the scope is expanded to cytosolic or organelle condensates. Even though the condensates are different from the conventional organelles, which are segregated by physical barrier of membrane, due to their long-term stability and unique functions, they could be considered as pseudo-organelles, providing new scope for the intracellular landscape.

3.13 Organelle Probes in Live Cells and Fixed Cells

Most of the discussion in this chapter for the organelle selectivity was for the live-cell applications. If cells are fixed and permeabilized, the organelle scaffolds may remain, but most of the lipid bilayer components are removed. The common fixing reagent is formaldehyde or glutaraldehyde (5-carbon dialdehyde) which makes crosslinked covalent boding between amine groups of proteins. So, with chemical fixing, the proteins are interconnected by crosslinking reactions, while membrane structures are less influenced. During permeabilization by adding a detergent, most of the phospholipid components of the membrane are washed out, leaving the protein scaffold of the cells in an aqueous environment. The fixing and permeabilization process helps to provide a stable sample condition over relatively

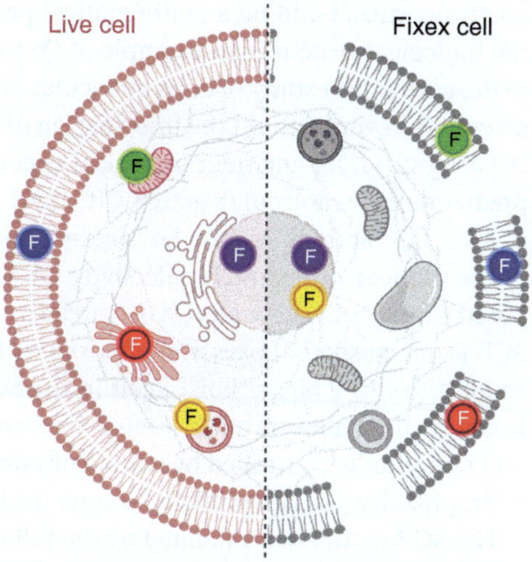

Figure 3.52 Change of the localization of organelle-selective probes from live to fixed cells. F, fluorescence.

long-term handling. Most of the historical cell/tissue staining were performed on the fixed samples. Antibody-based intracellular imaging process also requires the fix and permeabilization, as antibodies cannot penetrate the intact membranes of cell and organelles. As most of the antibody targets are proteins, proper sample preparation could provide quite vivid images for the specific protein localization and assembled structure.

In contrast, if small-molecule organelle selective probes are added to fixed and permeabilized cells, the probe distribution will become dramatically simpler. There is no more physical barrier among organelles and the small-molecule probes can access any area of the cells. Then, the localization of the probes is determined by mainly electric force between probes and organelle components. The obviously dominant charged component of cells would be DNA in the nucleus, so most of the positively charged dyes may be localized in the nucleus. Depending on the permeabilization degree (concentration and treatment time of the detergent), some of the membrane lipids may remain to some extent. Then, highly hydrophobic dyes may stain such leftover membrane traits (Figure 3.52).

3.14 Modeling for the Organelle-Selective Probes

It would be great if a chemist can predict the organelle localization of a dye in live cells. Cells are composed of multiple organelles, such as endosomes, lysosomes, peroxisomes, ERs, Golgi body, mitochondria, nuclei, and nucleoli. Plasma membrane also can be considered as a location. Strictly speaking, by the same analogy, each organelle-specific probe may be further classified into membrane or lumen as the detailed localizations. However, the refraction limit of visible light makes it difficult to distinguish the sub-organellar localization, unless special techniques such as super-resolution microscopy are employed. No entry to the cell or cytoplasmic localization is another category of the probe localization. By looking at the probe structure and properties, building a mathematical prediction model of the probe behavior in the biological system is an example of QSAR study. QSAR has been widely applied to the correlation study of drug molecules and their biological activities. In QSAR of drug study, even without the information of the biological target, by simply utilizing the drug structure and their biological effect, a mathematical model is generated for predicting new molecule's activity. If QSAR is applied to organelle selectivity model of dyes, the "activity" would be "organelle staining property" of a dye.

The pioneer of organelle selectivity QSAR is Richard Horobin. In 1990, he proposed simplest Chinese box (SCB) model as a simplified model of cell structure [87]. SCB is composed of boxes and of boxes within boxes. Organelles are considered as small boxes in a bigger box of cell in the SCB model. Each box is compartmentalized with lipid bilayers and contains different contents. He proposed the probe entry and retention is controlled by simple physicochemical properties of probes, such as hydrophilicity and lipophilicity (Figure 3.53).

The SCB model was validated by the following experiments. Utilizing 41 cationic fluorescent probes and rat fibroblast as the cells, Horobin categorized the dyes into four classes: no entry to the cell, plasma membrane, mitochondria, and other intracellular regions [88]. By simple analysis, the dominating factors were obtained. Hydrophilic dyes do not enter the cells, while extremely hydrophobic dyes stain the cell membrane. Only slightly hydrophobic dyes ($0 < \log P < 5$) stain mitochondria. Mitochondrial selectivity can be induced by membrane potential or by the lipophilic phospholipid cardiolipin of the mitochondrial inner membrane. The latter is independent of the metabolic activity of mitochondria, and may be an important clue for the nonclassical mitochondria probe (membrane potential independent) (Figure 3.54).

Similar approach and analysis have been applied to different organelles with introduction of new physicochemical property parameters for QSAR model for each organelle: lysosomes [89], ER [90], nuclei [91], DNA vs. RNA (chromatin

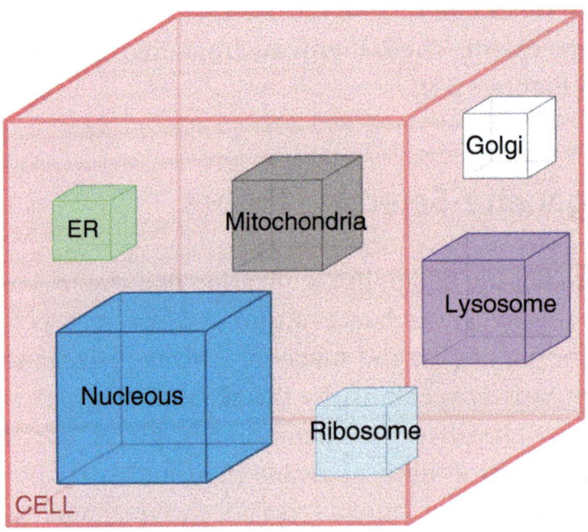

Figure 3.53 SCB model.

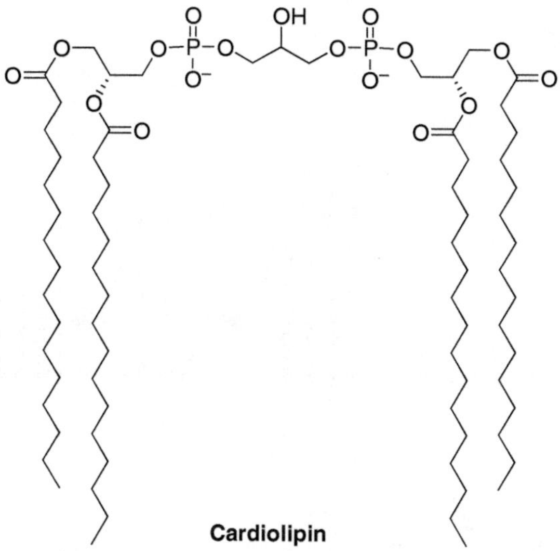

Figure 3.54 Structure of cardiolipin.

vs. nucleolar) [92–94], membranes [95], and multiple organelles [96, 97]. The developed parameters to describe the probes are *Z* (electric charge calculated by pK_a), CBN (conjugated bond number), LCF (the largest conjugated fragment), MW (molecular weight), log*P* (octanol–water partition coefficient), etc. These studies provided clues on what to consider while designing organelle-specific probes in the live cells (Figure 3.55).

When these parameters were applied to 119 BODIPY dyes to construct multi-organelle prediction model by decision tree format, the predictive success rate for each organelle was 92% (LD), 45% (cytosol), and 41% (lysosome). While usable to some extent, the model is not sophisticated enough to give high success rate [98] (Figure 3.56).

Fluorescent styryl dyes can be built by conjugating two building blocks combinatorially. In this case, each dye molecule can be encoded by $Ai + Bj$. Then, a mathematical model is constructed through training, and the model could predict the mitochondrial localization with 72% succeed rate [99] (Figure 3.57).

For a more information-rich approach, modern molecular descriptors are used for the organelle selectivity model. Molecular descriptors are mathematical representations of molecules' properties that are systematically generated by algorithms. More than 300 known organelle-selective probes were collected and categorized into nine organelle locations (nucleus, plasma membrane, Golgi body, ER, cytosol, lysosome, mitochondria, LD, no entry into the cell). Then 786 molecular descriptors are generated and 80% of the training set was subjected to random forest modeling. The resulting prediction model (with an accuracy of 100% for the training set) was applied to 20% of the test set providing 75% accuracy. This is the largest scale organelle selectivity modeling in terms of probe number and organelle subclassification, along with the highest overall prediction accuracy [100].

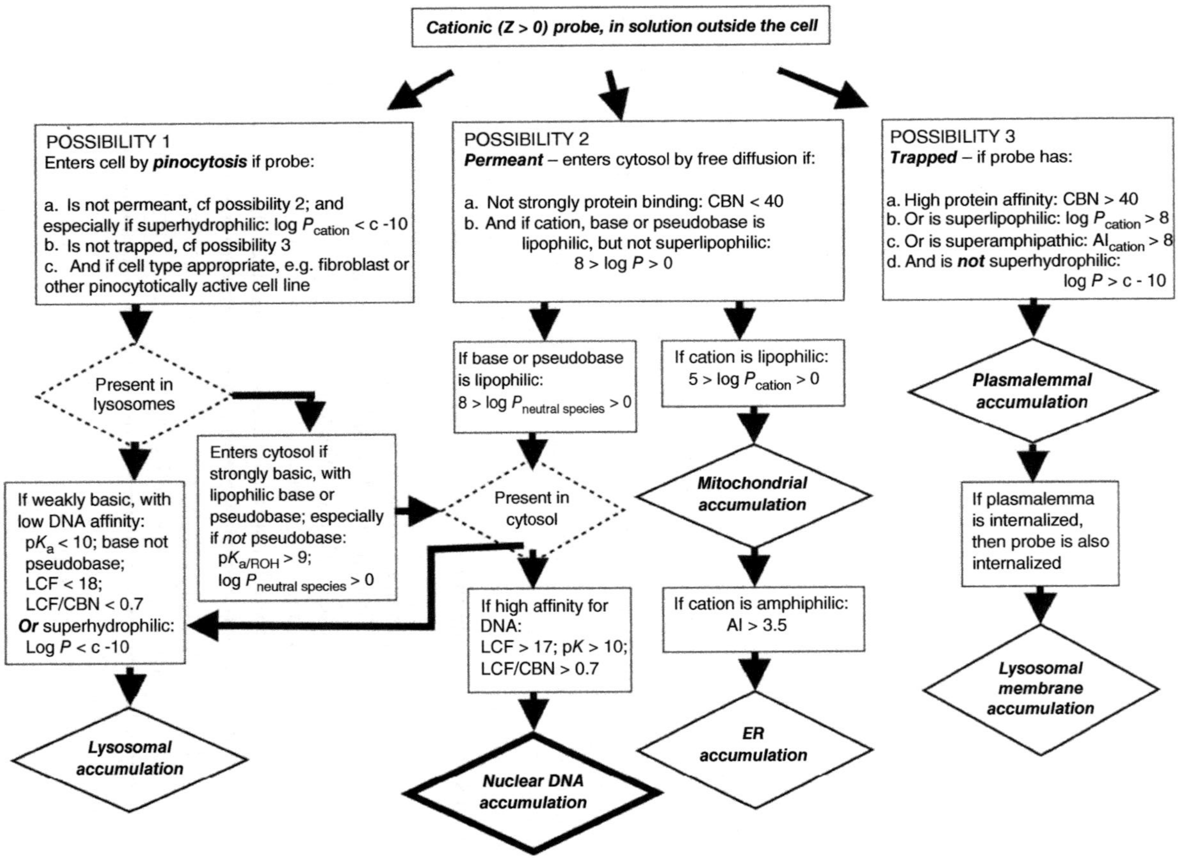

Figure 3.55 Cell picture with summarized parameters. Source: Horobin et al. (2006)/Springer Nature.

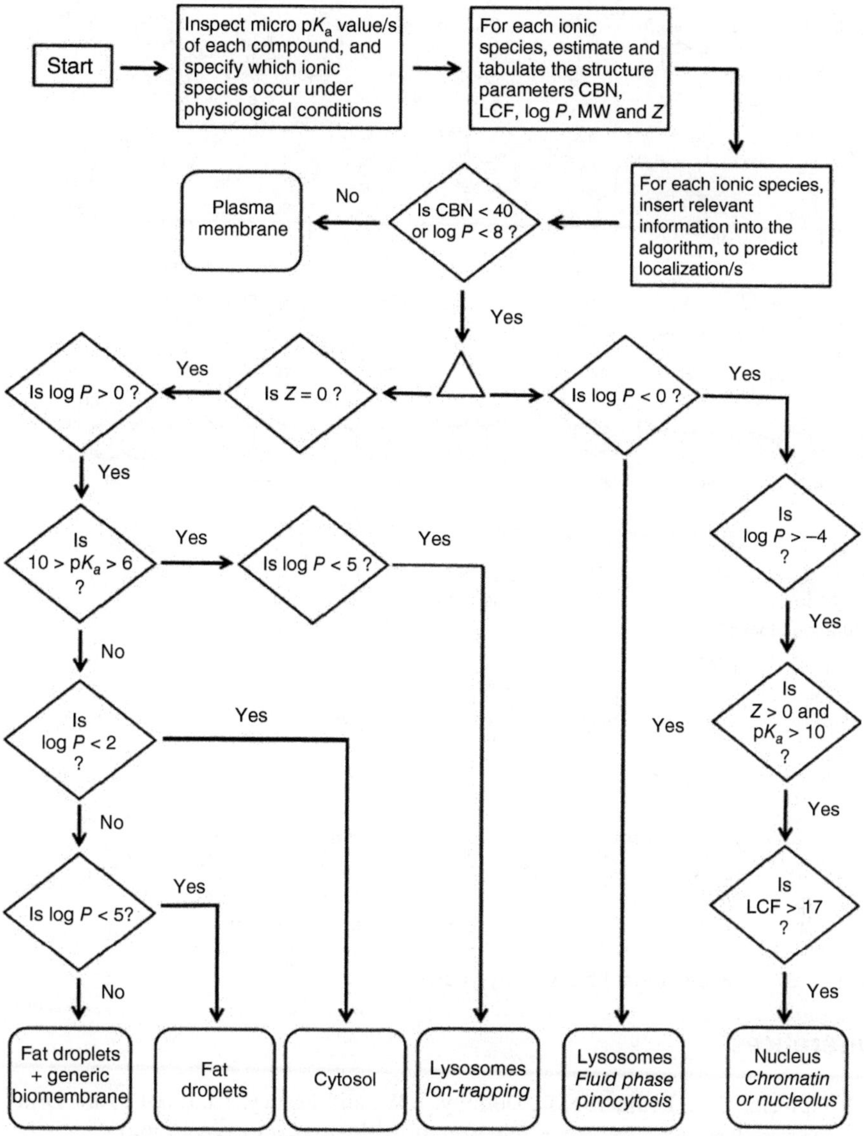

Figure 3.56 Simplified algorithm of QSAR model of organelle prediction. Source: Shohei Uchinomiya et al. (2016)/Bentham Science Publishers.

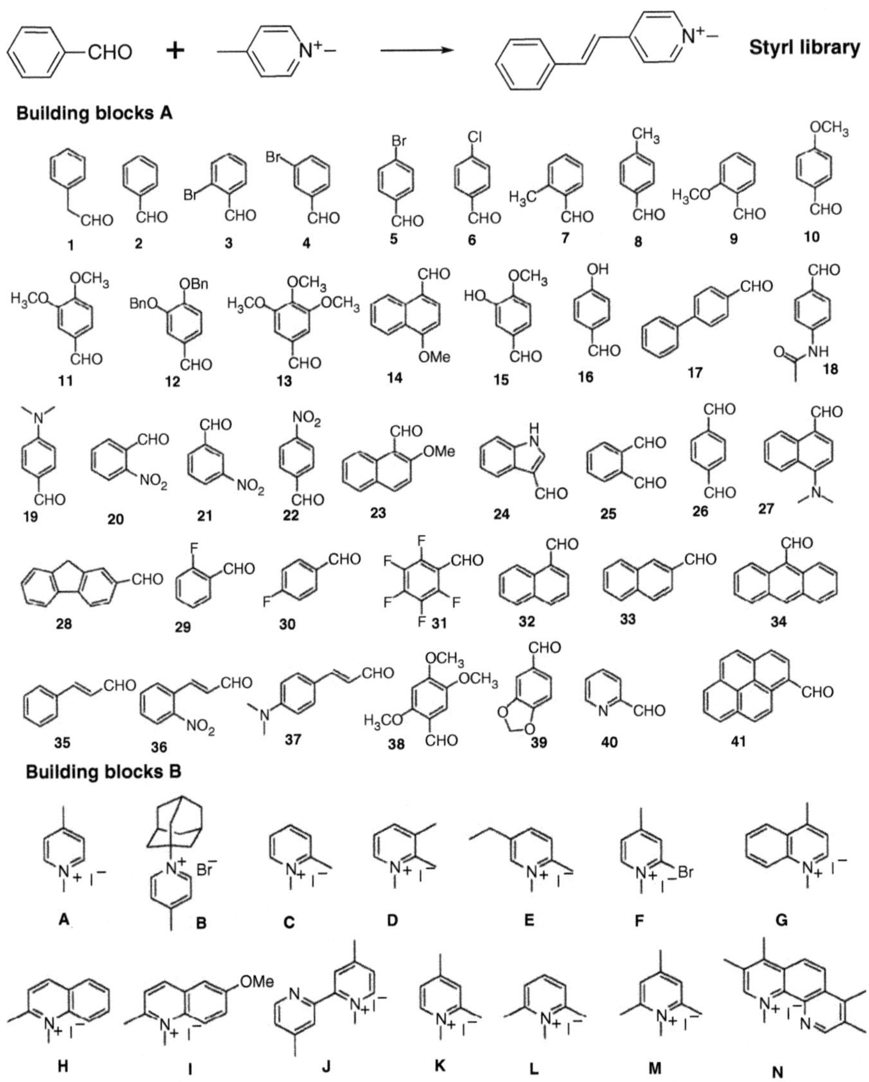

Figure 3.57 Combinatorial library of styryl dyes.

References

1 Lipinski, C.A., Lombardo, F., Dominy, B.W., and Feeney, P.J. (2001). *Adv. Drug Delivery Rev.* 46: 3–26. https://doi.org/10.1016/s0169-409x(00)00129-0.

2 Minton, A.P. (2006). *J. Cell Sci.* 119: 2863–2869. https://doi.org/10.1242/jcs.03063.

3 Schrödinger, E. (1944). *What Is Life? The Physical Aspect of the Living Cell.* Cambridge: Cambridge University Press.

4 Fujimoto, T. and Parmryd, I. (2017). *Front Cell Dev. Biol* 4: 155. https://doi.org/10.3389/fcell.2016.00155.

5 Cheng, C., Trzcinski, O., and Doering, L.C. (2014). *Front. Neuroanat.* 8: 30. https://doi.org/10.3389/fnana.2014.00030.

6 Liu, C., Gao, X., Yuan, J., and Zhang, R. (2020). *Trend Anal. Chem.* 133: 116092. https://doi.org/10.1016/j.trac.2020.116092.

7 Zhang, C., Jin, S., Yang, K. et al. (2014). *ACS Appl. Mater. Interfaces* 6: 8971–8975. https://doi.org/10.1021/am5025897.

8 Collot, M., Kreder, R., Tatarets, A.L. et al. (2015). *Chem. Commun.* 51: 17136–17139. https://doi.org/10.1039/C5CC06094J.

9 Collot, M., Ashokkumar, P., Anton, H. et al. (2019). *Cell Chem. Biol.* 26: 600–614.e7. https://doi.org/10.1016/j.chembiol.2019.01.009.

10 Klymchenko, A.S. and Kreder, R. (2014). *Chem. Biol.* 21: 97–113. https://doi.org/10.1016/j.chembiol.2013.11.009.

11 Sezgin, E., Levental, I., Grzybek, M. et al. (2012). *Biochim. Biophys. Acta* 1818: 1777–1784. https://doi.org/10.1016/j.bbamem.2012.03.007.

12 Jiang, Y.-W., Guo, H.-Y., Chen, Z. et al. (2016). *Langmuir* 32: 6739–6745. https://doi.org/10.1021/acs.langmuir.6b00193.

13 Platt, F.M., d'Azzo, A., Davidson, B.L. et al. (2018). *Nat. Rev. Dis. Primers* 4: 27. https://doi.org/10.1038/s41572-018-0025-4.

14 Wemhöner, A., Frick, M., Dietl, P. et al. (2006). *J. Biomol. Screening* 11: 286–295. https://doi.org/10.1177/1087057105285284.

15 Johnson, I. and Spence, M. (ed.) (2010). *The Molecular Probes Handbook*, 11e. Life Technologies.

16 Lu, S., Sung, T., Lin, N. et al. (2017). *PLoS ONE* 12: e0173771. https://doi.org/10.1371/journal.pone.0173771.

17 Braulke, T. and Bonifacino, J.S. (2009). *Biochim. Biophys. Acta* 1793: 605–614. https://doi.org/10.1016/j.bbamcr.2008.10.016.

18 Yapici, N.B., Bi, Y., Li, P. et al. (2015). *Sci. Rep.* 5: 8576. https://doi.org/10.1038/srep08576.

19 Zhu, H., Fan, J., Xu, Q. et al. (2012). *Chem. Commun.* 48: 11766–11768. https://doi.org/10.1039/C2CC36785H.

20 Piazzolla, F., Mercier, V., Assies, L. et al. (2021). *Angew. Chem. Int. Ed.* 60: 12258–12263. https://doi.org/10.1002/anie.202016105.

21 Miao, F., Uchinomiya, S., Ni, Y. et al. (2016). *ChemPlusChem* 81: 1209–1215. https://doi.org/10.1002/cplu.201600249.

22 Watson, J.D. and Crick, F.H.C. (1953). *Nature* 171: 737–738. https://doi.org/10.1038/171737a0.

23 Dragan, A.I., Pavlovic, R., McGivney, J.B. et al. (2012). *J. Fluoresc.* 22: 1189–1199. https://doi.org/10.1007/s10895-012-1059-8.

24 Lee, J.W., Jung, M., Rosania, G.R., and Chang, Y.-T. (2003). *Chem. Commun.* 1852–1853. https://doi.org/10.1039/B303960A.

25 Smith, P.J., Blunt, N., Wiltshire, M. et al. (2000). *Cytometry* 40: 280–291. https://doi.org/10.1002/1097-0320(20000801)40:4%3C280::aid-cyto4%3E3.0.co;2-7.

26 Feng, S., Kim, Y.K., Yang, S., and Chang, Y.-T. (2010). *Chem. Commun.* 46: 436–438. https://doi.org/10.1039/B916858C.

27 Peng, X., Wu, T., Fan, J. et al. (2011). *Angew. Chem. Int. Ed.* 50: 4180–4183. https://doi.org/10.1002/anie.201007386.

28 Nakamura, A., Takigawa, K., Kurishita, Y. et al. (2014). *Chem. Commun.* 50: 6149–6152. https://doi.org/10.1039/C4CC01753F.

29 Eames, C., Frost, J.M., Barnes, P.R.F. et al. (2015). *Nat. Commun.* 6: 8497. https://doi.org/10.1038/ncomms9497.

30 Ghosh, K.K., Jeong, Y.-M., Kang, N.-Y. et al. (2015). *Chem. Commun.* 51: 9336–9338. https://doi.org/10.1039/C5CC02295A.

31 Li, Q., Kim, Y., Namm, J. et al. (2006). *Chem. Biol.* 13: 615–623. https://doi.org/10.1016/j.chembiol.2006.04.007.

32 Li, Q. and Chang, Y.T. (2006). *Nat. Protoc.* 1: 2922–2932. https://doi.org/10.1038/nprot.2006.484.

33 Speese, S.D., Ashley, J., Jokhi, V. et al. (2012). *Cell* 149: 832–846. https://doi.org/10.1016/j.cell.2012.03.032.

34 Maharana, S., Wang, J., Papadopoulos, D.K. et al. (2018). *Science* 360: 918–921. https://doi.org/10.1126/science.aar7366.

35 Guillén-Boixet, J., Kopach, A., Holehouse, A.S. et al. (2020). *Cell* 181: 346–361.e17. https://doi.org/10.1016/j.cell.2020.03.049.

36 Lu, Y.-J., Deng, Q., Hu, D.-P. et al. (2015). *Chem. Commun.* 51: 15241–15244. https://doi.org/10.1039/C5CC05551B.

37 Sharma, S., Acharya, J., Banjara, M.R. et al. (2020). *BMC Res. Notes* 13: 29. https://doi.org/10.1186/s13104-020-4895-7.

38 Wang, K.-N., Chao, X.-J., Liu, B. et al. (2018). *Chem. Commun.* 54: 2635–2638. https://doi.org/10.1039/C8CC00256H.

39 Baniecki, M.L., Wirth, D.F., and Clardy, J. (2007). *Antimicrob. Agents Chemother.* 51: 716–723. https://doi.org/10.1128/aac.01144-06.

40 Cervantes, S., Prudhomme, J., Carter, D. et al. (2009). *BMC Cell Biol.* 10: 45. https://doi.org/10.1186/1471-2121-10-45.

41 Cervantes, S., Stout, P.E., Prudhomme, J. et al. (2012). *BMC Infect. Dis.* 12: 1. https://doi.org/10.1186/1471-2334-12-1.

42 Mitra, K., Ubarretxena-Belandia, I., Taguchi, T. et al. (2004). *Proc. Natl. Acad. Sci. U.S.A.* 101: 4083–4088. https://doi.org/10.1073/pnas.0307332101.

43 Terasaki, M. (1989). *Methods Cell Biol.* 29: 125–135. https://doi.org/10.1016/S0091-679X(08)60191-0.

44 Kessel, D., Conley, M., Graça, M. et al. (2005). *Photochem. Photobiol.* 81: 569–572. https://doi.org/10.1562/2004-12-16-RA-403.

45 Vila-Carriles, W.H., Zhao, G., and Bryan, J. (2007). *FASEB J.* 21: 18–21. https://doi.org/10.1096/fj.06-6730hyp.

46 McDonald, L., Liu, B., Taraboletti, A. et al. (2016). *J. Mater. Chem. B* 4: 7902–7908. https://doi.org/10.1039/C6TB02456D.

47 Zhang, H., Fan, J., Dong, H. et al. (2013). *J. Mater. Chem. B* 1: 5450–5455. https://doi.org/10.1039/C3TB20646G.

48 Danylchuk, D.I., Jouard, P.-H., and Klymchenko, A.S. (2021). *J. Am. Chem. Soc.* 143: 912–924. https://doi.org/10.1021/jacs.0c10972.

49 Li, R.S., Gao, P.F., Zhang, H.Z. et al. (2017). *Chem. Sci.* 8: 6829–6835. https://doi.org/10.1039/C7SC01316G.

50 Zhang, H., Fan, J., Wang, J. et al. (2013). *J. Am. Chem. Soc.* 135: 11663–11669. https://doi.org/10.1021/ja4056905.

51 Liberman, E.A., Topaly, V.P., Tsofina, L.M. et al. (1969). *Nature* 222: 1076–1078. https://doi.org/10.1038/2221076a0.

52 Santo-Domingo, J. and Demaurex, N. (2012). *J. Gen. Physiol.* 139: 415–423. https://doi.org/10.1085/jgp.201110767.

53 Zielonka, J., Joseph, J., Sikora, A. et al. (2017). *Chem. Rev.* 117: 10043–10120. https://doi.org/10.1021/acs.chemrev.7b00042.

54 Schnitzler, J.G., Bernelot Moens, S.J., Tiessens, F. et al. (2017). *J. Lipid Res.* 58: 2210–2219. https://doi.org/10.1194/jlr.d073197.

55 Koopman, R., Schaart, G., and Hesselink, M.K. (2001). *Histochem. Cell Biol.* 116: 63–68. https://doi.org/10.1007/s004180100297.

56 Greenspan, P., Mayer, E.P., and Fowler, S.D. (1985). *J. Cell Biol.* 100: 965–973. https://doi.org/10.1083/jcb.100.3.965.

57 Spangenburg, E.E., Pratt, S.J.P., Wohlers, L.M., and Lovering, R.M. (2011). *J. Biomed. Biotechnol.* 2011: 598358. https://doi.org/10.1155/2011/598358.

58 Collot, M., Fam, T.K., Ashokkumar, P. et al. (2018). *J. Am. Chem. Soc.* 140: 5401–5411. https://doi.org/10.1021/jacs.7b12817.

59 Wanders, R.J.A., Waterham, H.R., and Ferdinandusse, S. (2016). *Front. Cell Dev. Biol.* 3: 83. https://doi.org/10.3389/fcell.2015.00083.

60 Beckman, K.B. and Ames, B.N. (1998). *Physiol. Rev.* 78: 547–581. https://doi.org/10.1152/physrev.1998.78.2.547.

61 Hettema, E.H., Distel, B., and Tabak, H.F. (1999). *Biochim. Biophys. Acta* 1451: 17–34. https://doi.org/10.1016/s0167-4889(99)00087-7.

62 Dansen, T.B., Pap, E.H.W., Wanders, R.J.A., and Wirtz, K.W.A. (2001). *Histochem. J.* 33: 65–69. https://doi.org/10.1023/A:1017927728892.

63 Zhou, Y., Li, P., Fan, N. et al. (2019). *Chem. Commun.* 55: 6767–6770. https://doi.org/10.1039/C9CC02483B.

64 Zhou, Y., Li, P., Wang, X. et al. (2020). *Chem. Sci.* 11: 12149–12156. https://doi.org/10.1039/D0SC02922J.

65 Zhou, Y., Wu, C., Wang, X. et al. (2021). *Anal. Chem.* 93: 9609–9620. https://doi.org/10.1021/acs.analchem.1c01776.

66 Izumi, S., Urano, Y., Hanaoka, K. et al. (2009). *J. Am. Chem. Soc.* 131: 10189–10200. https://doi.org/10.1021/ja902511p.

67 Veziroglu, E.M. and Mias, G.I. (2020). *Front. Genet.* 11: 700. https://doi.org/10.3389/fgene.2020.00700.

68 Trams, E.G., Lauter, C.J., Salem, N. Jr., and Heine, U. (1981). *Biochim. Biophys. Acta* 645: 63–70. https://doi.org/10.1016/0005-2736(81)90512-5.

69 Rudraprasad, D., Rawat, A., and Joseph, J. (2022). *Exp. Eye Res.* 214: 108892. https://doi.org/10.1016/j.exer.2021.108892.

70 van der Pol, E., Böing, A.N., Harrison, P. et al. (2012). *Pharmacol. Rev.* 64: 676–705. https://doi.org/10.1124/pr.112.005983.

71 Barros, F.M., Carneiro, F., Machado, J.C., and Melo, S.A. (2018). *Front. Immunol.* 9: 730. https://doi.org/10.3389/fimmu.2018.00730.

72 Morello, M., Minciacchi, V., de Candia, P. et al. (2013). *Cell Cycle* 12: 3526–3536. https://doi.org/10.4161/cc.26539.

73 Valadi, H., Ekström, K., Bossios, A. et al. (2007). *Nat. Cell Biol.* 9: 654–659. https://doi.org/10.1038/ncb1596.

74 Théry, C., Witwer, K.W., Aikawa, E. et al. (2018). *J. Extracell. Vesicles* 7: 1535750. https://doi.org/10.1080/20013078.2018.1535750.

75 Shedden, K., Xie, X.T., Chandaroy, P. et al. (2003). *Cancer Res.* 63: 4331–4337. https://doi.org/10.3390/cancers13040749.

76 Feng, D., Zhao, W.-L., Ye, Y.-Y. et al. (2010). *Traffic* 11: 675–687. https://doi.org/10.1111/j.1600-0854.2010.01041.x.

77 Dominkuš, P.P., Stenovec, M., Sitar, S. et al. (2018). *Biochim. Biophys. Acta, Biomembr.* 1860: 1350–1361. https://doi.org/10.1016/j.bbamem.2018.03.013.

78 Chuo, S.T.-Y., Chien, J.C.-Y., and Lai, C.P.-K. (2018). *J. Biomed. Sci.* 25: 91. https://doi.org/10.1186/s12929-018-0494-5.

79 Gray, W.D., Mitchell, A.J., and Searles, C.D. (2015). *MethodsX* 10: 360–367. https://doi.org/10.1016/j.mex.2015.08.002.

80 Morales-Kastresana, A., Telford, B., Musich, T.A. et al. (2017). *Sci. Rep.* 7: 1878. https://doi.org/10.1038/s41598-017-01731-2.

81 Liao, Z., Jaular, L.M., Soueidi, E. et al. (2019). *J. Extracell. Vesicles* 8: 1628592. https://doi.org/10.1080/20013078.2019.1628592.

82 Feric, M., Vaidya, N., Harmon, T.S. et al. (2016). *Cell* 165: 1686–1697. https://doi.org/10.1016/j.cell.2016.04.047.

83 Sabari, B.R., Dall'agnese, A., Boija, A. et al. (2018). *Science* 361: eaar3958. https://doi.org/10.1126/science.aar3958.

84 Guo, Y.E., Manteiga, J.C., Henninger, J.E. et al. (2019). *Nature* 572: 543–548. https://doi.org/10.1038/s41586-019-1464-0.

85 Larson, A.G., Elnatan, D., Keenen, M.M. et al. (2017). *Nature* 547: 236–240. https://doi.org/10.1038/nature22822.

86 Klein, I.A., Boija, A., Afeyan, L.K. et al. (2020). *Science* 368: 1386–1392. https://doi.org/10.1126/science.aaz4427.

87 Horobin, R.W. and Rashid, F. (1990). *Histochemistry* 94: 205–209. https://doi.org/10.1007/bf02440189.

88 Rashid, F. and Horobin, R.W. (1990). *Histochemistry* 94: 303–308. https://doi.org/10.1007/bf00266632.

89 Rashid, F., Horobin, R.W., and Williams, M.A. (1991). *Histochem. J.* 23: 450–459. https://doi.org/10.1007/BF01041375.

90 Colston, J., Horobin, R.W., Rashid-Doubell, F. et al. (2003). *Biotech. Histochem.* 78: 323–332. https://doi.org/10.1080/10520290310001646659.

91 Horobin, R.W., Stockert, J.C., and Rashid-Doubell, F. (2006). *Histochem. Cell Biol.* 126: 165–175. https://doi.org/10.1007/s00418-006-0156-7.

92 Horobin, R.W., Stockert, J.C., and Rashid-Doubell, F. (2013). *Histochem. Cell Biol.* 139: 623–637. https://doi.org/10.1007/s00418-013-1090-0.

93 Horobin, R.W., Stockert, J.C., and Rashid-Doubell, F. (2015). *Biotech. Histochem.* 90: 241–254. https://doi.org/10.3109/10520295.2015.1005129.

94 Horobin, R.W., Stockert, J.C., and Rashid-Doubell, F. (2015). *Biotech. Histochem.* 90: 255–263. https://doi.org/10.3109/10520295.2015.1006680.

95 Horobin, R.W. and Stockert, J.C. (2011). *Biotech. Histochem.* 86: 379–393. https://doi.org/10.3109/10520295.2010.515489.

96 Horobin, R.W., Rashid-Doubell, F., Pediani, J.D., and Milligan, G. (2013). *Biotech. Histochem.* 88: 440–460. https://doi.org/10.3109/10520295.2013.780634.

97 Horobin, R.W. and Rashid-Doubell, F. (2013). *Biotech. Histochem.* 88: 461–476. https://doi.org/10.3109/10520295.2013.780635.

98 Uchinomiya, S., Horobin, R.W., Alvarado-Martínez, E. et al. (2016). *Comb. Chem. High Throughput Screening* 19: 378–383. https://doi.org/10.2174/1386207319666160408150528.

99 Shedden, K., Brumer, J., Chang, Y.T., and Rosania, G.R. (2003). *J. Chem. Inf. Comput. Sci.* 43: 2068–2080. https://doi.org/10.1021/ci0341215.

100 Park, S.-H., Lee, H.-G., Liu, X., Lee, S.K., and Chang, Y.-T. (2023). *Quantitative Structure-Activity Relationship of Fluorescent Probes and Their Intracellular Localizations. Chemosensors.* 11: 310. https://doi.org/10.3390/chemosensors11050310.

4

Live-Cell-Selective Probes

The first form of cell appeared on Earth most likely emerged from seawater. Cell needs to keep the identity of self, and the first step would be separation between "self" and environment. Water and oils do not mix. Therefore, it would have been a good choice for the cell to segregate the "self" by a lipid bilayer membrane from the aqueous environment. The cells accept nutrients and excrete wastes, maintaining their identity. The cells may compete each other and larger cells may have swallowed small ones, increasing their size. Similar to a soap bubble, when the size of cells becomes bigger and bigger, it would get more and more difficult to keep the size of the cell by a thin membrane. Once they reach the critical size, they may divide to reduce the structural instability. At this point, the meaning of growth may have been changed from size growth to number growth. To keep the identity of each cell, genetic material could be introduced, and the regulation of growth and division would have become sophisticated. In a good environment with rich resources, they could proliferate fast, competing with other cells. The basic materials for cells are "organic molecules," i.e. molecules of life. On the Earth, all life forms are carbon based, and the organic molecules are composed of carbon and hydrogen as the backbone. Cells may have swallowed other cells and absorbed the highly condensed organic molecules as nutrients. Some cells may have developed a cell wall outside of the plasma membrane to increase the physical stability and also as a defense against other competing cells.

Some of the cells acquired special skills to obtain energy from sunlight and synthesize organic molecules by themselves, called photosynthesis. These cells make glucose using carbon dioxide as the carbon source and sunlight as the energy. As a result, the cells no longer need to eat other competing cells for obtaining organic molecules. This ability of self-sustaining photosynthesis would be desirable to all the cells, but at the same time, the sophisticated photosynthetic system is expensive to develop and maintain. Thus, some cells chose the strategy of inviting the photosynthetic cells as a partner of living together (symbiosis). In the beginning, the host cells may have swallowed the photosynthetic cells and majority of them were digested. Some of the swallowed photosynthetic cells survived and secured their own living space in host cells. They may have provided the photosynthetic product, glucose, to the host cells. As a return, they may have relied on the host cell's biochemical machineries for other functions, focusing on their own photosynthetic

Sensors and Probes for Bioimaging, First Edition. Young-Tae Chang and Nam-Young Kang.

role. These photosynthetic cells may have evolved into an organelle in host cells, called chloroplast.

In the ancient Earth atmosphere, oxygen concentration was low. Without oxygen, the digestion and energy extraction from glucose was not efficient. The intermediate of glycolysis, pyruvate or lactate, still contains high energy, though not fully utilized. In the meantime, the photosynthesis made a big change in air composition by generating oxygen as a byproduct. With the increase in photosynthetic organisms, i.e. plants, the oxygen concentration in the air increased dramatically, reaching up to 35% in the Permian period. Current oxygen concentration in the air is about 21%. Oxygen attracts electrons from other materials (reduction), generating highly reactive oxygen species (ROS) intermediate that are potentially harmful to the cells. With the increased oxygen concentration, some cells acquired special skill of using oxygen to efficiently burn the nutrient molecules to maximally extract out the energy. The high efficiency of the cell metabolism is a highly desirable trait for the cell's survival in competition with other cells. These specialized cells are also adopted into other cells and have evolved into a specialized organelle, mitochondrion. The full oxidation of organic molecules performed by mitochondria utilizing oxygen is called cellular respiration. This is another big example of symbiosis with a similar example of chloroplasts. The mitochondrial respiration and chloroplast photosynthesis share similar molecular mechanisms, with opposite direction. And they share common molecular machineries in general. Though they started as independent lives and merged with other cells later, chloroplasts and mitochondria still keep the traits of original life, such as their own genetic information and special double-layer membrane structures.

Cell-level fusion or symbiosis was one strategy, and other cells chose cell colonies by contacting each other for better survival in difficult situations. The early form may have been a loose cell colony with a common protecting structure, such as a biofilm of bacteria. More advanced form is tightly interconnected organization by taking part of different roles in the complex. For example, as the cell number increases in the complex, the inner location cells may have limited resources such as oxygen and nutrients compared to the outer layer cells. Instead, inner cells may benefit by the better protection from the shock given to the outer layer. If the overall benefits of the complex are greater than the sum of individual cells, they may evolve into a permanent symbiotic system by making an exchange system of information and material among the cells. The simplest life form of the multicellular symbiosis would be sponges, which have only two types of cells. Figure 4.1 shows our skin cells and inner intestinal cells (Figure 4.1).

The multicellular organisms developed into many different structures and shapes, e.g. the nematode worm, *Caenorhabditis elegans* has about a thousand of well-defined cells. For humans, it was estimated an adult has about 10 trillion cells in a 1977 paper [1]. Interestingly, in the same paper, the parasitic or symbiotic bacteria in our body was estimated 100 trillion, 10 times of our own cells! Do we have more bacteria than human cells? If we assume our cell has an average 10 μm size, and bacteria has 1 μm, the volume or weight ratio would be 1000 to 1. Still, total bacteria weight would be 1% of our body. An adult of 60 kg weight may have

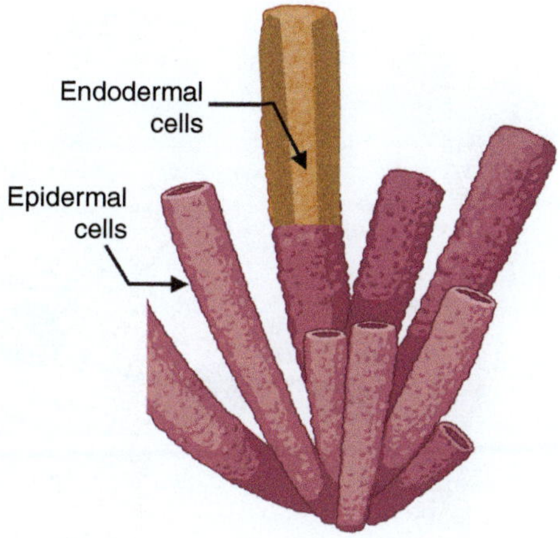

Figure 4.1 Structure of sponge.

600 g of bacteria, which look quite a high number, even considering the microbiome in our gastrointestinal tract. In an improved measurement in 2016, human cells were estimated as 30 trillion with red blood cells (RBCs) as the most abundant cells as of 70%. RBCs are relatively small cells and their total weight is only 2.5% of the whole body [2]. In this chapter, the bacteria number was estimated as 38 trillion, so the ratio of human and bacteria is about 1 : 1.

When two bacteria interact, to properly understand their action and response, identification of each bacterium would be the very basic first step for observation. If the dynamic interaction is the goal to observe, the identification method should be achieved in real time in live bacteria. The situation is similar in multicellular organisms. To understand the organized activity of a cell complex or tissue, identification of the target cells without interfering with their biological function is necessary. That is why a live-cell-selective probe is so important.

Conventionally, the standard method for cell discrimination is using a cell surface maker, CD (cluster of differentiation or designation) and its antibody. When cells differentiate from stem cells, various plasma membrane proteins are differentially expressed depending on the differentiation pathways. The differential proteins were identified and proposed for the first time to use to distinguish different immune cells in Human Leukocyte Differentiation Antigen Workshops in 1982 [3]. Starting 15 CDs from the first meeting, the number of CD increased to 371 in the latest 10th meeting in 2014 [4]. The CD system was originally designed for immune cells, but currently it is used for all human cells. For example, CD4 is glycoprotein that is expressed in T helper cells. Therefore, fluorescently labeled CD4 antibody or anti-CD4 can recognize and visualize T helper cells. In this case, T helper cell is defined as $CD4^+$ cells. In a similar way, CD8 is a glycoprotein selectively expressed in T killer cells. So, $CD8^+$ cells mean T killer cells. Both CD4 and CD8 are co-receptors for T-cell receptor. CD4 binds to major histocompatibility complex (MHC) Class II and CD8 binds to MHC Class I. MHCs are antigen-presenting proteins. In T cell maturation, the early T cells are $CD4^-CD8^-$ (DN: double negative) status, and change into $CD4^+CD8^+$cells. Then, through further selection, $CD4^-CD8^+$ and $CD4^+CD8^-$ are formed as matured cells. As seen in the example, defining one cell

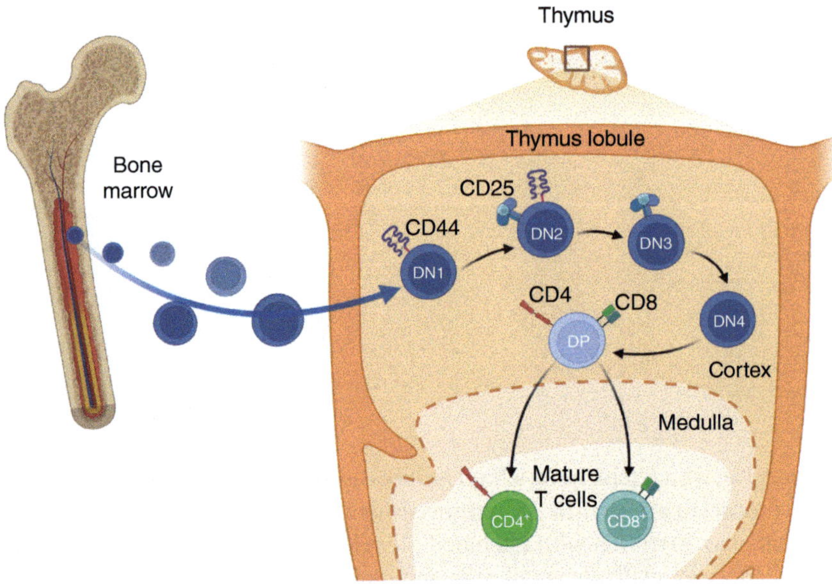

Figure 4.2 T-cell maturation.

type with a single CD is difficult, and combination of multiple CDs with + or − status may make a call for the cell identity (Figure 4.2).

As shown in the CD4 and CD8 examples, most CDs are proteins or glycoproteins with rare cases of carbohydrates or lipids. Also, most CDs are functional receptors or enzymes, so binding CD with its selective antibody may block or stimulate the function. Depending on the cases, the antibody treatment is significantly toxic and difficult to use in live cells. Therefore, it is important to monitor and check if the cell status is normal or has changed after antibody treatment.

4.1 Protein-Oriented Live-Cell Distinction (POLD)

If there is a cell-selective small-molecule probe, it is most common to assume there is a binding target in the cell. This reminds us the "magic bullet" of Paul Ehrlich. Magic bullet is an ideal drug molecule, which acts by binding only to the target protein, without bothering with any other proteins in the body. Considering the huge numbers (tens of thousands) of proteins in our body, magic bullet would be a dream drug with no side effect at all [5]. By the same token, an ideal probe should bind specifically to a single protein, only existing in the target cell. Such a protein is called as a "biomarker." If protein X is expressed only in cancer, then protein X can be a biomarker of cancer. Therefore, a specific probe for protein X would be an ideal imaging agent for cancer! Then, how could we find such a biomarker and design a specific probe for it? That is the question.

If there is already a known target protein and its magic bullet drug molecule, one could make a selective probe by simply attaching a fluorescent moiety to the

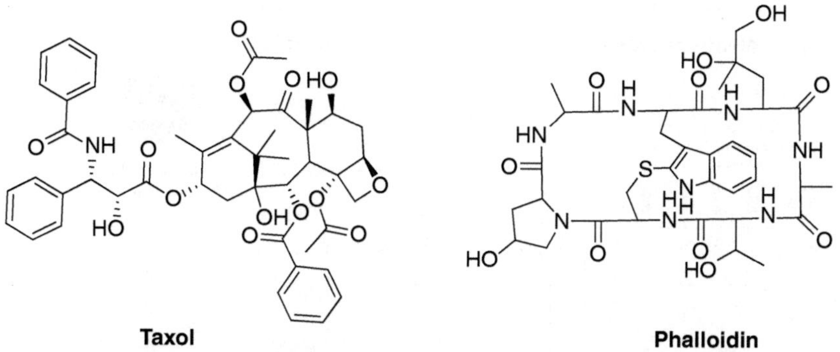

Figure 4.3 Structure of taxol & phalloidin.

drug molecule. For example, famous anticancer drug taxol binds to microtubule composed of tubulins, so taxol can be modified by attaching a green fluorescent molecule, fluorescein, to yield tubulin probe. In a similar way, actin-binding molecule phalloidin can be labeled with a fluorescent molecule to provide actin probe. If two different color fluorescent molecules are used, the cocktail probes may even show microtubule and actin structure at the same time (Figure 4.3).

It looks straightforward. However, there are problems. First, following Lipinski's rule of 5, the molecular weight of a drug molecule is preferred to be less than 500 for the entry into the live cells. These molecules are commonly called as "small molecules" or "drug-like molecules." Although taxol and phalloidin are biologically active molecules, which imply their efficient entry into the live cells, the conjugated molecules with a fluorescent moiety will have a molecular weight easily higher than 1000. The conjugated molecules are not small molecules, anymore. As a result, there is a high possibility that the conjugated probe molecules may not be able to enter the live cells, limiting their application only to fixed and permeabilized cells. Then, the situation is similar to antibody probes which target the intracellular proteins. Secondly, even though the conjugated molecules indeed enter the live cell, there is another possibility that the localization of the probe is driven by the fluorescent molecule rather than the original drug molecule, depending on the relative size of each molecule and their physical and chemical properties. Thirdly, if the target proteins are universal and exist in most of cells, it is difficult to expect the cell selectivity.

Therefore, for the cell-selective probe development, it is essential to discover a cell-specific biomarker first. For the protein biomarker discovery, proteomic analysis using two-dimensional (2D) gel or 2D high-performance liquid chromatography (HPLC) could be an option. Usually, 2D gel utilizes the intact proteins and 2D HPLC uses protease-digested peptide fragment for the analysis. Human genome was estimated to contain 30 000–35 000 genes from the first published papers in 2001 [6], but the updated paper in 2003 has further lowered the estimate to 20 000–25 000 genes [7]. When the genes are expressed as proteins, various post-translational modifications (glycosylation, phosphorylation, lipidation, ubiquitination, acetylation, methylation, etc. [8]) occur and result in presumably more than millions of different proteins depending on the cell types and conditions. The 2D gel

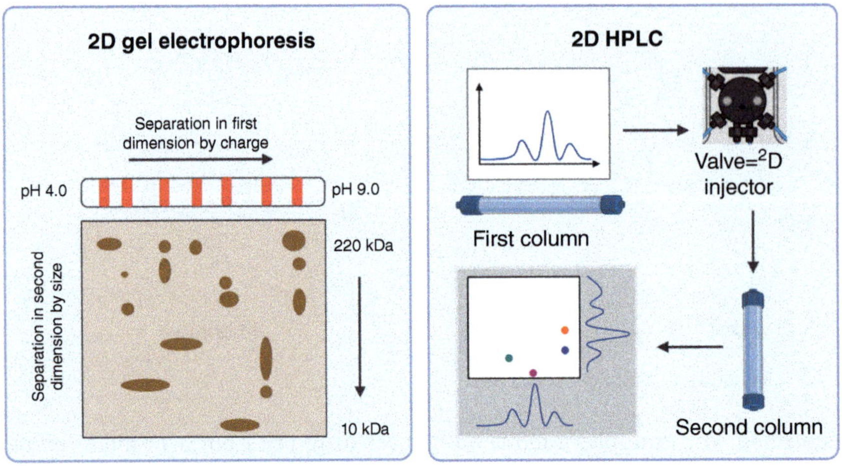

Figure 4.4 2D gel and 2D HPLC methods.

method usually uses two different protein separation principles: sodium dodecyl sulfate–polyacrylamide gel electrophoresis (SDS-PAGE) for molecular-weight-based separation and isoelectric focusing for pK_a-based separation. The distinguishable protein numbers from 2D gel may be hundreds to thousands at best. Still, comparing the proteome (total collection of all the proteins of a cell) of two cells and identifying one cell-specific protein are extremely challenging (Figure 4.4). Even the same proteome may generate frame-shifted protein patterns by batch-to-batch variation.

One trick to make the protein comparison easy is 2D **Di**fferential **G**el **E**lectrophoresis (DIGE). One proteome was labeled with green fluorescence and the other with red fluorescence, and the two proteomes were run in one gel to remove the batch variation. The labeling is usually through amine modification by succinimidyl ester or thiol modification by maleimide of the proteins. The majority of the proteins may be common for essential cellular function and overlap to show yellow (green + red). The outstanding green or red proteins could be the candidates for one-cell-specific biomarkers. They could be either unique proteins in one cell or unique variation of post-translational modification in the given status of a cell. The protein spots could be cut and analyzed by mass spectrometry to identify the protein (Figure 4.5).

2D HPLC connected with on-line mass spectrometry for high-throughput protein/peptide analysis is called **Mu**lti**d**imensional **P**rotein **I**dentification **T**echnology (MudPIT). To mimic the 2D gel separation, two HPLC columns are serially connected. Most common design is combining Strong Cation eXchange (SCX) as the first column and Reverse Phase of C18 (RP-18) as the second column serially. The two columns have different solvent power properties. In the SCX column, different ionic strengths aqueous solvents are used, and the higher the ionic strength, the more the bound proteins are released. In RP-18, water–organic solvent mixture is commonly used, and the higher the organic solvent concentration, the more the releasing of the sample. Therefore, SCX and RP-18 have opposite solvent powers.

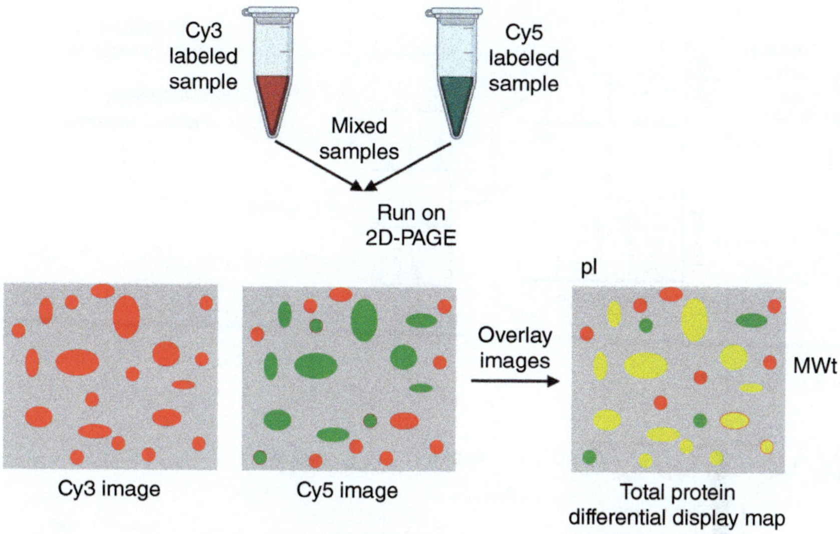

Figure 4.5 2D DIGE (through succinimidyl ester and maleimide reaction).

First, all the protein/peptide samples are trapped in the first SCX column. Then, a portion of the protein is released by a low ionic strength solvent and the released proteins enter into the RP-18 column. Then, the solvent is changed into water–organic solvent gradient system to isolate the proteins in RP-18 columns. During the procedure, all other proteins in SCX are stuck and wait for the next solvent wave. The second round starts with a higher ionic strength solvent to release the second portion of the proteins from the SCX column and then continues to the PR separation with water–organic solvent gradient system. This procedure is repeated with an increasing ionic strength of the solvent for each round. Usually, 5–10 steps of ion strength solvents are used; water–organic solvent gradient is same for each round for PR-18 separation. Therefore, SCX separation is stepwise, and RP-18 separation is rather continuous. The mosaic assembly of each chromatogram could mimic the 2D gel separation. MudPIT is easy to automate and has a much higher number of proteins that could be identified compared with the 2D gel approach (Figure 4.6).

An alternative approach of proteomics for biomarker discovery is transcriptomics using mRNA expression analysis. mRNA is the intermediate template for protein synthesis and transcriptome is the total mRNA expressed in a cell. mRNA expression change is correlated with the amount of protein synthesis, though the correlation is not always linear [9]. The quantitative analysis of mRNA has been achieved by DNA microarray technology, after an amplification through reverse transcription polymerase chain reaction (RT-PCR) and in vitro transcription of mRNA. The throughput of transcriptomics is much higher than proteomics, and overall procedure is also much simpler for transcriptomics (Figure 4.7).

Gene sequencing is another emerging technique for quantitative DNA or RNA analysis. Especially, automated gene sequencing coupled with massive clonal amplification, called next-generation sequencing (NGS), accelerated the gene analysis

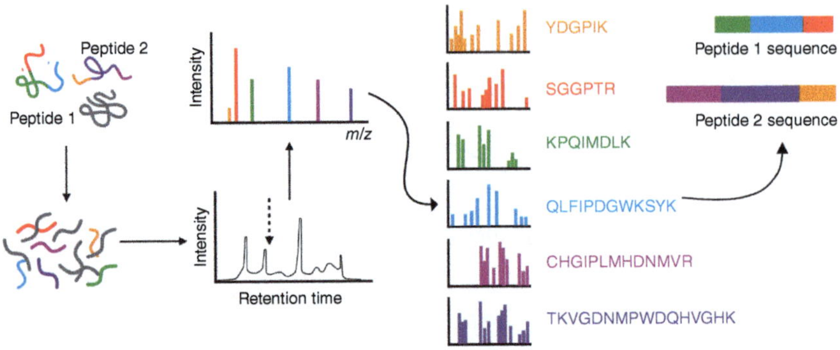

Figure 4.6 MudPIT.

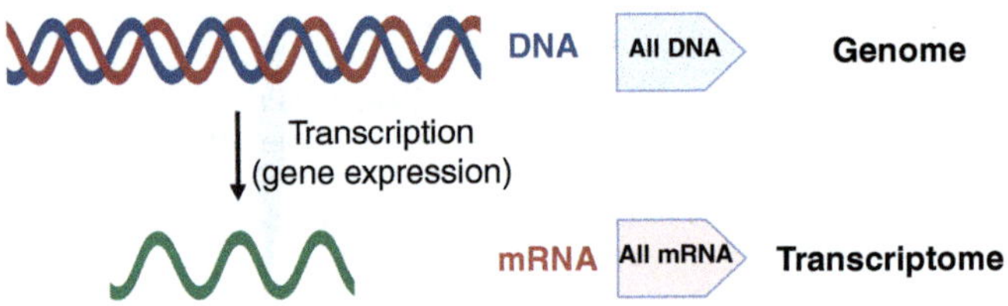

Figure 4.7 Transcriptomics.

dramatically, enabling quantitative deep sequencing and metagenome sequencing [10]. Even without an amplification step, a direct single-chain sequencing for DNA or RNA is another revolutionary technique introduced recently [11].

If the elucidated biomarker is a cell membrane protein, specific antibody could be raised and utilized in line with the CD system. But, if the biomarker is an intracellular protein, the antibodies could not be used for the live-cell imaging. A drug-like small molecule is an ideal option in that case. With the fixed-target protein, small-molecule probes could be either designed through a bottom-up approach of molecular modeling or screened in silico using a virtual candidate molecular library. In either case, they are based on the known structure of the biomarker protein. If the protein could be produced and purified, in vitro screening with candidate probe molecules is another possible option. Once the promising candidate molecule is identified, through protein–probe co-crystallization, the binding mode is confirmed, and further optimization is also possible. The final optimized small molecule needs to be labeled with a reporter motif such as a fluorescent moiety, and then the abovementioned problem with taxol or phalloidin, for example, appears again. In this case, if the candidate molecules are already labeled with a fluorescent moiety or are fluorogenic themselves, the extra step of labeling and its potential problem could be removed from the beginning. Diversity-oriented fluorescence library approach (DOFLA) is introduced for this purpose. In this case, DOFL is similar to a fluorescently labeled antibody set, but can be used for intracellular target protein (Figure 4.8).

Even with DOFL in hand, sometimes a specific biomarker is not available or known to the target cell. Under this situation, still it is possible to use the target

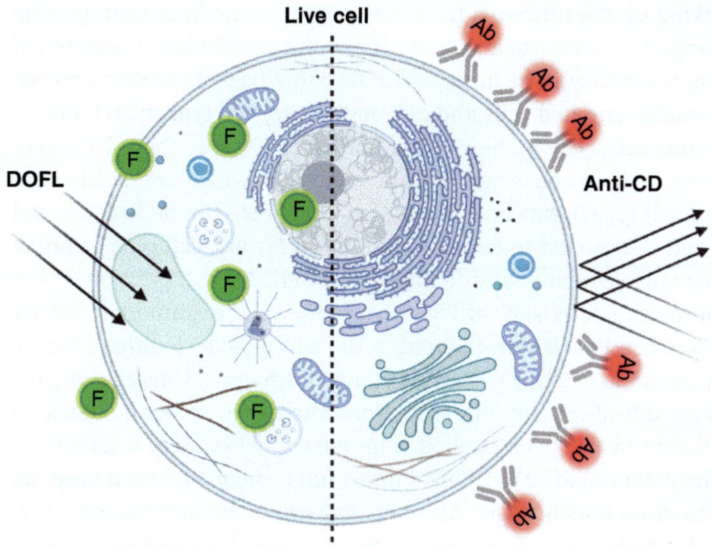

Figure 4.8 DOFL and anti-CD.

cell itself with negative control cells for the screening system, called phenotypic screening. The same amount of DOFL can be treated to the cells with various conditions and time points, and high-throughput or high-content imaging-based screening could be performed to identify the target-cell-selective probe candidate. As the phenotypic screening is a kind of blind screening performed without any hypothesis for the selectivity, there remains a burden to elucidate the selectivity mechanism afterwards. At the same time, there is a good chance to encounter a joy of learning totally unexpected new mechanism and novel biomarkers. In this chapter, examples of protein-binding targets will be introduced.

4.1.1 Embryonic Stem Cell Probe: CDy1

All the cells in our body originate from a single cell, a fertilized egg cell! When a lucky sperm fuses with the egg, two nuclei fuse together, and the genetic information from the father and the mother work together to start the development of a new life. In humans, the full genetic information is stored in 23 chapters (called chromosome), and sperm and egg carry each one set (haploid: n) and a fertilized cell has one pair of them (diploid: $2n$). One gene is conventionally defined as the information for making one protein, and the matching two genes are allele to each other. They can be exactly same, or slightly different in sequence, resulting in two slightly different proteins. The inherited trait is defined by the balance of the two allele products.

The fertilized egg cell divides into hundreds of cells and differentiates into an inner cell mass and placenta structure. Placenta is the connecting tissue between an embryo and a mother, and interestingly it is not originated from the mother's tissue, but from the embryo. The embryo and the mother are in a competing situation for sharing common resources, and the placenta will be on the embryo side for maximally sucking up the nutrients from the mother, sometimes even putting the mother in a dangerous state, such as gestational diabetes (induced high blood glucose state during pregnancy). The inner cell mass contains embryonic stem cells (ESC), which are undifferentiated cells and can make every cell type in an embryo, called pluripotent stem cells (PSC)! The ESC are known to last only about 10 days of embryo development, and are converted to adult stem cells, which are multipotent (limited number of cell types), but not pluripotent cells. PSC can be isolated and cultured in vitro, and is expected to be potentially used for any cell type or organ regeneration for the replacement of any damaged part of body.

However, one of the characteristics of PSC is self-renewal, i.e. immortal, similar to tumors. Actually, when PSC is transplanted in the adulty body, teratoma can be formed. Although teratoma is a benign tumor, it may form eyes or internal organs on the skin, inducing out-of-control differentiation. Therefore, proper guidance of stem cell differentiation in the given context is important. Also, there is an ethical issue as PSC has the potential to be developed into a full human. Should it keep the human right? If yes, from which stage? Another problem is the immune rejection possibility, when allogeneic stem cell (originated from another human being) is used for treatment. Autologous or patient-specific stem cells could be an option, but it requires the key step of PSC generation from adult cells. This is similar to converting old cells into young cells, which has been believed to be impossible for a long time, especially in mammals.

The first breakthrough came from the first cloned sheep, Dolly, in 1996. The nucleus of the somatic cell (which has $2n$ DNA) was transferred to the egg cell after removal of the nucleus (which had $1n$ DNA), mimicking fertilization process or rejuvenating the somatic cell. The egg cell with $2n$ DNA was implanted in a surrogate mother sheep, and it started cell division and grew to give birth to Dolly. Dolly has the same genetic information with her mother who provided the nucleus of somatic cell. Technically speaking, it is closer to a "twin elder sister" rather than a mother, though. This was the first demonstration that mammal embryo could be generated without sperm. While it was a shocking event as a new way of baby generation, the early aging and unexpected lung disease of Dolly cast concerns that the cloned animal is not a perfectly new baby, carrying the history of twin sister's life.

In 2004, a similar technique of nuclear transplantation of somatic cell had been applied to human egg, and ESC was isolated and cultured [12]. The next year, the same research group reported multiple cases of patient-specific stem cell generation [13], casting the hope that regeneration medicine is nearby. After the first paper was published, however, there was an accusation of illegal egg cell trading issue for the research. More seriously, the papers turned out to be fake without clear evidence of successful stem cell generation. It was a serious scientific scandal, and the nuclear transplantation approach had been discarded at least for human stem cell research afterwards.

In 2006, a new breakthrough hit the world. Without using either egg or sperm cell, Yamanaka group reported a method of changing somatic cells into PSC only by four genetic factors [14]. This artificial PSC was called as **i**nduced **P**luripotent

Stem cells (iPS). The approach was simple and even seemed an absurd dream. They studied which genes are highly activated in the stem cell stage compared to somatic cells. They selected 24 candidates and observed that when they transfected all 24 genes into somatic cells, they indeed changed their character into stem-like cells. Encouraged by the results, by omitting one by one, they carefully tested which genes are essential for the activity, and finally narrowed down to the so-called 4 Yamanaka magic factors. All the cells and life forms were thought to change only in one direction: from young to old, no way back, period! However, this report reverted the long-believed common sense forever! The 4 Yamanaka factors (Oct4, Sox2, Klf4, c-Myc) turned out to be either transcription factors or growth factors which affect the epigenetic modification of the genome. Whatever they are, only are genes can change the old cells into young, even youngest cell in the embryo! The change of genome structure into younger cells is called retro-differentiation or reprogramming. This should be a revolutionary discovery in modern biology, and even better the observation was reproduced by many research groups in due course. It really works and Nobel prize in Physiology or Medicine was given to Yamanaka in 2012!

While exciting, there are some technical problems. The efficiency of reprogramming was usually low, though variation of Yamanaka factors was continuously developed for higher efficiency and better quality of the stem cells. Retro viral vector was used in the beginning for high efficiency (about 0.01% to 0.1%; 1 out of 1000 or 10000 cells converts into a stem cell), and it inserted the 4 factor genes into the genome of the host cells. Then, the generated stem cells were called genetically modified (GM) cells. GM products are not commonly welcomed even for food, so GM cells for clinical application may raise lots of concerns. To avoid the genome changes, non-retroviruses such as adeno virus or liposome vectors were introduced, but usually the efficiency was not as good as the retrovirus method. Rather than gene transfection, purified proteins were also tried and proved to be active for the reprogramming [15]. Eventual goal may be utilizing small molecules to replace all the biological materials for the reprogramming. While not perfect yet, the promise of the drug cocktail method is emerging [16] with room for improvement in the efficiency and safety of stem cell generation.

With the advent of various methods of PSC generation, the need for selective fluorescent probe for good-quality stem cell was also rising. Without a small-molecule probe, green fluorescent protein (GFP)-based genetic reporter conjugated to Oct4 or Sox2 was commonly used, but the cells were already GM products. While useful, the GM cells were difficult to be used for clinical samples, such as patient-specific stem cell generation. Detecting the activated Oct4 or Sox2 as the marker of stem cells was possible with antibodies, but it is only for autopsy, i.e. detection by killing the cells, not for the live cells.

In 2010, the first small-molecule fluorescent probe for ESC was developed with the name **C**ompound of **D**esignation **y**ellow **1** (CDy1), following the notation of a cell surface maker, **C**luster of **D**esignation or **D**ifferentiation (CD) [17]. To distinguish the probe name with CD system, the fluorescence color code (yellow in CDy1) was added (Figure 4.9).

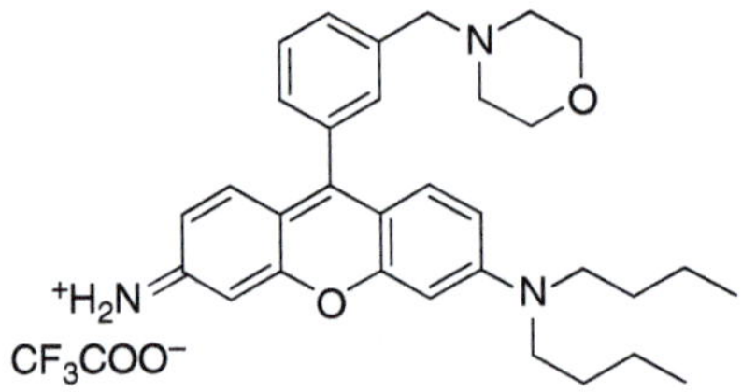

CDy1

Figure 4.9 The structure of CDy1.

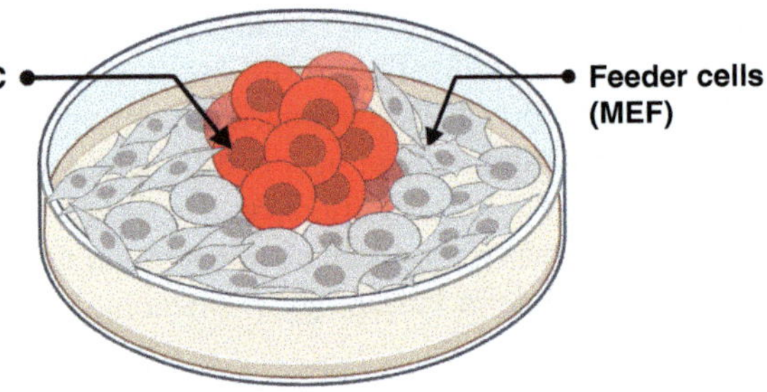

Co-culture of mESC with MEF

Figure 4.10 Imaging of CDy1 in mESC and MEF co-culture.

CDy1 was discovered from unbiased high-throughput phenotypic screening of mouse Embryonic Stem Cell (mESC) and mouse embryonic fibroblast (MEF) through automated fluorescence microscope. CDy1 stains mESC and its colony strongly, with minimal background in MEF (Figure 4.10).

Without CDy1, mESC and MEF are not separable in flow cytometry, and only with CDy1, the two cell populations are clearly separated. Not only analysis, fluorescence activated cell sorter technique could also physically separate mESC and the isolated mESC cells were still healthy and functional. This nontoxic property of CDy1 is important for further biological experiments of ESC in live status.

Once mESC differentiates into three germ layers (endoderm, mesoderm, and ectoderm), the CDy1 staining disappeared showing its selectivity only to mESC. Therefore, CDy1 can be used for monitoring the process of stem cell differentiation in real time. The selectivity of CDy1 to ESC is not only for mouse cells, but also for human ESC (hESC)[18]. CDy1 stains not only mESC, but also iPS generated by reprogramming of somatic cells. In mouse, the reprogramming of fibroblasts into iPS with Yamanaka factors takes about two weeks. If the efficiency is 0.1%, one out of one thousand cells are properly reprogrammed and the reporter genes usually start to show up at 12 days or later. Then, before 12 days, there is no way to monitor which cells are undergoing reprogramming process. In the first report, CDy1 showed several days earlier detection of reprogramming cells than the reporter gene. The ability of CDy1 to recognize stem cells was applied to the rapid chemical effect evaluation on cellular reprogramming. Compared to genetic reporter, the CDy1-based screening was robust and cost effective [19] (Figure 4.11).

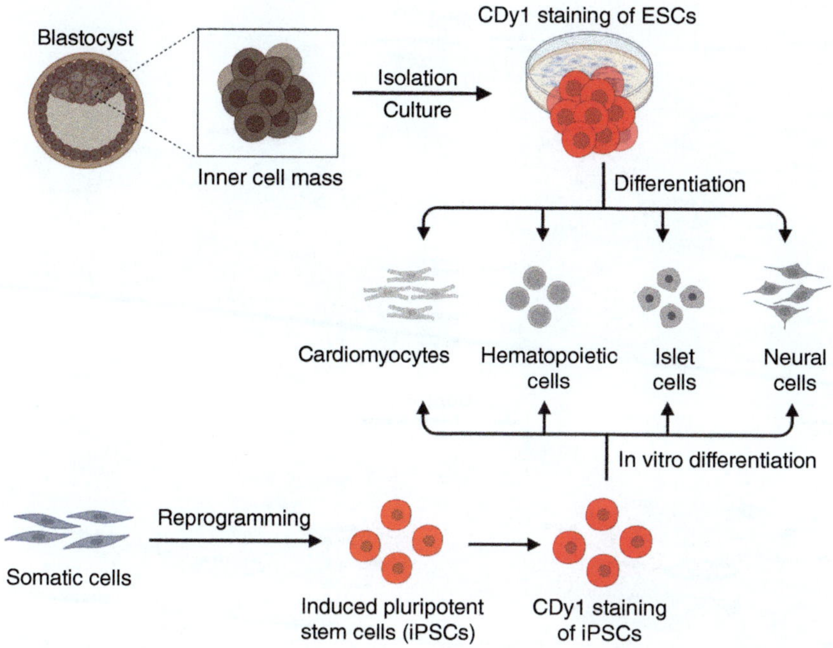

Figure 4.11 Selective staining of ESC during the differentiation and reprogramming.

A long-term monitoring of reprogramming revealed that there are CDy1-positive cells occurring as early as three days of Yamanaka factor treatment. The early reprogramming cells from 3, 5, 7, and 9 days were collected and the gene expression was analyzed systematically. The global gene cluster analysis elucidated platelet-derived growth factor-BB (PDGF-BB) as one of the essential switch genes in the reprogramming [20] (Figure 4.12).

ES or iPS are promising for regeneration therapy, but teratoma formation is one of the critical problems for safe clinical application. CDy1 was investigated for the generation of ROS and demonstrated to induce selective death of stem cells upon visible light irradiation. Importantly, the CDy1 and/or light irradiation did not negatively affect differentiated endothelial cells. The photo dynamic therapy (PDT) of ES/iPS with CDy1 and visible light irradiation blocked the teratoma formation in mice, suggesting a promising new approach for safe cell therapy [21] (Figure 4.13).

In separate studies, in addition to ES, CDy1 was proven to be useful to enrich mouse neural stem cell from murine brain, expanding the scope of CDy1 into other adult stem cell selectivity [22]. CDy1 also showed selectivity for mouse spermatogonial stem cells, both in germline stem cell culture and mature adult testes [23]. These studies cast the possible common aspects of various stem cells along the differentiation pathways (Figure 4.14).

Despite the high potential and wide usage, the target of CDy1 has not been identified over 10 years since its first discovery. The binding target protein of CDy1 was finally elucidated by covalent bond forming derivative of CDy1. The derivative (CDy1CA) contains the chloroacetyl (CA) motif which can form a covalent bond

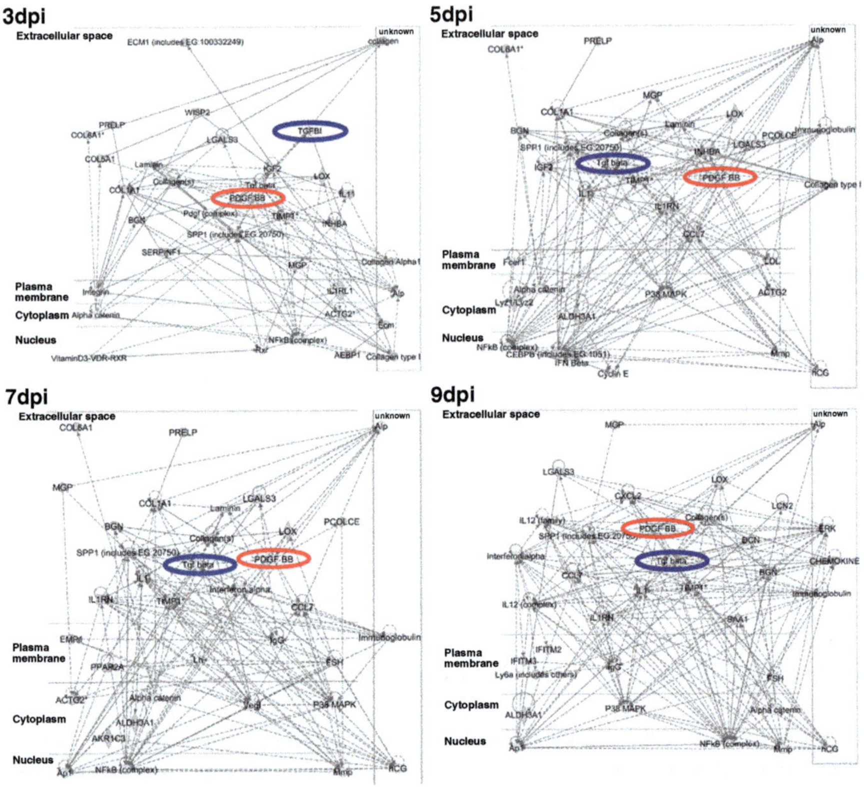

Figure 4.12 Global gene network for reprogramming. Source: Sung-JinPark et al. [20]/With permission of Elsevier.

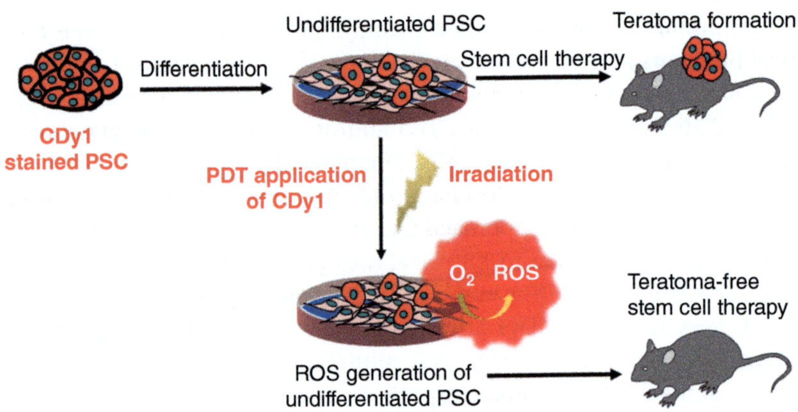

Figure 4.13 PDT application of CDy1.

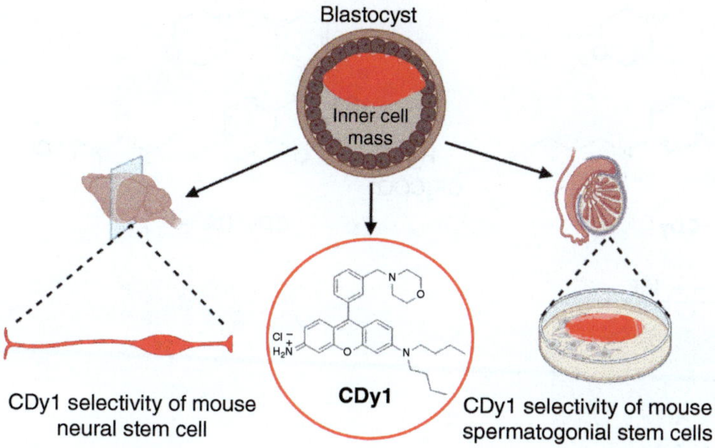

Figure 4.14 Scope of CDy1 in stem cell selectivity.

with the thiol group of the cysteine side chain that may exist nearby the binding site of the target protein. CDy1CA itself is cell permeable and a mESC-selective probe, so the covalent binding could be induced in situ inside the live mESC. The fluorescent band analysis in SDS gel provided the target protein as mitochondrial ALDH2, which is matching to the localization of CDy1 in ESC. ALDH2 is also highly expressed in ESC compared to feeder cells (MEF), and KD (Knock Down) of ALDH2 decreased the staining of CDy1 in ESC, validating ALDH2 as the relevant binding target of CDy1. ALDH2 is a detoxification enzyme to break down reactive aldehyde and may play an important role in ESC. Through a systematic transporter screening using ABC CRISPRa library, ABCB1 emerged as another player in CDy1 selectivity. The details are discussed in Section 4.4.10 [24] (Figure 4.15).

In 2014, a shocking paper was published in *Nature*. According to this paper, iPS could be generated by a mild acidic stimulus and such a new iPS was named as stimulus-triggered acquisition of pluripotency (STAP) cells [25]. If proven, this simple method will revolutionize the iPS field again and facilitate the clinical application of stem cell dramatically. Unfortunately, this report was fabricated and ended as another big scandal in the stem cell field and the paper was retracted. Still, the efforts to make high-quality iPS in a stable procedure are on the way.

4.1.2 Neural Stem Cell Probes

4.1.2.1 CDr3

Embryonic stem cells differentiate into three germ layers: endoderm, mesoderm, and ectoderm. Then, each layer generates multipotent stem cells. For example, mesoderm generates mesenchymal stem cells and hematopoietic stem cells. Mesenchymal stem cells can differentiate into osteoblast, myoblast, adipocyte, chondrocyte, and fibroblast. Hematopoietic stem cells make all kinds of blood cells, i.e. erythrocytes (RBCs), lymphocytes, monocytes (precursor of macrophage), and granulocytes. Multipotent stem cells possess the ability of differentiating

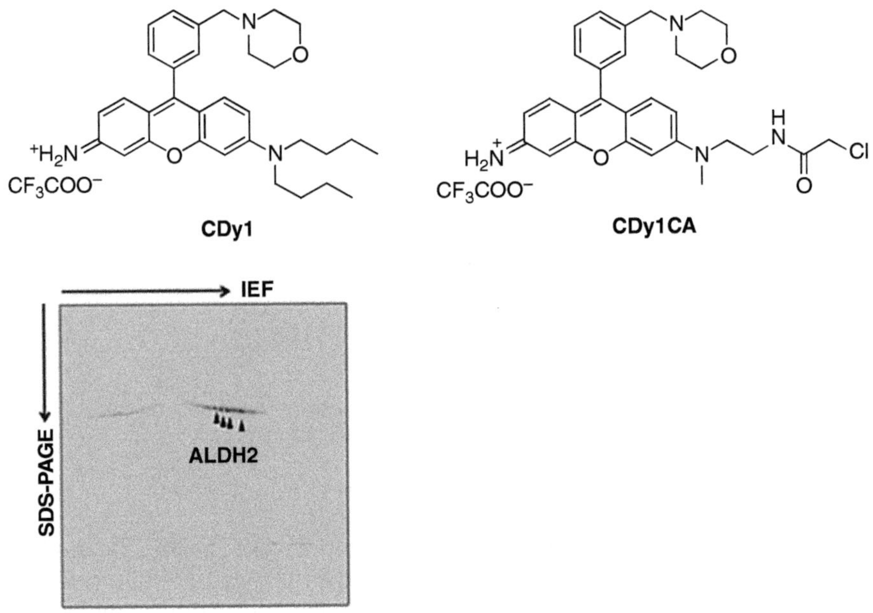

Figure 4.15 Structure of CDy1CA and SDS gel image of bound target. Source: Ref. [24], from Royal Society of Chemistry.

into limited number of different types of cells in the given lineage. Unlike ESCs, multipotent stem cells survive for a long period up to later stages of lifespan, so they are also called adult stem cells. While ESCs are relatively easy to grow under in vitro conditions such as on a coated plate or on the feeder cells (usually fibroblasts), adult stem cells are more difficult to culture and maintain. ESCs are the main population of cells in the inner spear of blastocyst, so the collection of cells is rather straightforward. Adult stem cells are quite rare and scattered in adult tissue, so it is difficult to detect and collect them. Furthermore, they are believed to reside in a specially protected environment, called niche area. Their growth was controlled in the resting period, and only when a serious damage happens in the tissue, adult stem cells are activated and differentiate to replenish the wound area. Therefore, without constructing a proper niche environment, one may easily expect the difficulties of harvesting and maintaining adult stem cells (Figure 4.16).

One of the rare adult stem cells is neural stem cell, which is originated from the ectoderm. Neural stem cells differentiate to either neurons or neural glia cells, such as oligodendrocytes and astrocytes. It has been thought that our neural system development is finished within several years after birth. The basic structure of neuronal connection is basically completed, so neural damage is believed to be irreversible with minimal renewal activity. The discovery of neural stem cells cast a hope of regeneration of damaged neural systems and attracted high attention due to the therapeutic potential. Similar to other adult stem cells, neural stem cells are also difficult to enrich and culture in vitro. Even worse, the definition of neural stem cell is also vague compared to ESC.

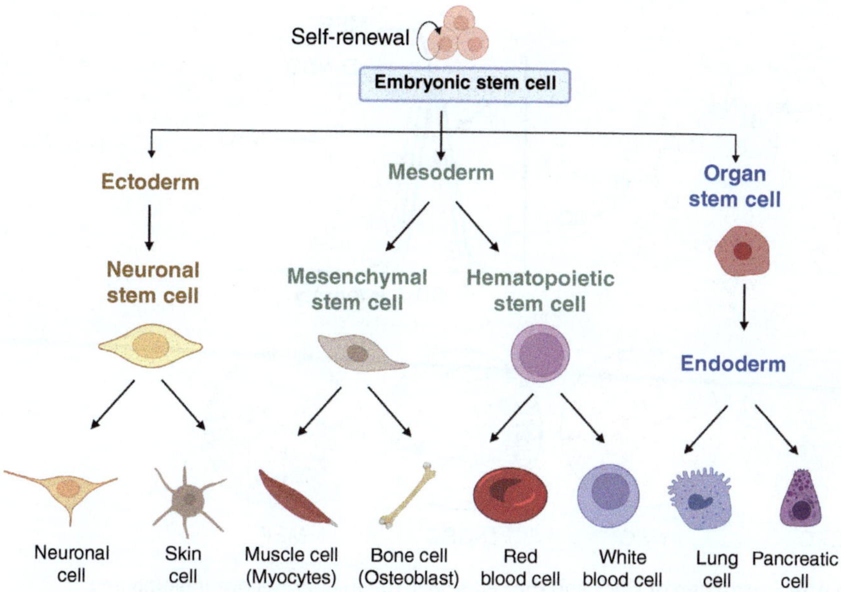

Figure 4.16 ESC differentiation into three germ layers and adult stem cells.

The first developed neural stem-cell-selective probe was CDr3 (**C**ompound of **D**esignation **r**ed **3**). Similar to CDy1 discovery, CDr3 was also elucidated by phenotypic cell screening. In this case, neural stem cells (NS5 mouse cell line) were the positive control, and ESC, astrocytes (differentiated from NS5), and fibroblasts as negative controls. CDr3 stained neural stem cells much stronger than any other control cells, and the selectivity was also confirmed in human neural stem cells (Figure 4.17).

The mechanism study for CDr3 through 2D gel of neural stem cell extract and fluorescence imaging elucidated FABP7 (Fatty-Acid-Binding Protein 7) as the binding target of CDr3. The complex of CDr3–FABP7 was stable enough to survive hot SDS treatment for the 2D gel, giving a bright single spot of fluorescence. This is quite a unique event as CDr3 does not have any covalent bond forming moiety, and the strong binding may be due to the hydrophobic property of CDr3 and also the inner space of FABP7 (Figure 4.18).

FABP7 is highly expressed in brain tissue, especially in embryonic or fetal brain, implying a connection to neural development. FABP7 is also the binding protein of docosahexaenoic acid (DHA), which is an important omega-3 fatty acid for developing fetus and newborn. While FABP7 is a reasonable target of CDr3 in the context, further molecular target validation was performed. First, the knockdown of FABP7 of neural stem cell discarded CDr3 staining and also induced further differentiation, losing the neural stemness. This implies FABP7 is important for the maintenance of neural stemness. For further target validation, CDr3-negative HEK-293 (Human embryonic kidney 293) cells were transfected with FABP7 and the CDr3 staining was examined. Both mouse and human FABP7 expression fully acquired the selective

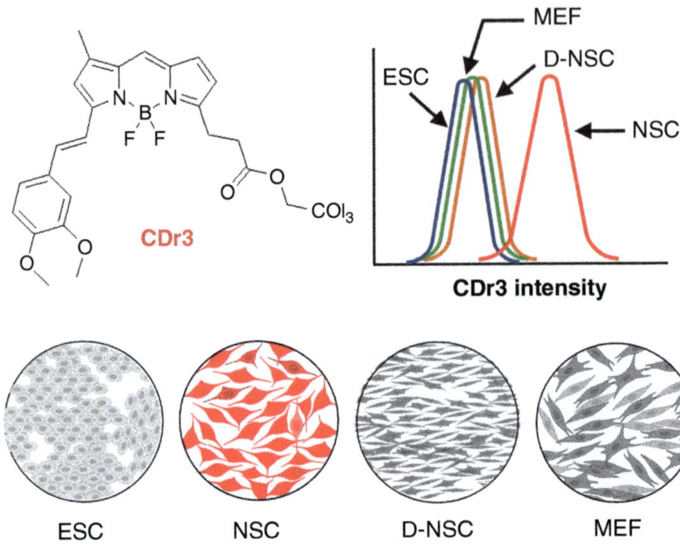

Figure 4.17 Structure of CDr3 and the cell selectivity in fluorescence imaging and flow cytometry.

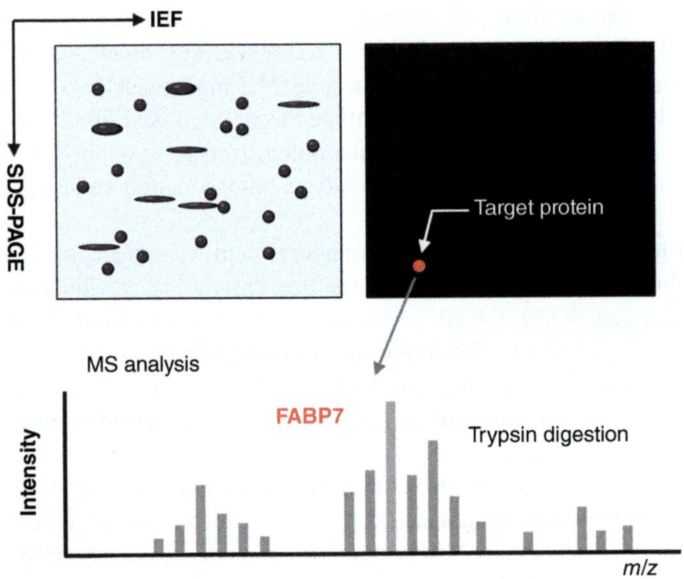

Figure 4.18 2D gel image of CDr3-bound FABP7.

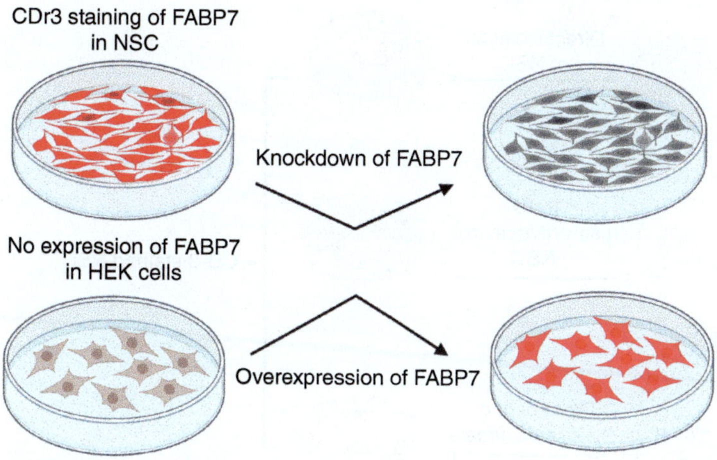

Figure 4.19 Target validation of CDr3: FABP7.

staining of CDr3 in HEK-293 cells, confirming the molecular mechanism of CDr3 binding to FABP7 [26] (Figure 4.19).

In mouse fetal brain, with the high expression of CD133 or SSEA-1, an antibody-based isolation of neural stem cells was possible. Interestingly, CD133 and SSEA-1 are general stem cell markers (first found from ESC) and they disappear after birth. Therefore, mouse adult brain could not be used for antibody-based neural stem cell isolation. Aldefluor, a fluorescent neural stem probe that targets ALDH1, is also working only for the fetal brain, but not for the adult brain [27]. CDr3 works in both the fetal and adult brain, providing universal applicability of functional neural stem cell isolation. FABP7 may be a more specific biomarker than CD133 or SSEA-1 for the neuronal stem cells. It is noteworthy that FABP7 is an intracellular protein, and anti-FABP7 cannot be used for live-cell detection or isolation [28]. In later studies, CDr3 was also successfully used for the monitoring of trans-differentiation of mesenchymal stem cell (MSC) into neural stem cell (NSC) [29] and enhanced NSC detection from human gingival cells treated with *Gynura divaricata* (alternative medicine herb for type 2 diabetic treatment) treated [30] (Figure 4.20).

4.1.2.2 CDy5 for Neural Stem Cell Division Monitoring

CDy1CA, a covalent-binding version of CDy1, was tested in different stem cells and showed a high selectivity in NSC. A new name of CDy5 (**C**ompound of **D**esignation **y**ellow **5**) was dubbed to the molecule and systematically studied for the probe scope. Unlike CDr3, CDy5 contains a CA moiety, which can form a covalent bond with the thiol group. If CDy5 binds to the target protein in the cell, there is a chance that CA reacts with the cysteine group nearby the binding site in the target protein. Once reacted, the target protein will be labeled with a fluorescent motif, and it would be easy to detect it on SDS gel. CDy5 was the lucky case, and the labeled protein was identified as acid ceramidase as the binding target of CDy5. It is noteworthy that

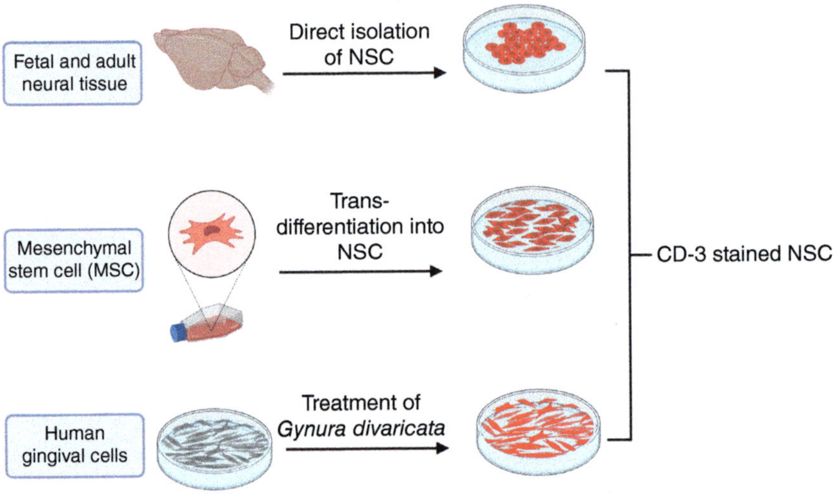

Figure 4.20 Application of CDr3 for neural stem cell detection and isolation.

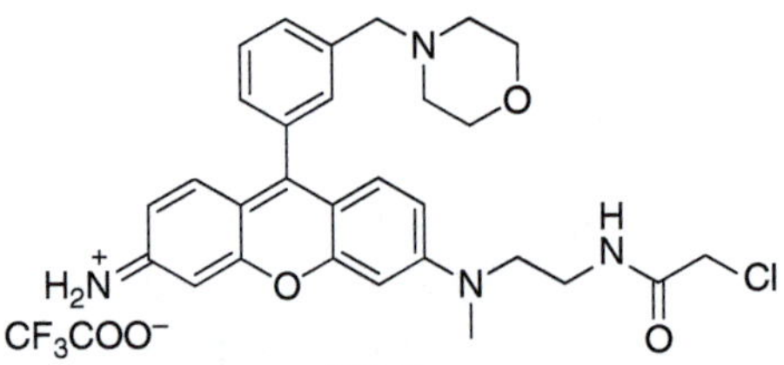

CDy5

Figure 4.21 Structure of CDy5.

CDy5 elucidated ALDH2 as the target protein from ESC, implying the main target difference between the two cell types. With the covalent-binding property, CDy5 was expected to stain NSC stably and thus a long-term monitoring of cell division was attempted (Figure 4.21).

Stem cells may have two different modes of division. The first is symmetric division to maintain the self-renewal activity of stemness, and the other is asymmetric division to generate one stem cell and one differentiated cell. When CDy5 was applied to neurosphere, the staining pattern showed heterogeneous mixed populations of NSC and differentiated cells. The long-term monitoring of the cells clearly showed both symmetric and asymmetric divisions. In symmetric division, same size of two daughter cells was evenly stained by CDy5 and divided. In asymmetric division, one small stained stem cell and the other bigger unstained cell were divided into two cells. The result clearly showed that the destiny of the two cells is determined right before or at the cell division event. Such a fluorescent film was reported previously in amphibian cells using genetic reporters, but this was the first example of mammalian stem cells for the distinguishable symmetric/asymmetric cell division [31]. Compared to the extra efforts required for the preparation of genetically modified cells or animals, the convenience of using CDy5 for the application demonstrates the advantage of small-molecule probes (Figure 4.22).

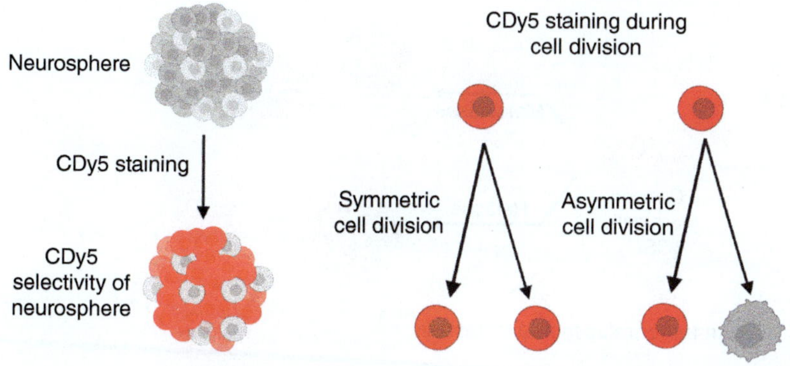

Figure 4.22 CDy5 staining of neurosphere and symmetric/asymmetric cell division.

4.1.3 Tumor-Initiating Cell Probes

4.1.3.1 TiY

To keep the harmony of our body, it is essential for the cells to communicate to each other and regulate the proper growth rate by not allowing one type of cell or tissue to overgrow. When the control is lost, and one kind of cells exhibit uncontrolled growth, we call such an abnormal tissue as a tumor. If the increased tumor still keeps itself within its boundary (called basal lamina or basement membrane), it is a benign tumor. Once the tumor invades other tissue's boundary, breaking up the whole body's harmony, it becomes a malignant tumor or cancer. Usually, the cancer development proceeds with accumulation of multiple mutations, generating heterogeneity of the cancer cells. The unbalanced overgrowth of tumor results in the deficiency of oxygen and nutrients especially in the inner side area. The low oxygen condition (hypoxia) increases ROS by incomplete oxidation–reduction reaction and induces necrosis. To overcome the shortage of supply, cancer cells recruits new blood vessels into the tumor tissue by secreting growth factors. Still, compared to the fast-growing outside of the tumor, the cell growth is slower in the inner side, with limited resources and room for expanding.

Through surgery, chemotherapy, or radiotherapy, massive part of tumor could be eliminated, but tumor relapse or metastasis (migration of tumor from the primary site to a different site) often occurs. The usual suspect is cancer stem cell (CSC), which is believed to be a hidden minor cancer population with a slow growth rate but resistance to chemotherapy or radiotherapy. Unlike the fast-growing tumor cells, CSC may have evolved to survive in hypoxic and low-nutrient condition. As the first generation of chemotherapy and radiotherapy mainly targets the fast-growing cells by targeting DNA synthesis or cell division machinery (such as microtubules), CSC may be less sensitive to them and have better chance to survive. The survived CSC from the cancer therapy may restart to proliferate and part of it differentiate into fast-growing tumor cells for relapse (Figure 4.23).

Interesting question here is the origin of CSC. It is not clear whether CSC is formed first and develops into tumor, or CSC is just adapted and survived cells from already-formed tumor in harsh conditions. Considering the normal development of

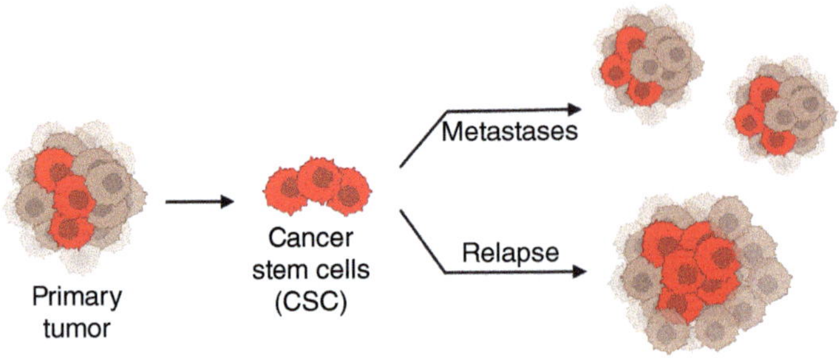

Figure 4.23 CSC in tumor relapse and metastasis.

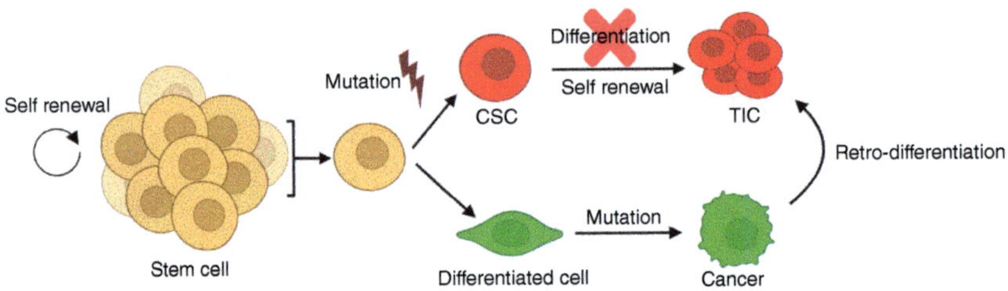

Figure 4.24 Possible mechanism of TIC formation.

an embryo with ESC, it sounds natural that CSC is originated from ESC and hidden somewhere in the body, waiting until tumorigenic conditions occur. But, if iPS can be formed from somatic cells by reprogramming or retro-differentiation, it may be also possible for various cancer cells to undergo reprogramming to form CSC. The major part of tumor cells are fast growing, but they do not have the ability to form a new tumor when transplanted in a non-tumor site or another body. The ability of new tumor formation of CSC has some common features with ESC, but there are clear differences. With the accumulated mutations, there may be many kinds of CSC, and it would be impossible for CSC to change back to normal ESC. While CSC can form different kinds of tumor cells, they are not pluripotent like ESC. So, to avoid such a possible misconception, another name, tumor-initiating cell (TIC) may be more neutral and technically correct term than CSC. A qualified TIC should have the ability to form a new tumor even by a single cell, regardless the origin or mechanisms (Figure 4.24).

Similar to the ESC probe, if there is a TIC probe available, it would be quite useful for cancer studies. However, there were technical difficulties. Usually, the number of TIC is minor in tumors. Even after TIC are somehow enriched by antibodies, culturing of TIC may quickly generate diverse cell populations and the TIC ratio drops in the culture. So, maintenance of pure and good quality of TIC in culture was quite challenging. Fortunately, in 2012, a TIC cell line TS32 was established from a patient-derived non-small cell lung carcinoma (NSCLC) through anti-CD166 selection and expansion in immunodeficient mice [32]. Using TS32 and its differentiated cell 32 A as negative controls, the first TIC probe TiY (**T**umor-**i**nitiating cell probe **Y**ellow) was developed in 2018 [33] (Figure 4.25).

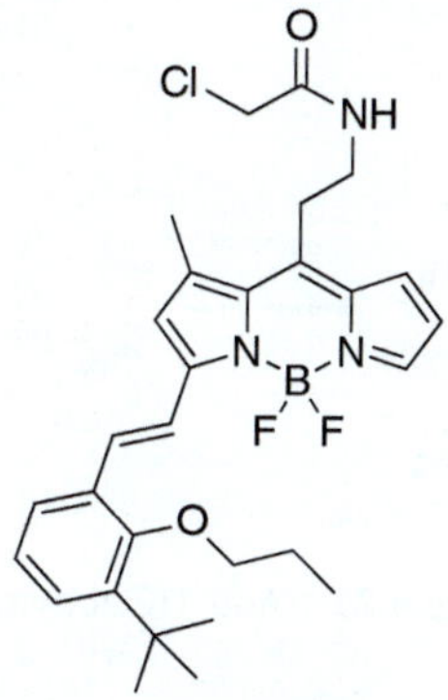

Figure 4.25 Structure of TiY.

Impressively, the small-molecule probe, TiY, showed a higher selectivity to TIC over anti-CD166. The TIC activity is usually tested by tumorsphere formation ability in vitro, but the ultimate proof of TIC is the formation of a tumor in another body. TiY-positive cells were functionally active TIC, and even single cell could induce a tumor in animals. TiY contains CA in the molecule and could form covalent bond with the target protein in TIC. The fluorescently labeled protein was cut and analyzed by mass spectrometry to give vimentin as the target protein. Through in vitro-binding test, TiY binds to tetrameric vimentin stronger than monomeric vimentin. Vimentin is a kind of intermediate filament and essential for the cell motility. Vimentin is also known to be a marker of epithelial–mesenchymal transition (EMT) and plays an important role in cancer metastasis. The high signal of TiY is correlated with the expression level of vimentin in cancer cells. As further validation of the target, overexpression of vimentin increased the TiY signal in the cell, and the knockdown of vimentin decreased the TiY signal (Figure 4.26).

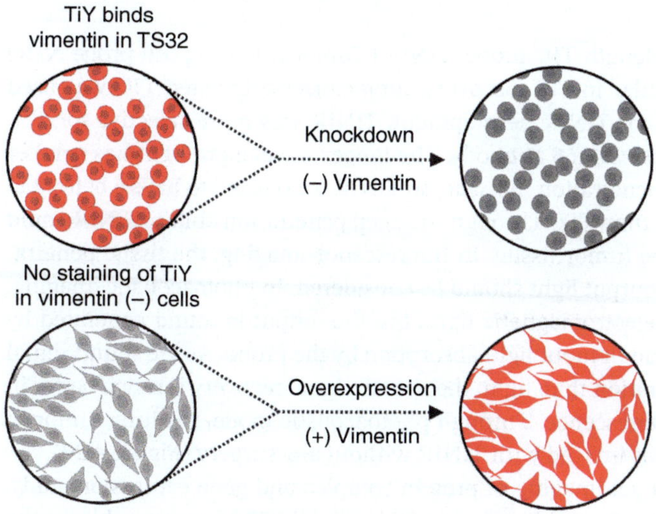

Figure 4.26 Target validation of TiY.

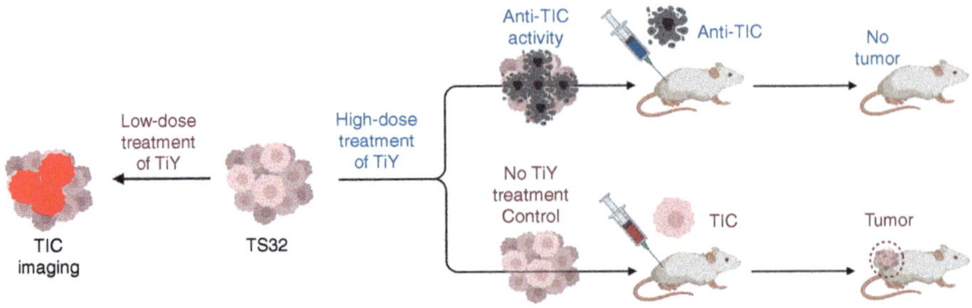

Figure 4.27 Anti-TIC activity of TiY.

Interestingly, it was observed that high vimentin increased tumorsphere formation and the knockdown decreased the size and number of tumorspheres. The result implies the functional importance of vimentin in TIC activity. As expected, the vimentin inhibitor withaferin was reported to have anti-cancer activity, but with high side effect of non-specific cell toxicity. Withaferin also competes with TiY for the binding of vimentin. Interestingly, TiY also showed an anti-TIC effect at a high concentration without side effects to normal cells or differentiated 32A cells. So, TiY could be used as an imaging probe for TIC at a low concentration and therapeutic agent at a high dose (Figure 4.27).

As there has been no common CD biomarker for TIC known, multiple antibodies have been used to identify different TICs depending on the origin of the tissues. In contrast, TiY showed the broadest spectrum of TIC selectivity over differentiated tumor cells in all the tested cancer cell lines. This may be due to the fact that vimentin is a common machinery protein and can act as a universal biomarker for TIC. It is noteworthy that anti-vimentin cannot enter the cell, but the small-molecule probe TiY can, underlining the advantage of a chemical approach (Figure 4.28).

4.1.3.2 TiNIR

In 2019, a longer wavelength TIC probe, TiNIR (**T**umor-**i**nitiating cell probe **NIR**) was reported [34]. Similar to TiY discovery, lung-cancer-originated TIC was used as a screening system for TiNIR development. TiNIR was more selective for TIC distinction than the anti-CD166 antibody. The longer wavelength light has smaller scattering and deeper penetration in tissue, so TiNIR is expected to have a better tissue imaging capability than TiY. Through the deep penetration ability, TiNIR could image the TIC from the tumor tissue. In fluorescence imaging, the tissue penetration of both input and output light should be considered. In photoacoustic imaging, the input is the same electromagnetic light, but the output is sound generated by thermal expansion through input light absorption by the probe. As the sound signal has much better penetration than light, the overall signal recovery is much better in photoacoustic than fluorescence. Through photoacoustic mode, the lung tumor in intact animal model was imaged with TiNIR without any surgery (Figure 4.29).

Through fluorescent gel imaging of protein complex and gene expression study, the binding target of TiNIR was identified as HMOX2. HMOX2 is responsible for the elimination of ROS in the body. TIC is formed in a hypoxic environment where ROS

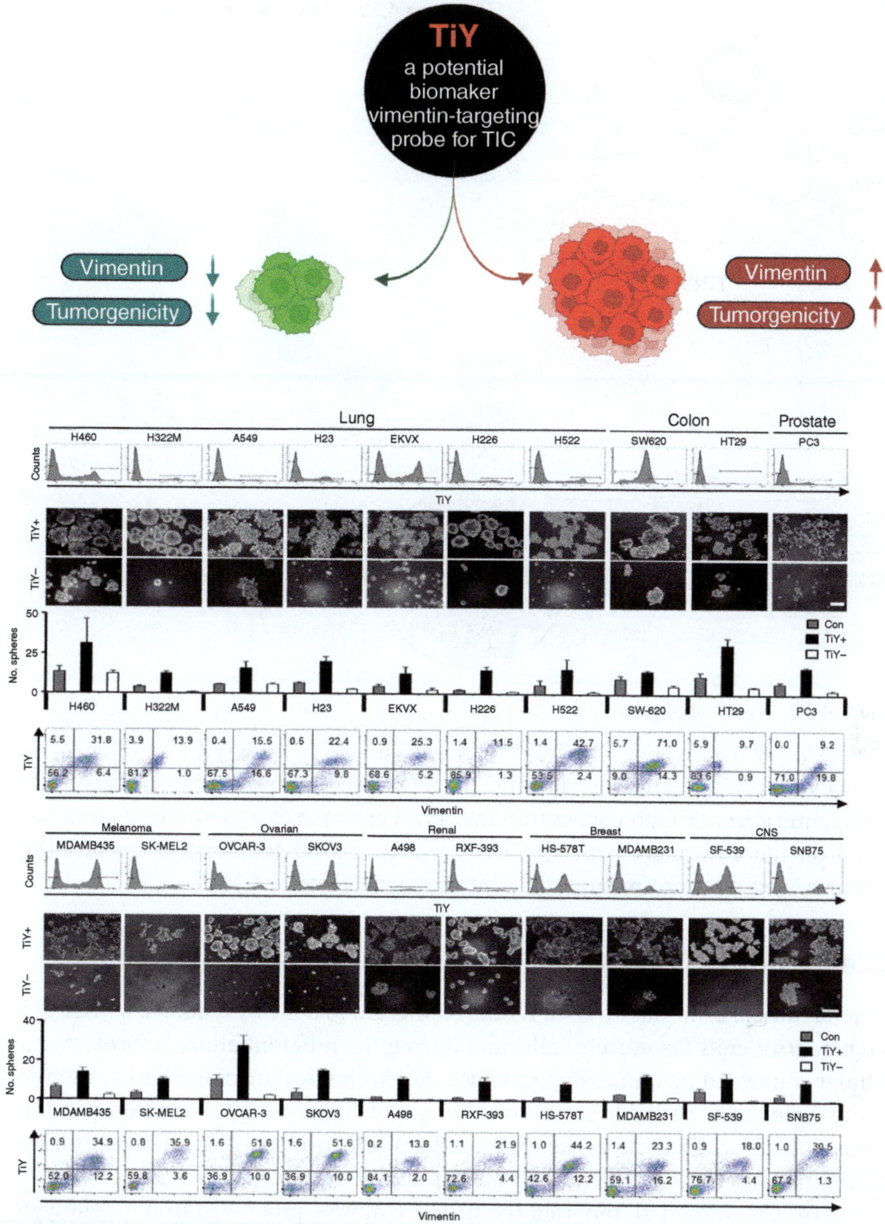

Figure 4.28 The universal selectivity of TiY. Source: Ref. [33], from John Wiley & Sons.

is high. While ROS may be functionally important for TIC formation and maintenance, the high ROS should be a burden to TIC. Therefore, removal or suppressing ROS by HMOX2 seems to be essential for TIC survival. TiNIR binds to HMOX2 and also inhibits the function as a competitive inhibitor for cofactor heme-binding site. Inhibition of HMOX2 by TiNIR increases the ROS in TIC and eventually kills

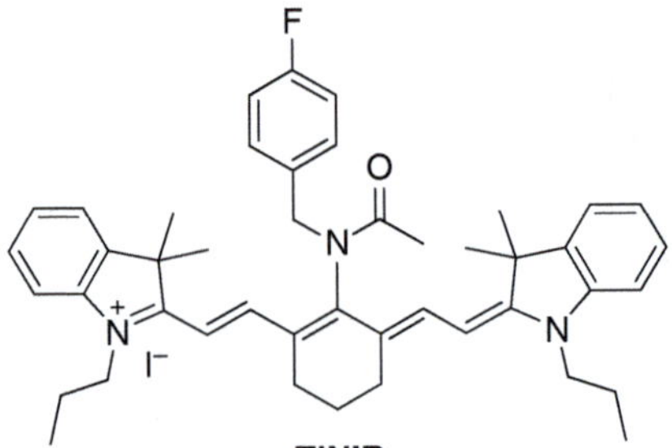

Figure 4.29 Structure of TiNIR.

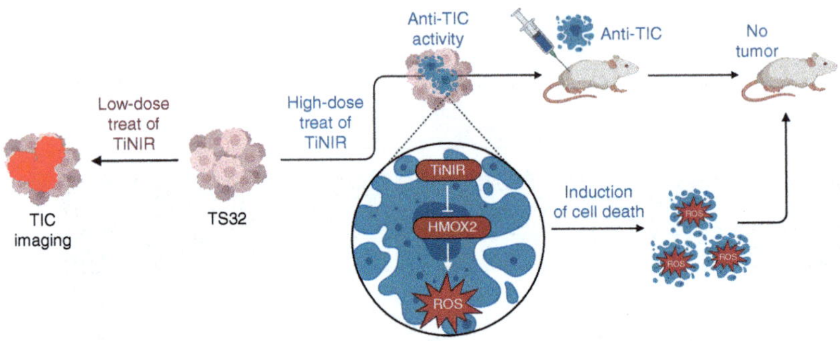

Figure 4.30 TiNIR application.

the cells. Therefore, TiNIR is also an imaging probe at low concentrations and a therapeutic agent at high concentrations, as an example of a theranostic molecule. Based on the anticancer activity, TiNIR was proved to be effective for blocking cancer relapse after tumor surgery in mouse model (Figure 4.30).

4.1.4 Muscle Cell Probes

Skeletal muscle is fibrous striated tissue, connected to bone by tendons. Myoblast is the precursor cells for muscle cells, and during the differentiation, it forms multi-cellular connected myotubes and provides the mechanical machine for the motion. While there are remarkable structural changes during the myotube differentiation, it was difficult to distinguish the differentiated cells from myoblast or monitor the differentiation process, especially in the live status.

In 2008, Clemons et al. reported the first live muscle probe E26 from phenotypic screening. Green probe E26 selectively stains differentiated muscle cells over undifferentiated myoblast, reporting the live cell status of muscle cells [35]. While E26 could be used for various drug effect screening for the muscle differentiation, the probe was rather weak for the light exposure, fading out quickly. For a robust screening system with long-term light exposure, a better photostability probe would be desirable. Through extensive screening of fluorescence library, new probes I25 and I31 were elucidated as red fluorescence and more photostable than E26 (Figure 4.31).

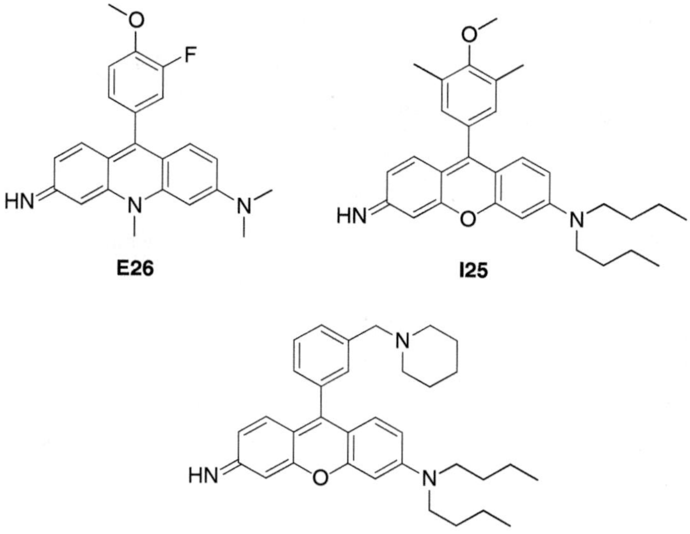

E26 I25 I31

Figure 4.31 Structure of E26, I25, and I31.

To figure out the biomarker of differentiated muscle cells, binding target proteins of the new probes were systematically studied. The conventional approach for target identification of small molecules is the affinity matrix, by attaching the molecule to a hydrophilic carbohydrate polymer through a linker. For making the efficient affinity matrix, structure–activity relationships (SAR) study is necessary to find a proper position of the linker connection on the molecule without losing the original selectivity. Assuming that the probe binds to the target protein (but without knowing the binding mode), the linker site is expected to be exposed to the media direction, without disturbing the binding between the probe and the protein. The SAR often requires multiple derivative syntheses and activity test iterations, which easily take months of time and effort. If the binding mode employs full wrap-up of the small molecule by the target protein, leaving no exposure position, all SAR studies would be useless from the beginning. Even worse, it is impossible to predict the bad possibility due to the lack of binding mode information. Therefore, the SAR study is often a notorious rate-determining step at best, with a low success probability. One possible solution for avoiding the SAR is to use a prebuilt-in linked library for the screening. If the probe works with the built-in linker, at least the linker does not bother with the target binding. The worst scenario may be the linker which is actually part of the active binding structure, but figuring out the fact is straightforward. In either case, the linker could be used for connection to the matrix and tested without SAR (Figure 4.32).

Alternative approach would be comparing various active candidates and extract the important binding motif information as the surrogate of SAR studies. In muscle probes, I25 and I31 shared a common structure, which may be important for the binding to the target protein. Then, modifying the variable part of the molecule

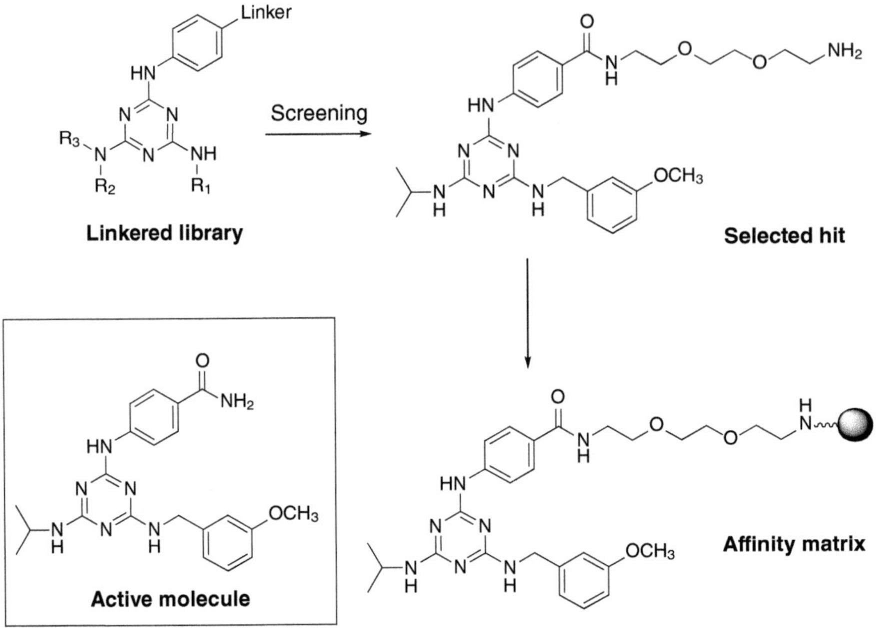

Figure 4.32 Built-in linked library approach.

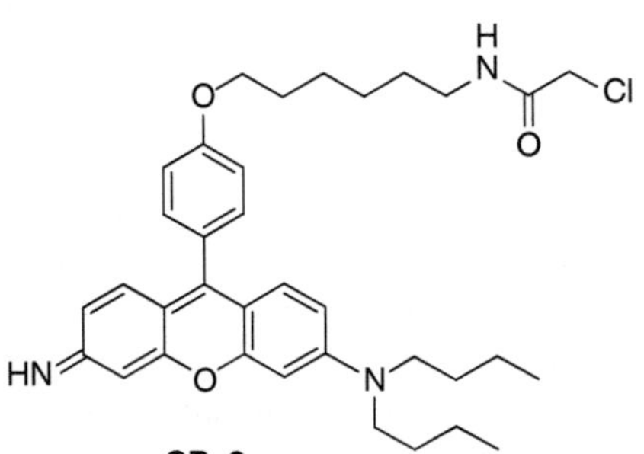

Figure 4.33 Structure of CDy2.

would be a reasonable option to make an efficient affinity matrix (shown in Figure 4.33). Conventional affinity matrix is a chromatography using cell lysate made by detergent extraction of cellular proteins and isolating strongly binding proteins as the potential target of the affinity molecule. The procedure contains detergent treatment, which demolishes all the intracellular compartmentalization, and multiple washing procedures. Some of the proteins may be lost during the lysate preparation or denatured by environmental changes or washed out by extensive washing steps if the binding affinity is low. We prepared CDy2, a CA derivative of I25 or I31, attached to the linker as an alternative affinity molecule and systematically compared with the conventional affinity matrix. The CA derivative could be used as a probe by itself and can reach to the target protein in the live cells and induce the covalent binding through thiol reaction of the cysteine side chain under native conditions of the cells (Figure 4.33).

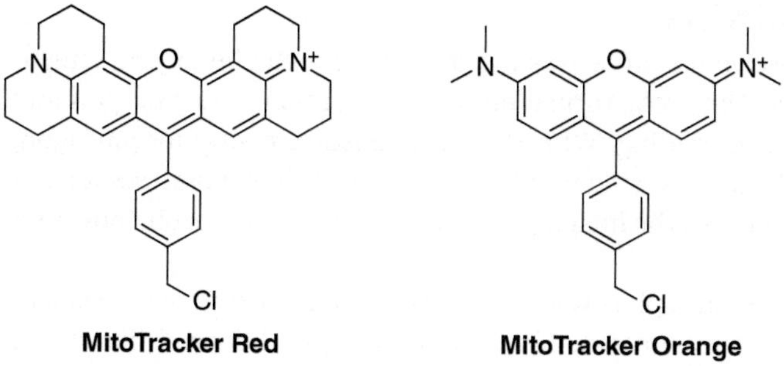

Figure 4.34 Structure of mitotracker Red and Orange.

Surprisingly, the two comparable methods generated totally different results. Conventional affinity matrix showed tubulin as the most abundant binding protein, but CDy2-binding protein was identified as mitochondrial protein ALDH2, which is abundant in muscle cells. Considering the cellular location of CDy2 as well as that I25 and I31 is mitochondria, ALDH2 would be more appropriate target protein in live-cell environment. This interesting study underlines the common artifact of conventional affinity matrix procedure [36].

Through the follow-up study of exact binding site on the target protein, Cys302 of ALDH2 was identified as the binding site of CDy2 by proteomic mass spectrometry analysis. With the exact binding site information, the close-up binding mode of CDy2 to ALDH2 could be modeled in high resolution. When compared to other covalent-bond-inducing mitotrackers such as MitoTracker Red and MitoTracker Orange, CDy2 showed much higher selectivity to ALDH2 over other proteins [37] (Figure 4.34).

4.1.5 Pancreatic Cell Probes

Pancreas is a huge gland organ. Acinar cells, which are the major cells of pancreas, produce exocrine digestion enzymes. In addition to acinar cells, pancreas has a very unique micro-organ, called Langerhans islets or pancreatic islets, which occupy about 1% of total pancreatic cells. Pancreatic islets are composed of various gland cells, and most important cells are α- (minor) and β- (major) cells. α-Cell secretes glucagon and β-cell secretes insulin into the blood stream. In contrast to the exocrine system, glucagon and insulin secretions are classified as the endocrine system. The notation reflects the concept that the blood system is the inner space of the body, and the gastrointestinal tract is the outer space from the body. Glucagon is a hormone which increases blood glucose, and insulin has the opposite function by glucose uptake by muscle cells or adipocytes. Glucose is the primary energy source of our body, and a low glucose condition is dangerous, especially for the brain. But, abnormally high glucose concentration is also problematic, and the uncontrolled high glucose induces a disease, diabetes. Therefore, α- and β-cells are important to maintain the homeostasis of glucose concentration in our blood stream. However, currently, there is no clinical contrast agent which can detect the status of either of the cell type.

4.1.5.1 Pancreatic α-Cell Probes

In 2009, the first α-cell-selective probe was discovered through phenotypic screening of α- and β-cell lines. The probe turned out to selectively bind to glucagon and increase fluorescence upon binding. With the clear sensor property for glucagon, the probe was dubbed Glucagon Yellow. While Glucagon Yellow possesses a good α-cell selectivity in cell line test, the imaging of pancreatic islet was not obvious from pancreas tissue [38] (Figure 4.35).

In 2015, TP-α (Two Photon alpha) was developed through a systematic screening of two-photon fluorescent molecule library in α- and β-cell lines. When tested with various proteins and biological materials in vitro, TP-α selectively responded to glucagon with increased fluorescence. Mouse pancreatic islets are known to be composed of 70–80% of β-cells and 10–25% of α-cells. When the islet cells were analyzed by glucagon antibody and TP-α, a good correlation was confirmed through a two-dimensional flow cytometry and in fluorescent islet images. It is noteworthy that the compact structure of the islet hampers the access of probes, especially macromolecular antibody probes, so the islet has to be loosened and attached to the plate bottom by in vitro culturing for a couple of days. Therefore, it seems TP-α enters live α-cells and binds to glucagon, illuminating strongly in the α-cells. Here the abundant hormone glucagon acts as the biomarker and target of α-cells [39] (Figure 4.36).

As a partner of TP-α, TP-β was developed for concurrent imaging of α- and β-cells in multicolor mode. As the mechanism of TP-β is different from POLD, the details are discussed in Section 4.4.3.

4.1.5.2 Pancreatic β-Cell Probes

As the counterpart of alpha cells, and more importantly as the insulin-producing cells, the interest for pancreatic β-cell-selective probe has been high. Dithizone

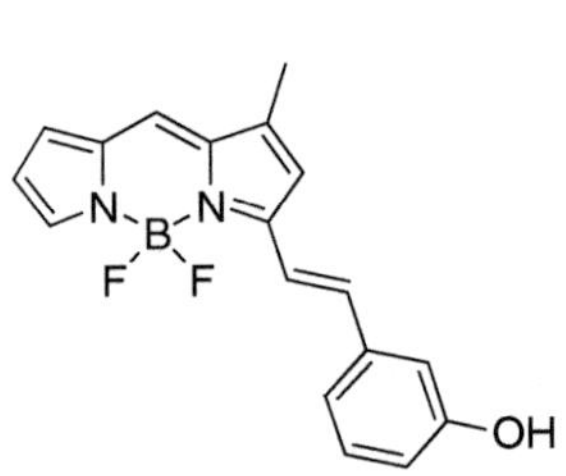

Glucagon Yellow

Figure 4.35 Structure of Glucagon Yellow.

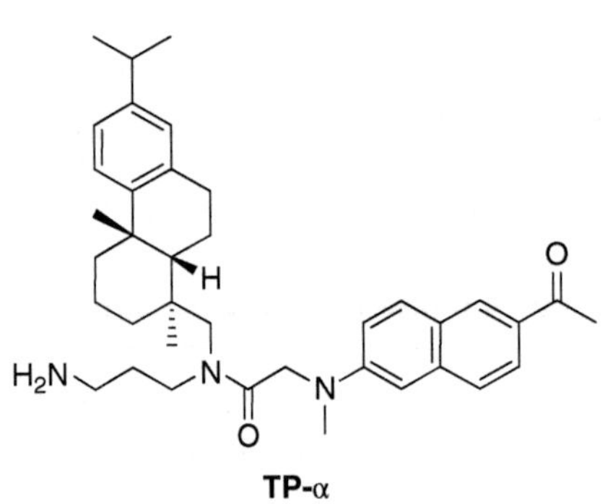

TP-α

Figure 4.36 Structure of TP-α and alpha cell staining image and flow cytometry data of pancreatic islet cells.

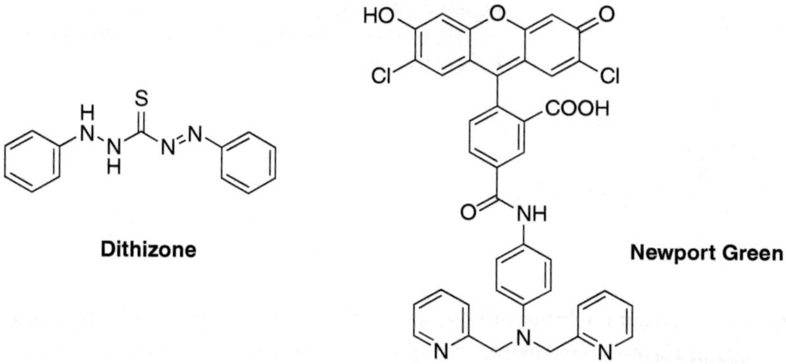

Figure 4.37 Structure of dithizone and Newport green.

was the first β-cell probe targeting the high Zn ion concentration of the cells [40], but deleterious effect to the β-cell was also reported [41]. Another Zn ion sensor, Newport green, has been used for functional islet separation and quality control in case of islet transplantation [42] (Figure 4.37).

In 2013, a new fluorescent β-cell probe PiY (**P**ancreatic **i**slet **Y**ellow) was developed through an unbiased imaging-based screening with α-, β-, and also acinar cell lines [43]. Acinar cells are dominant cells in pancreas surrounding the islets, so for the good islet imaging, the selectivity over acinar cell is very important. PiY exhibited high selectivity only for β-cells against α-cell or acinar cells. As expected, PiY showed a good islet imaging in the pancreas tissue section and the name of the probe was given considering the islet imaging ability (Section 5.3.1). From flow cytometry analysis, PiY-positive cells were clearly insulin-expressing cells. Even with the apparent β-cell selectivity, the molecular target of PiY was not elucidated (Figure 4.38).

In 2020, a multi-modality β-cell probe, PiF (**P**ancreatic **i**slet **F**luorinated probe), was designed and developed for β-cell-selective probe [44]. First, through an in vitro screening of insulin response, a series of fluorescent compounds were collected as the candidate probes for β-cell. β-Cells produce insulin as the functional hormone and the secretion vesicles of β-cells contain super high concentration of insulin.

Figure 4.38 Structure of PiY and cell selectivity/ flow cytometry (one mountain).

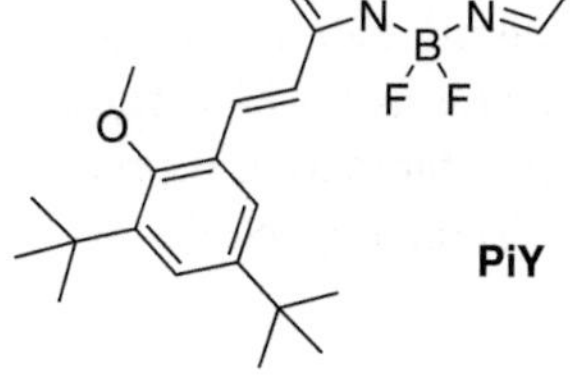

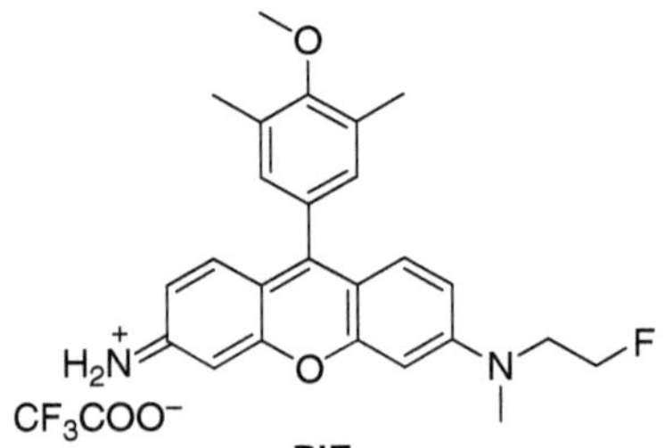

Figure 4.39 Structure of PiF and shuffling of structure of candidates (insulin response?) + flow cytometry data.

For the highly condensed packing and storage of the insulin, Zn^{2+} ion is involved. The most common form of Zn–insulin complex is a hexamer containing two zinc ions, but zinc ion is not necessary for the complex formation as a similar insulin hexamer can also be formed without zinc ion [45]. While dithizone and Newport Green were developed as β-cell probes targeting the high concentration of zinc ion, there was no β-cell probe directly targeting insulin.

The collected insulin sensor molecules were rosamine-based fluorophores. By shuffling the candidate structure and biological test, a ^{19}F incorporated probe was selected and dubbed PiF. PiF was superior to PiY in β-cell selectivity in flow cytometry with two clear cell populations. This may be due to the inundated amount of target of PiF in β-cells in comparison to α-cells. This is a unique example in DOFLA as a target was predefined, and a sensor molecule selected by an in vitro screening was applied to the cell-selective probe. Therefore, the mechanism of PiF is POLD with insulin as the binding target. PiF also showed superior performance to PiY in pancreatic islet tissue imaging (Section 5.3.1) and demonstrated a unique whole-body imaging by modality conversion (Section 6.7) (Figure 4.39).

4.1.6 Amyloid Probe: CDy11

Amyloid is an abnormal protein aggregate characterized by a nano-sized fibril with an enriched β-sheet secondary structure. Of the 40–42 peptides discovered in Alzheimer's disease, β-amyloid would be the most studied. The precursor peptide is formed inside cells, but after cleavage by β- and γ-secretase, the β-amyloid is secreted to the extracellular domain. There are many sensors and imaging probes for β-amyloid target, but the application is usually for the brain tissue rather than specific cells (Figure 4.40).

When bacteria form colonies, through quorum sensing, they make biofilm as the protecting barrier. One of the components of biofilm is Fab amyloid. In 2016, the first biofilm probe CDy11 (**C**ompound of **D**esignation **y**ellow **11**) was developed targeting Fab as the biomarker. The screening was performed on Fab-expressing bacteria as positive control, and Fab knock out bacteria as the negative control. As the biomarker is predefined (reverse genetic approach: genotype to phenotype), the screening could be performed on purified protein, and selected probe is

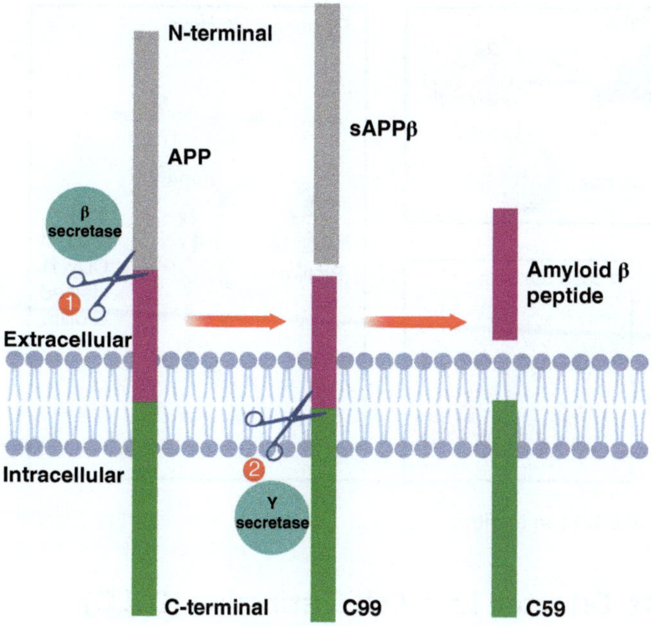

Figure 4.40 Amyloid generation by cleavage and aggregate formation.

Figure 4.41 Structure of CDy11 and its binding to Fab amyloid.

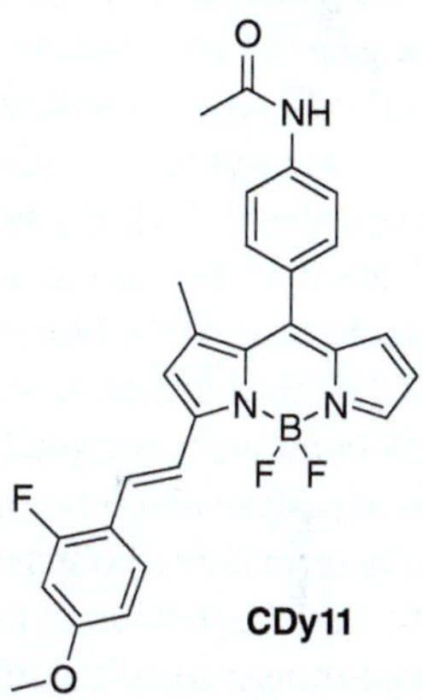

CDy11

applied to bacteria later. However, considering the complex biological situation, whenever possible, direct test in whole system (i.e. bacteria with amyloid rather than pure amyloid, one separate component) would increase the chance to rule out nonpractical probe from the screening (Figure 4.41).

CDy11 was demonstrated to detect bacterial infection in eyes and peritoneally infected bacteria colony [46]. CDy11 was also used for dynamics of innate immune response to *Staphylococcus aureus* biofilm in mouse ear skin model [47] (Figure 4.42).

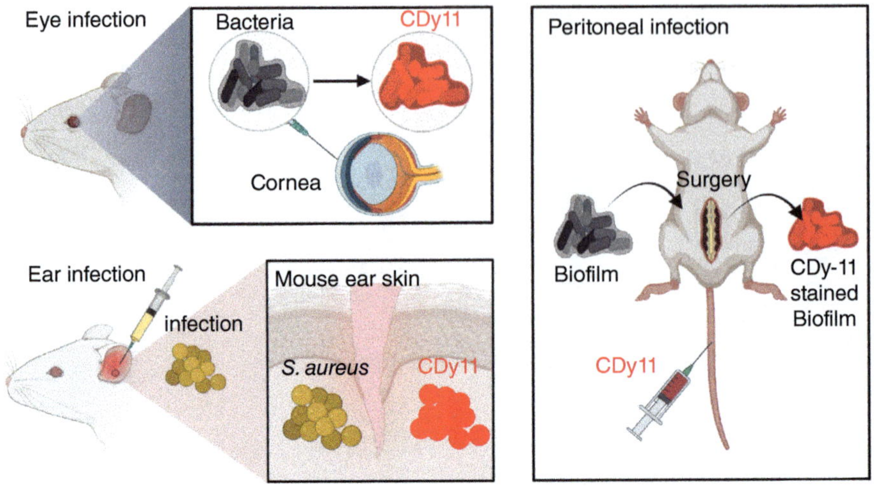

Figure 4.42 Application of CDy11 in biofilm.

4.2 Carbohydrate-Oriented Live-Cell Distinction (COLD)

The development of embryo starts with the expression of the inherited genetic information. The first direct products of gene information are proteins. Each cell has the same genetic information in DNA, but the kinds and quantity of expressed proteins are all different depending on the cells. The specific protein expression could be directly measured by antibody (western blot) or indirectly through mRNA quantitation (northern blot). The interesting name was given by Dr. Southern who developed for DNA detection using antisense oligonucleotide probe. Following the inventor of the technique, DNA blot is called as southern blot, and the name was expanded to RNA (as northern blot) and protein (as western blot). While the name of western blot is commonly accepted as an antibody-based protein detection, interestingly the term of eastern blot has not reached common consensus yet. Some scientists tried to use the term for carbohydrate detection, and others tried to use it for small-molecule probe-based detection of biomolecules. There is still different definition of the term as post-translational modification detection or even lipid detection with the name of far-eastern blot. It would be fun to see in the future which approach finally acquires the terminology as the survivor of the evolution process. Anyway, CDs, the cell surface markers, are mainly proteins, and antibodies are the common tools to discriminate them.

Next to proteins, the second biggest diversity in biomolecules are from carbohydrates. Proteins are composed of 20 amino acids and the side chain of amino acids contains a broad range of chemical and structural diversity, including + and − charges, hydrophilic and hydrophobic groups, and also aromatic motives. Based on the diversity of the building blocks, the structure and properties of proteins cover an extremely broad range. Property of carbohydrates is relatively limited. The molecular formula of glucose, the representative carbohydrate, is $C_6H_{12}O_6$. In abbreviated form, it is $6(CH_2O)$: carbon + water! Carbohydrate also means hydrated carbon, which reflects its molecular formula. Many other carbohydrates, such as galactose, mannose, and fructose, have exactly same molecular formula. Five-carbon carbohydrates, such as ribose, have the formula of $5(CH_2O)$, basically same structure of hydrated carbon. Unlike proteins with various functional groups, most likely the only functional group in carbohydrates is the hydroxyl group. Therefore, the general property of all carbohydrates is being hydrophilic and water miscible (Figure 4.43).

Glucose is the most common and the basic materialized energy source for all the living things, but at the same time quite toxic material at high concentration. The chemical toxicity of glucose is due to the reactive reducing end structure. Glucose is in a dynamic equilibrium of at least 5 isomers; one open form contains chemically reactive aldehyde group, and pairs of 5- and 6- membered cyclic hemi-acetal structures. The main isomers of glucose are 6-membered ring structures with different ratios of α- and β-isomers. The 1-position of glucose is called anomeric center, and the resulting stereoisomers around 1-position are uniquely called as anomers. Due to the dynamic equilibrium of the anomeric center, the one position stereochemistry is often drawn as a wiggled bond as in Figure 4.43. The aldehyde group of the open form can be oxidized by reducing other materials and is called reducing end. The reducing end can induce condensation reactions with amine groups and leads to caramelization or glycation reactions (Figure 4.44).

Proteins are modified after translation with various functional groups, and glycosylation (addition of with carbohydrate) is one of the most common modifications.

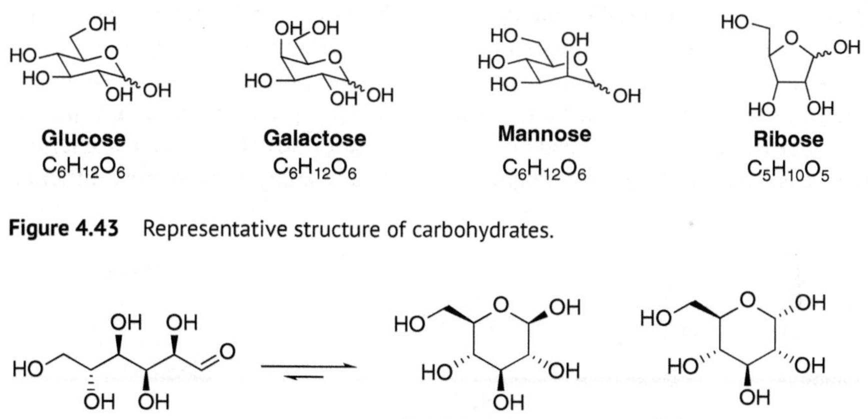

Figure 4.43 Representative structure of carbohydrates.

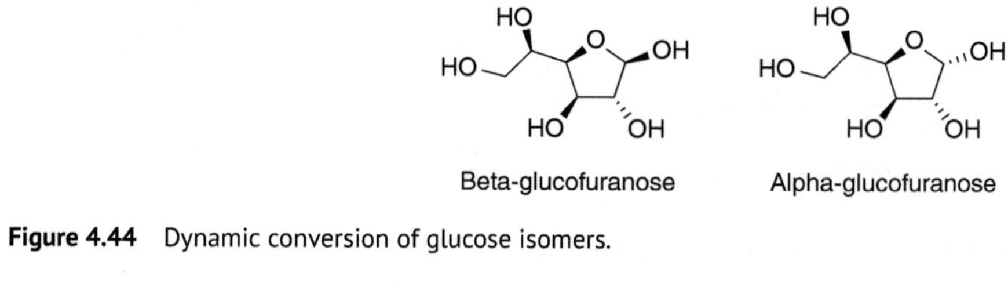

Figure 4.44 Dynamic conversion of glucose isomers.

Glycosylation of protein often occurs not only by a single carbohydrate, but also by carbohydrate polymers. When carbohydrate forms a polymer, one water molecule is removed between two hydroxyl groups, forming an ether linkage. The structure of carbohydrate polymers is still hydrated carbon with just a longer shape. Therefore, the glycosylation usually makes the protein surface more hydrophilic. The size of carbohydrate polymers attached to protein is sometimes even bigger than the protein itself. The ability of raising and binding to antibodies is called antigenicity. In general, proteins have a high antigenicity with various functional groups, but carbohydrates have usually a low antigenicity, i.e. boring molecules. So, glycosylation of cell surface proteins could be a good strategy to avoid immune attacks. That is why carbohydrate vaccine (antigen) is rare. By the same token, carbohydrate-based CD is relatively rare compared to protein CDs.

While the functional group is mainly limited to the hydroxyl group, carbohydrates still can generate huge structural diversity. In addition to the dynamic equilibrium of the anomeric center, the connection of carbohydrates in polymers generates another complex diversity. In proteins, the connection of amino acids is usually linear with a fixed direction (from N-terminal to C-terminal) between the amino and carboxylate groups. In carbohydrates, any hydroxyl pair (as ether) could be used for connection, even with stereochemical choices for anomeric positions. Furthermore, carbohydrate polymers allow branching of the monomer connection. Altogether, the structural diversity of carbohydrates is much more complicated than that of proteins (Figure 4.45).

In proteins, the linear amino acid information is directly from gene sequencing, so protein sequence is predictable by reading the DNA or mRNA sequence. Also, the products are usually homogeneous, unless alternative splicing or different post-translational modification occurs. Carbohydrate polymerization is catalyzed by enzymes without a genetic blue print. So, there is a preferred pattern or partial motifs, but there is no consensus of the sequence or structure. Even worse, the structure is quite heterogenous, making the structural analysis even more difficult.

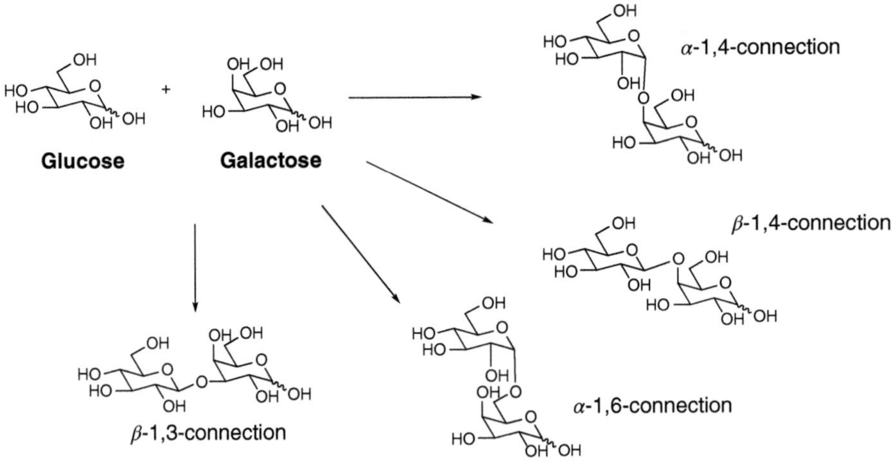

Figure 4.45 Examples of structural diversity of carbohydrate linkage.

4.2.1 Lectins

For protein recognition, antibodies are the most common choice. In contrast, carbohydrates have usually low antigenicity, and it is relatively difficult to obtain antibodies as the carbohydrate-binding tools. Instead, lectins are a special class of carbohydrate-binding proteins. Lectins are ubiquitous in nature, including foods, and play roles in biological recognition between cells, bacteria, and viruses through glycoprotein recognition. The binding of lectins to glycoproteins can be inhibited by monosaccharides or oligosaccharides, which can prevent the binding of lectins to the target cell. As expected, the binding mode of lectins with carbohydrates relies more on the stereochemical shape differences than on functional group recognition, such as amino or carboxylate.

Lectins are being used in various biochemical and medical researches. For example, lectins such as concanavalin A have been used widely for affinity matrix for purifying glycoproteins. While ABO blood typing is performed by antibodies, many other blood types are determined by lectins, by recognizing the carbohydrate modification on the RBC membrane [48]. The representative lectins are summarized in Table 4.1.

In addition to the individual study of lectin for the biochemical research, a systematic glycome study was also reported. In analogy to the genome or proteome, the whole collection of glycan structure is termed a glycome. As an example, lectin microarray was developed for the systematic analysis of the glycome on the T-cell membrane and HIV-1. HIV-1 is the virus responsible for AIDS and is known to hide in CD4 T helper cells. By lectin array (composed of ~70 lectins), it was shown that HIV-1 and native immunomodulatory microvesicles possess the common glycome structure. This result suggests that the virus and natural microvesicles may share the common exocytic pathway and may explain why the immune response is difficult against HIV. The fluorescently labeled lectin was also used for T-cell imaging in the study, but in fixed cells [49]. In another study, 96 lectins were prepared for microarray, and the glycomic analysis was applied to human-induced PSCs. The study showed ESCs and iPS shared a similar glycomic landscape, but uniquely different from somatic cells used for reprogramming. Interestingly, five of the somatic cells have distinct glycomic pattern to each other, representing their origin of the tissue history [50].

Table 4.1 Representative lectins and ligand motif.

Lectin symbol	Name	Ligand motif
ConA	Concanavalin A	α-D-mannosyl and α-D-glucosyl
LCH	Lentil lectin	Fucosylated core
RCA	Ricin	Galβ1-4GalNAcβ1
PNA	Peanut agglutinin	Galβ1-3GalNAcα1-Ser/Thr
WGA	Wheat germ agglutinin	Neu5Ac (sialic acid)
UEA	*Ulex europaeus* agglutinin	Fucα1-2Gal

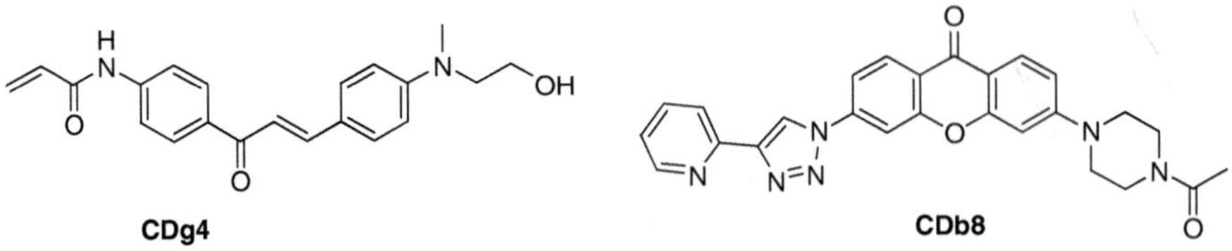

Figure 4.46 Structure of CDg4 and CDb8.

Although many lectins have toxicity, some lectins were also used for live-cell and whole-body imaging. For example, tomato lectin was fluorescently labeled and used for vascular structure imaging in mice through tail vein injection [51]. Recombinant lectin rBC2LNC was used for selective human PSC imaging both in fixed and live statuses and also analyzed by flow cytometry. Compared to antibodies which require mammalian cells for the production, the recombinant lectins could be generated in *Escherichia. coli* expression system and thus cost effective [52].

4.2.2 Embryonic Stem Cell Probes: CDg4 and CDb8

For fluorescent imaging of cells, multicolor staining is necessary in combination with other probes. As a complementary probe to CDy1, a green probe CDg4 (**C**ompound of **D**esignation **g**reen **4**) [53] and a blue mESC probe CDb8 (**C**ompound of **D**esignation **b**lue **8**) [54] were developed through phenotypic fluorescent image-based screening (Figure 4.46).

When ESC grow, they tend to form a sphere structure. The mouse stem cell sphere has a more three-dimensional globular shape than a rather flat human stem cell sphere. While CDy1 localized inside mESC, CDg4 and CDb8 tend to stain more strongly the surface of stem cell sphere than the cell inside. The surface of stem cell sphere is known to have a carbohydrate-rich glycocalyx. Among the components of glycocalyx, CDg4 was found to bind the glycogen structure. Amylase (glycogen digestion enzyme) treatment abolished CDg4 staining to the stem cell sphere, as evidence of the target. Furthermore, CDg4 actually increased its fluorescence upon binding glycogen, and thus CDg4 turned out to be a glycogen sensor. The presence of glycogen on stem cell sphere is unique only to ESC (not found in neurosphere), which may be due to the requirement of high glucose supply demand in ESC (Figure 4.47).

The multicolor imaging with CDy1, CDg4, and CDb8 on mESC was successfully performed using the complementary colors of the probes. While the molecular mechanism of CDb8 was not clearly identified, according to the similar staining pattern with CDg4, it is postulated that CDb8 may bind to one of the components of glycocalyx.

4.2.3 Gram-Positive Bacteria Probe

Boronic acid can form an ester with 1,2-diol or 1,3-diol, so it has been used for carbohydrate recognition motif for a long time. However, the ester formation and

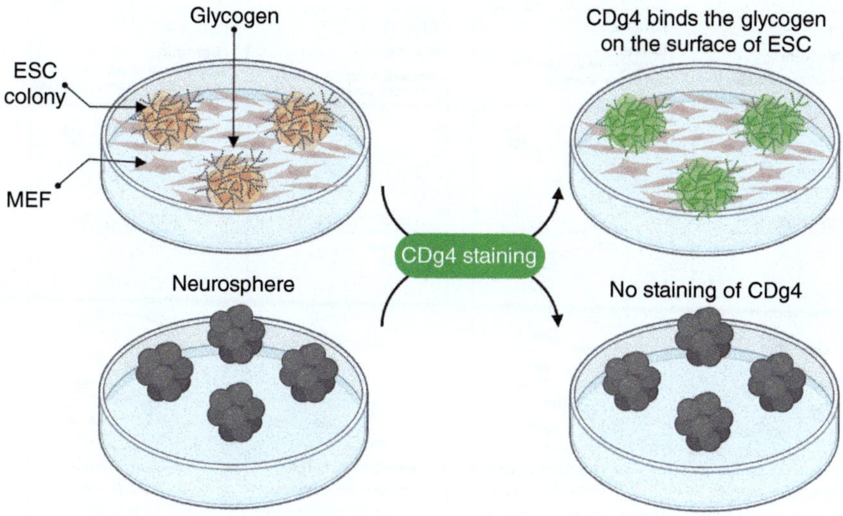

Figure 4.47 CDg4 image data.

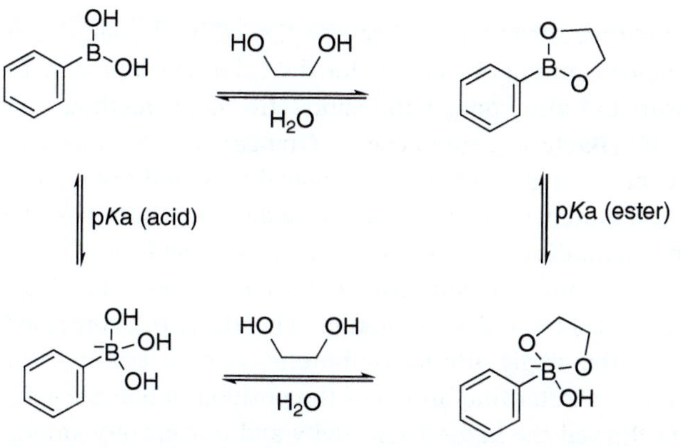

Figure 4.48 Reaction of boronic acid with carbohydrates.

hydrolysis are reversible, and the binding constant is not high. So, for the sensing of carbohydrates, a high concentration of boronic acid probe is required with relatively low selectivity among carbohydrates (Figure 4.48).

The clear difference between Gram-positive and Gram-negative bacteria lies in the cell wall. Gram-positive bacteria have a thick peptide glycan wall, outside of the cell membrane that protects the cell from physical attack or osmotic pressure. The cell wall of Gram-negative bacteria is much thinner than that of Gram-positive bacteria, and also has a second cell membrane outside of the cell wall. So, the peptidoglycan wall is not exposed to outside in Gram-negative bacteria. Dr. Gram took advantage of this difference for the discrimination of them, and Gram staining uses crystal violet that strongly binds to peptidoglycan. As a result, crystal-violet-binding cells are defined as Gram-positive bacteria. This method has been the gold standard in the classification of bacteria for a long time (Figure 4.49).

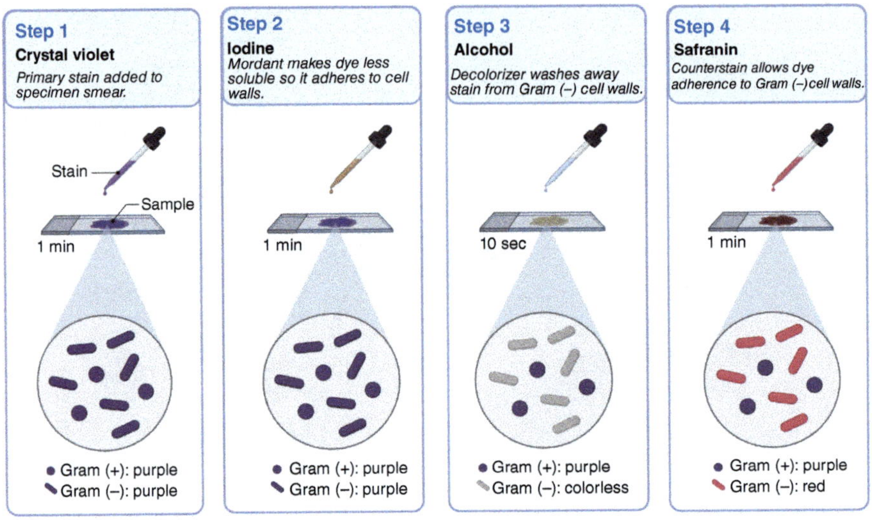

Figure 4.49 Gram method for the bacteria discrimination.

While useful, Gram method requires fixing of bacteria (i.e. killing bacteria) and takes multiple steps of time-consuming procedures for the color development. To overcome the viability issue and also convert the subtractive color method into fluorescence version, BacGO (**Bac**teria **G**ram-positive **O**range) was developed in 2019 [55]. Peptidoglycan contains high contents of carbohydrates, and fluorescent compounds with boronic acid motif were collected and tested for the Gram-positive bacteria selectivity. The results indicated that while the boronic acid motif takes important role in the recognition of Gram-positive bacteria, reasonably high hydrophobicity is also required for the discrimination. Therefore, the combined strength avidity, rather than the single affinity by boronic acid, is essential for the selective probe development. After fine tuning of the position of boronic acid, the finalized probe BacGO showed the highest selectivity and universality among all the published / commercialized fluorescent probe for Gram-positive bacteria (Figure 4.50).

Compared to Gram method, the BacGO method is faster and more sensitive. BacGO can be applied not only to single bacteria distinction, but also to the growing colony staining in real time. The quantitative analysis of sludge-treating bacteria was demonstrated, showing the possible application to a complex bacteria mixture (Figure 4.51).

4.2.4 Biofilm Probe: CDy14 and CDr15

When bacteria grow fast in a good environment, they actively divide and exist in a planktonic state. If the environment gets tough, they get together and form a colony. Once the bacteria number reaches some criteria, through quorum sensing, they modulate the gene expression to gain the maximum advantage for the whole colony.

Figure 4.50 Structure of BacGO (selectivity panel).

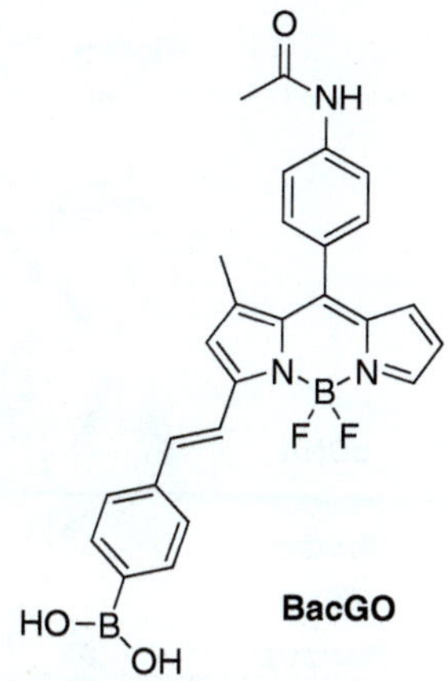

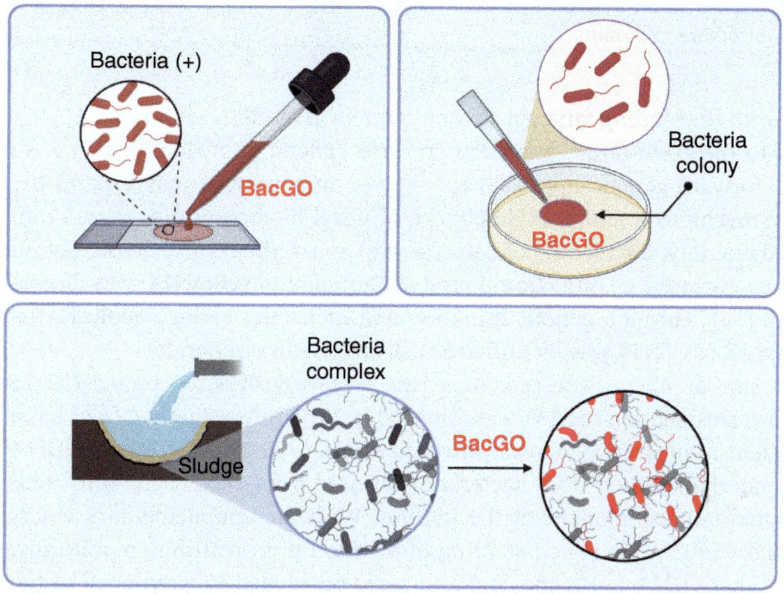

Figure 4.51 BacGO application.

One of the results is making a biofilm to protect the colony. Biofilm is composed of various peptidoglycans, amyloid proteins, and extracellular DNA. If bacteria hide inside the biofilm, antibiotics or bacteria probes may not reach the target bacteria and cannot perform properly. Using the biofilm, bacteria can evade antibiotic attack or immune surveillance, and can hide in the host body for a long time waiting for the next burst. In other words, the biofilm is like the camouflage of the bacterial colony. Therefore, effective antibiotics or probes should be able to penetrate the biofilm and reach the bacteria (Figure 4.52).

How about we change the approach? Forget about the bacteria themselves, but how does one detect the camouflage itself? The first probe developed from this idea

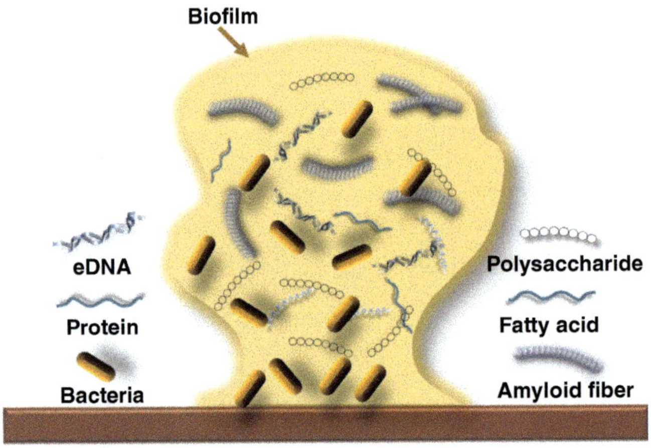

Figure 4.52 Structure of biofilm.

was CDy11 with the binding target of amyloid protein in biofilm (Section 4.1.6) [46]. In contrast to the fixed-target approach (reverse genetic approach: genotype to phenotype), forward genetic approach (phenotype to genotype) may expand the scope of the mechanism and provide chance of novel biomarker discovery. From the unbiased phenotype screening of bacteria with and without the general biofilm expression, a new probe CDy14 (**C**ompound of **D**esignation **y**ellow **14**) was discovered in 2018 [56]. Through genetic mutation and molecular biology analysis, the molecular target of CDy14 was identified as psl, an exopolysaccharide.

From the similar phenotypic screening, the discovery of a red probe, CDr15 (**C**ompound of **D**esignation **red** **15**), was followed in 2019. Interestingly, CDr15 has a dose-dependent fluorescence increase effect to eDNA (**e**xtracellular **DNA**). eDNA is a sticky material secreted from bacteria or released from dead bacteria to composite an important component of the biofilm. With the complementary colors of CDy14 and CDr15 allowing co-staining of the biofilm in real-time monitoring, it was found that eDNA forms the core and psl is positioned in outer shell of the biofilm. The sticky eDNA may act as a glue to attract the bacteria and components in the early biofilm stage, and psl appears as the decorating structure in the later stage [57] (Figure 4.53).

eDNA is not only generated in the biofilm, but also in neutrophil extracellular trap (NET) of neutrophils. Neutrophils are phagocytes and one of the granulocytes in our immune system with eosinophils and basophils. As the most abundant leukocytes in human blood, they act as the first defense line of the innate immune system. In addition to the release of granules including anti-microbials, neutrophils also secrete DNA to the extracellular matrix to form a trap to capture the invading bacteria. This sticky network structure is called as NET and eDNA is one of the major components. CDr15 also stains NET of neutrophils as a selective probe for eDNA [58] (Figure 4.54).

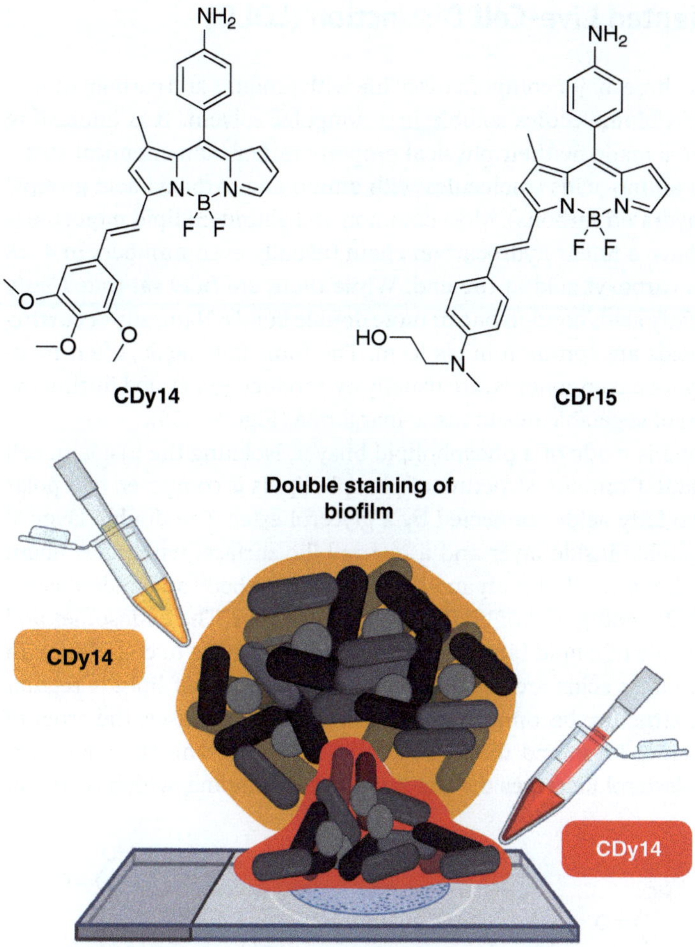

Figure 4.53 Structure of CDy14 and CDr15 and the double staining of biofilm.

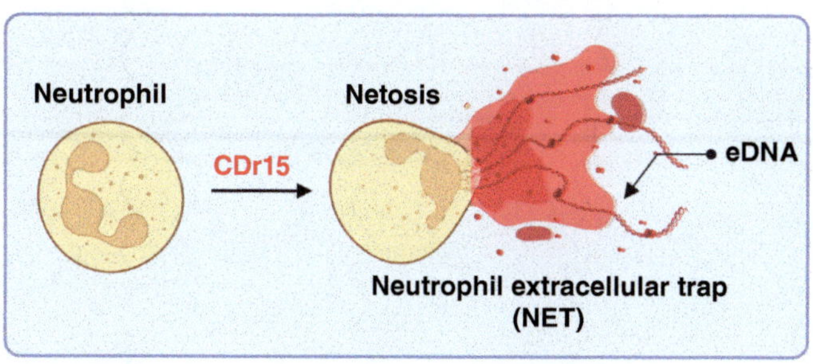

Figure 4.54 Neutrophil extracellular trap (NET).

4.3 Lipid-Oriented Live-Cell Distinction (LOLD)

Lipids are one of the three major components of life with proteins and carbohydrates. Lipids are defined as biomolecules soluble in a nonpolar solvent. It is interesting that lipid definition is made by their physical property rather than chemical structure as observed in amino acids (molecules with amino and carboxyl acid groups) or carbohydrates (hydrated carbons). Most common and abundant lipid molecule is fatty acids, which have a linear hydrocarbon chain (usually even numbers in 4–28 carbon long) and a carboxyl acid at the end. While there are fully saturated fatty acids, unsaturated fatty acids contain one or more double bonds. Naturally occurring unsaturated fatty acids are common in cis form. The trans-fatty acids, often mentioned as unhealthy food components, are usually by-products generated during the industrial reduction of vegetable oils to make margarine (Figure 4.55).

The cell membrane is made of a phospholipid bilayer, isolating the inside of cell from the environment. Common structure of phospholipids is composed of a polar head group and two fatty acids connected by a glycerol ester. The double layer of lipids has a hydrophobic inside layer and a hydrophilic surface, which has about 3–4 nm thickness. The intracellular organelles such as Golgi body and endoplasmic reticulum (ER) are also composed of lipid bilayer membranes. The endosomes and exosomes are also made of a lipid bilayer, so their shape is similar to cells, but with small size. If all the fatty acids are fully saturated, the stacking of lipid is regular and the membrane structure becomes rigid. If unsaturation increases, the order of the lipid stack becomes loose and the flexibility of the membrane also increases. The addition of cholesterol increases the rigidity of the membrane, which is similar

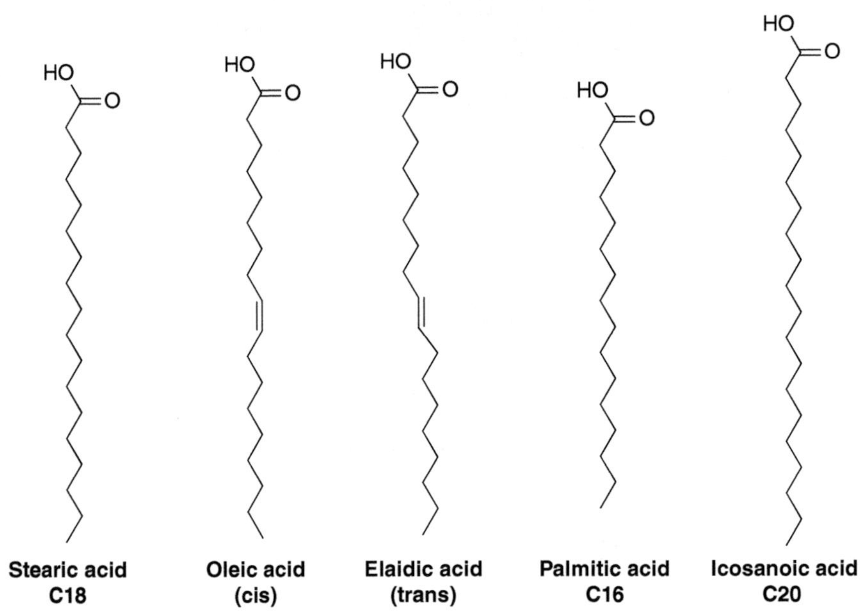

Figure 4.55 Fatty acid structures.

to stone embedding in a mud wall to make the structure more rigid. The cell membrane provides the anchoring basis for membrane proteins, which has a hydrophobic transmembrane domain.

4.3.1 Filipin as a Cholesterol Probe

Filipin is a natural product isolated from *Streptomyces filipinensis* and exists as a mixture of four compounds, with type III as the major component [59]. The extract has antifungal activity and inhibitory activity for caveolae-dependent endocytosis (Figure 4.56).

As filipin is highly fluorescent and binds to cholesterol, it has been used as a histochemical stain for cholesterol. Although there has been no direct application of Filipin for live-cell discrimination so far, it would be an interesting small-molecule candidate for distinguishing cells by targeting the different contents of cholesterol in the cell membrane.

4.3.2 Lipid Droplet Probes

Lipid droplet is a cellular energy storage organelle. Hydrophobic dyes such as Oil Red O, Nile Red, and BODIPY 494/503 have been used for lipid droplet staining. For industrial application, lipid droplet formation in algae or yeast cells has been used as a biofuel production process. BODIPY 494/503 was used to image lipid droplets to monitor the efficacy of biofuel production in oleaginous yeast. Through a high-throughput imaging platform, a small-molecule effector for enhanced lipid droplet production was elucidated by a triazine library screening [60]. This is an example of cell discrimination in different status or function using a small-molecule probe (Figure 4.57).

Through screening of Seoul Fluorophore (SF) library, a fluorescence turn-on lipid droplet probe FD13 was developed. As an interesting application, FD13 has been used for autophagy and lipid biogenesis monitoring in live cells. Autophagy (meaning self-eating) is a natural process of removing unnecessary or dysfunctional materials in the cell through a lysosomal digestion. Under autophagy conditions, the lipid

Figure 4.56 Structure of Filipin.

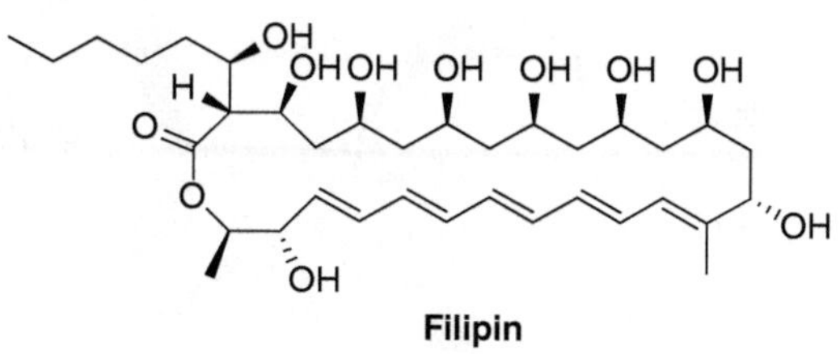

Figure 4.57 Structure of BODIPY 494/503.

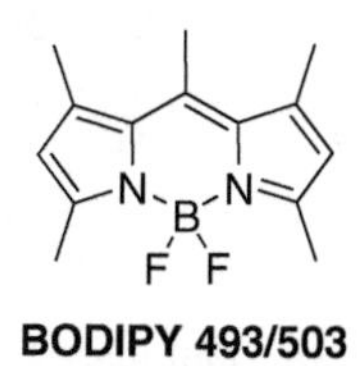

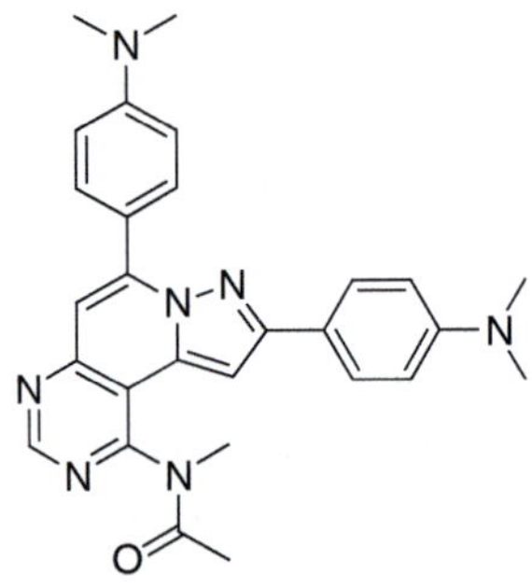

FD13

Figure 4.58 Structure of FD13.

droplets are reduced in the cells. During the lipid biogenesis in 3T3-L1 differentiation to mature adipocytes, the new lipid droplet formation could be imaged by FD13 in live cells [61] (Figure 4.58).

4.3.3 Neuron Probes

In the brain, there may be hundreds of different kinds of cell types, but classified into four main cell types: neurons, astrocytes, oligodendrocytes, and microglial cells. Neuronal cells have a long axon and dendrites, receiving and giving out signals. Astrocytes support neurons, and oligodendrocytes form myelin and insulate axon by wrapping. These cells are differentiated from neural stem cells. Microglia is a resident macrophage-like cells, migrating to the brain during the embryogenesis. Fluorescent probes for each cell type in the brain would be of great use (Figure 4.59).

Central nervous system (CNS) is a combined structure of brain and spinal cord, and connected to the peripheral nervous system (PNS) through neurons. Neuronal network is bidirectional. Signals from PNS to CNS are mainly through sensory

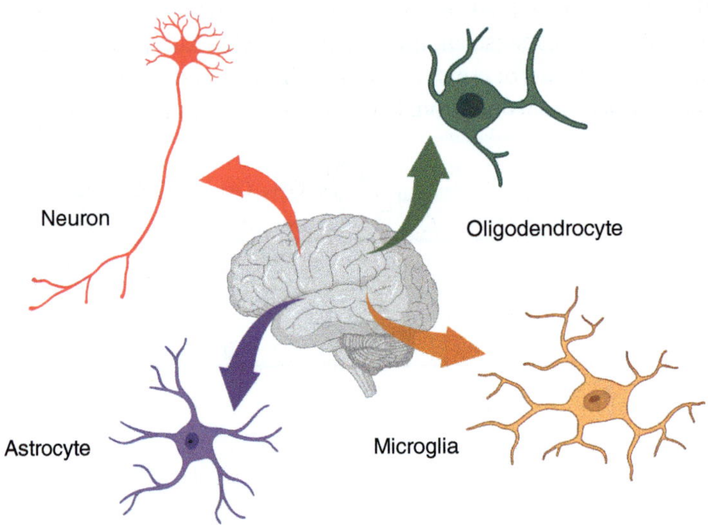

Figure 4.59 Cell types in the brain.

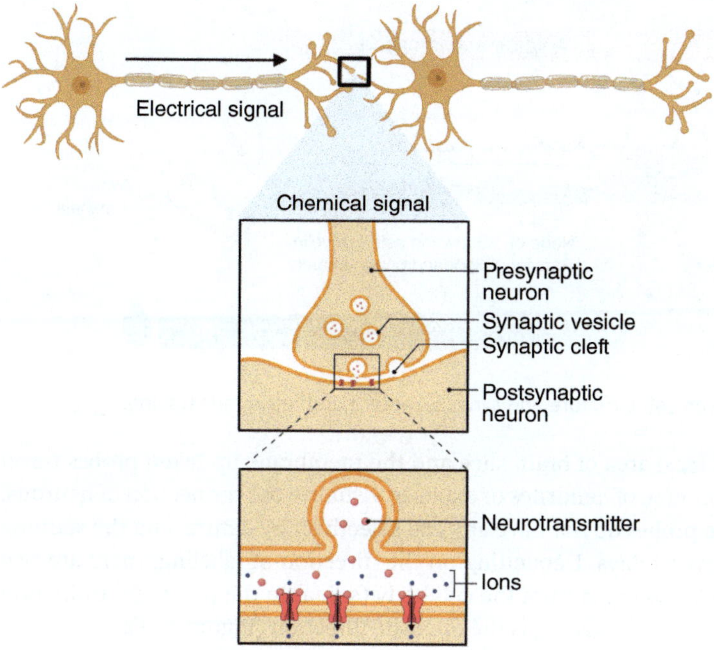

Figure 4.60 Neuronal signal transduction.

neurons and the opposite direction is through motor neurons. Neurons receive signals from multiple dendrites and give out signals through an axon. Some of the motor neuron axon is as long as 1 m. The neuronal signal travels through dendrites or axon as an electric signal via membrane potential wave. To prevent electric signal loss, axon is insulated by wrapped oligodendrocytes. The signal transduction between neurons is through synapse of each neuron, and chemical neurotransmitters are the signaling molecules. Therefore, neuronal signaling is a combined system of electric and chemical methods (Figure 4.60).

4.3.3.1 Nissl Stains as Neuron Body Probe

Historically Nissl dyes such as cresyl violet have been used for neuron staining in the tissue. The molecular target of Nissl dyes seems to be rRNA which is abundant in neurons. With the advent of fluorescent technologies, fluorescent version of Nissl staining dyes is introduced as NeuroTrace by ThermoFisher company. While various color probes are available and useful, the chemical information of the probes are limited, hampering the further improvement of the probes. Similar to the historical Nissl stain, the fluorescent NeuroTrance mainly stains the neuron body and the structural visualization of axon or dendrites is limited.

4.3.3.2 Plasma Membrane Dyes as Neuronal Network Probe

Utilizing the long membrane structure of dendrites or axons in neurons, hydrophobic plasma membrane probes, such as DiI or DiO series compounds, have been used to stain the network structure of neurons. These probes have been applied as a

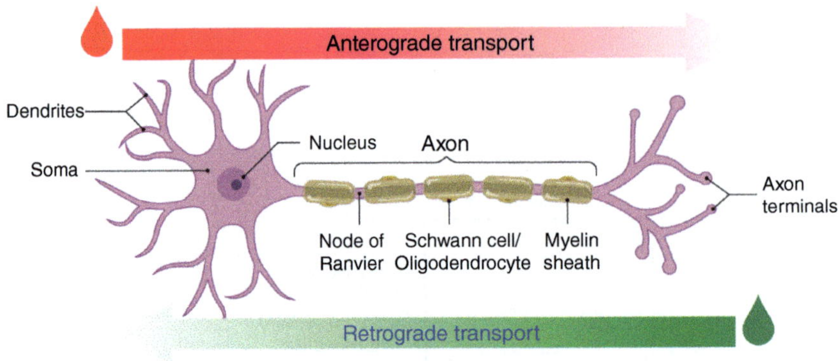

Figure 4.61 Neuron cell structure and synapse, antero and retrograde tracing.

solid chunk to a local area of brain slice and the membrane up-taken probes travel through the membrane of dendrites or axons to visualize the connection of neurons. While useful, the probes do not have any cell selectivity by nature, and the staining protocol takes several days. Depending on the direction of labeling, there are two ways. Anterograde tracing is from the cell body (soma) to the point of termination (synapse), and retrograde tracing is the opposite direction (Figure 4.61).

4.3.3.3 NeuO as a Universal Neuron Probe

In 2015, through an unbiased screening of mouse brain cells (neurons, astrocytes, and microglia) and systematic SAR studies, NeuO (**Neu**ron **O**range: $\lambda_{ex} = 468$ and $\lambda_{em} = 553$ nm) was developed as a universal neuron probe [62]. NeuO stains live neurons quickly and visualize the axons and dendrites in addition to the cell body of neuron. Neurons are classified based on the secretion of neurotransmitters, and the specificity of NeuO among the subtypes of neuron would define the applicability of NeuO. When tested, regardless of the neuron subtypes, NeuO stains most of the neurons including glutaminergic, GABAergic, cholinergic, dopaminergic, and serotonergic neurons. The cell selectivity of NeuO was also confirmed in human ES-derived neurons and in zebrafish neuromasts (Section 6.3), demonstrating the universal selectivity across species (Figure 4.62).

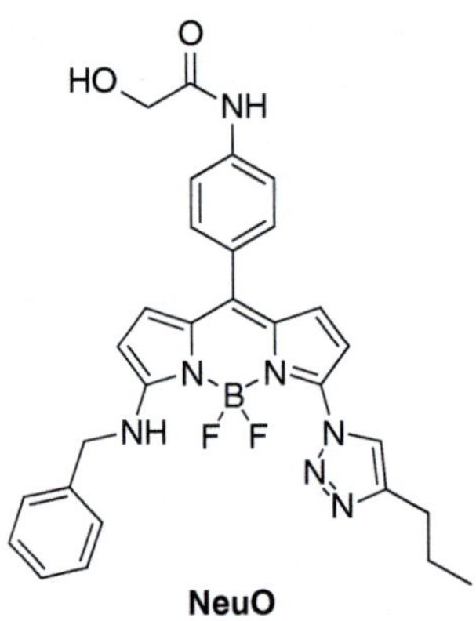

NeuO

Figure 4.62 Structure of NeuO and cell selectivity outline among brain cells.

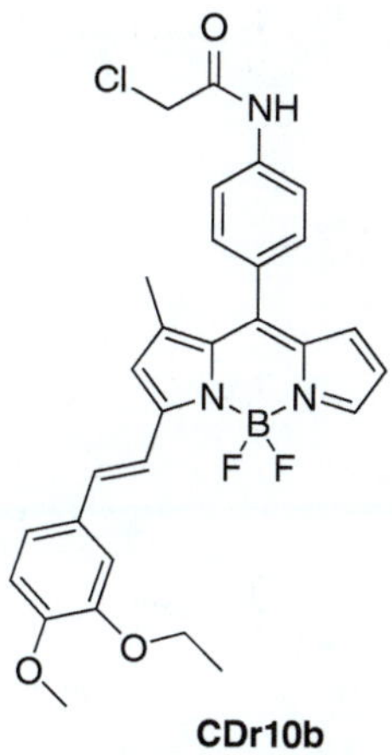

CDr10b

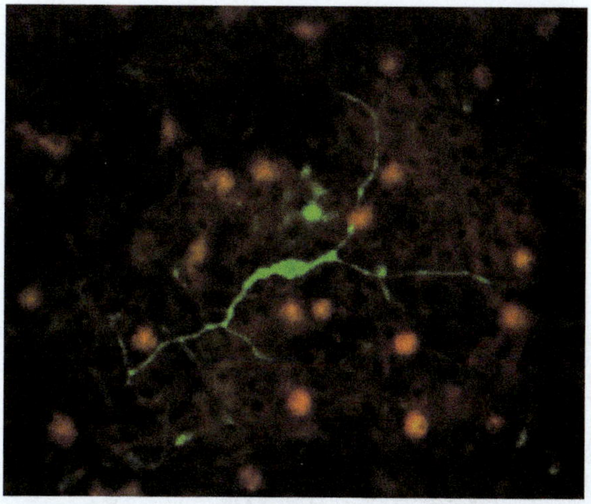

Figure 4.63 Structure of CDr10b and images (neuron and microglia). Source: Ref. [62], from John Wiley and Sons, Inc.

The neuron cell selectivity of NeuO could visualize microglial movement (labeled by red color CDr10b) in co-culture with neurons (labeled by orange color NeuO), demonstrating the multicolor labeling of cells. Compared to relatively static neuron structure, microglia dynamically move around the neurons in minutes time scale. In this condition, the neuron staining was stable up to 36 hours (Figure 4.63).

As neuronal growth and movement is much slower than that of dynamic microglia, multiple hours of long-term imaging would be necessary. For a long-term real-time fluorescence imaging, the photobleaching of the probe and the phototoxicity induced by the excited probe are main concerns. The low toxicity and high stability of NeuO permitted long-term monitoring of neuron growth and changes over a 20 hour time frame. From the inverted images (green fluorescence was converted into black color over white background), the fine structural changes of axons and dendrites could be visualized in detail (Figure 4.64).

In applications, NeuO showed the utility as a counter stain of astrocyte differentiation monitoring by natural product treatment [63]. In another study, the neurogenesis effect of curcurbitacin B was successfully visualized by NeuO, and the results were compared with memory protection in Alzheimer mouse model [64]. The axon genesis on vertical nanowire [65] and neuronal migration on silicon microcone arrays with different pitches were facilitated by live neuron visualization with NeuO [66]. Not only in cell cultures, NeuO could visualize also the neuron structure in whole brain slices (Section 5.3.2).

Despite the outstanding performance in neuron-selective staining, the selectivity mechanism of NeuO has been unknown for a long time. The NeuO signal in stained neurons survived paraformaldehyde fixing to some extent, but was fully washed out by methanol or Triton-X treatment. The observation indicates that NeuO does not make a covalent bonding and there is no strong binder in the cell. The NeuO staining in prefixed cells did not show any neuron selectivity. This is in contrast with the

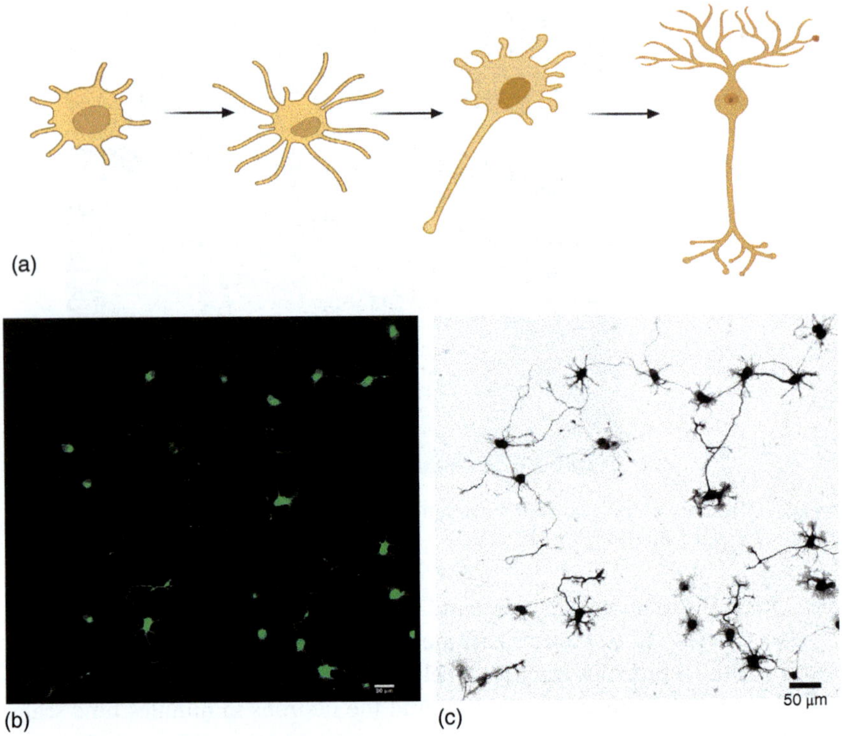

Figure 4.64 (a) Neuronal growth scheme (b) the fluorescence images of axon and dendrite with structural details. Source: (b) Ref. [62], from John Wiley and Sons, Inc. (c) The inverted fluorescence signal (green to black over white background) image of (b). Source: Created by Author.

conventional Nissel staining, which requires membrane permeabilization by detergent before the staining. Also, the behavior is different from membrane probes DiI or DiO, which works as far as the intact membrane structure is intact. In contrast, the neuron cell selectivity of NeuO only remained in live neurons, implying that an active biological function is involved in the selectivity mechanism. Further study showed that NeuO's selective uptake requires fast recycling of vesicles between plasma membrane and endosomes, casting the clue of the mechanism as LOLD.

4.3.4 B Lymphocyte Probe: CDgB

In our blood, there are main population of RBCs for oxygen supply to the body, and small numbers of white cells. White cells are responsible for immune response as our body's defense system. There are innate immune cells such as monocytes, granulocytes, and natural killer (NK) cells, and adapted immune cells such as B and T cells. B cells are responsible for antibody generation and T cells are responsible for cellular immune response using T-cell receptors. The lymphocytes B and T cells are very similar to each other by shape and size. In adults, the cells originate from the common hematopoietic stem cells in the bone marrow. Partially differentiated

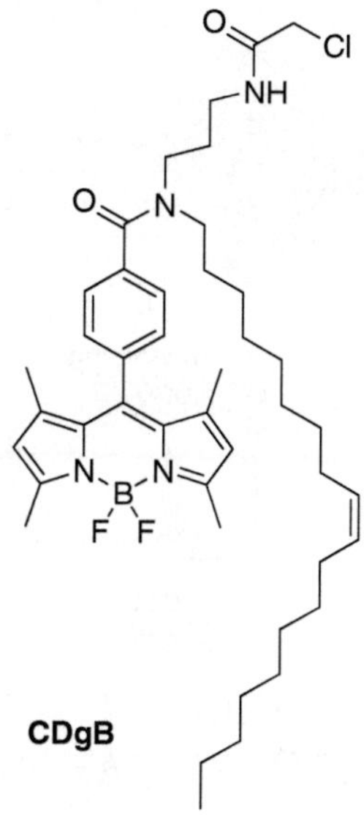

Figure 4.65 Structure of CDgB.

T cells migrate to thymus for further maturation and eventually move to lymph nodes and spleen. B cells partially mature in the bone marrow and then move to the lymph nodes and spleen. So, spleen is rich with T and B cells (1 : 3 ratio) and can be considered as a giant lymph node. Without antibody for CDs, it is almost impossible to discriminate them. So, it was expected to be extremely difficult to design or develop B- or T-cell-selective probe. Without a clue, more than 10 thousand DOFL have been screened in mouse B and T cells obtained from the spleen, and finally the first B-cell-selective small-molecule probe CDgB (**C**ompound of **D**esignation **g**reen **B** cell) was discovered in 2021 (Figure 4.65).

The structure of CDgB contains a long hydrocarbon chain (18-carbon long chain with an unsaturated double bond) and its staining in B cell was localized in the cell membrane. Therefore, the cell selectivity of CDgB may be originated from the cell membrane or its components such as membrane proteins. First of all, the carbon chain length of CDgB was systematically studied. Derivatives with shorter than 10 carbon chain showed no selectivity between T and B cells. The B-cell selectivity showed up when the chain length reached C12, and the stain index was highest at C14 and 16. When the chain length reached C20 and above, the separation of the cell populations became obscure, with a broader distribution of the cells. Interestingly, still, the brighter populations were always B cells, and there was no case of T-cell preferable recognition (Figure 4.66).

The lipidomic analysis of T- and B-cell membranes disclosed the difference of lipid composition of T-and B-cell membranes. The T cells have in general longer fatty acid chains than the B cells. Also, cholesterol content in the T-cell membrane was twice that of the B-cell' membrane. With longer fatty acid chains, the membrane would be more rigid. The higher cholesterol contents would further increase the membrane rigidity. Altogether the higher flexibility of the B-cell membrane than the T-cell membrane seemed the origin of the cell selectivity.

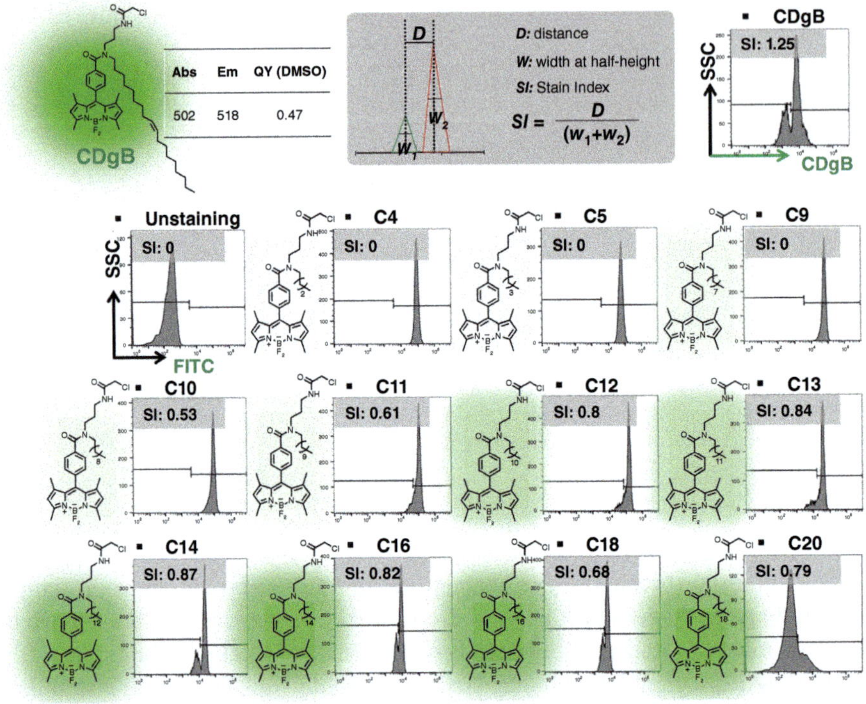

Figure 4.66 Carbon chain length effect of CDgB to the T- and B-cell selectivity. Source: Haw-Young Kwon et al. [67]/American Chemical Society.

To rule out the possible role of membrane protein for the cell selectivity, vesicles made of pure lipid molecules with different fatty acid chain lengths and cholesterol contents were prepared as T- and B-cell membranes. Through flow cytometry using SUV (Small Unilamellar Vesicle: 200 nm size) and fluorescent microscopy using GUV (Giant Unilamellar Vesicle: 20um size), it was clearly demonstrated that CDgB's selectivity could be achieved only by the lipid composition, without a need of other components such as a membrane protein. The hydrophobic nature of CDgB induced nanoparticle size aggregation of CDgB in aqueous media, quenching the fluorescence. The higher flexibility of the B-cell membrane takes up CDgB nanoparticle and fuses it to the membrane kinetically faster than the T-cell membrane. Once CDgB is fused in the cell membrane, the fluorescence was turned on based on disaggregation-induced emission (DIE) mechanism. The DIE mechanism suppressed the background signal of CdgB by self-quenching. Therefore, this is the first example of pure Lipid-Oriented Live-cell Distinction (LOLD) (Figure 4.67).

T cell and B cells are originally from the common precursor cells. So, if the T-cell membrane is thicker and more rigid than B-cell membrane, how about the common ancestor cell? Is it like B or T cell? CDgB would be a unique tool to monitor the change of the membrane during the differentiation procedure. For the study, the immature bone marrow cells were analyzed by CDgB and compared to spleen cell results. Spleen has CDgB+ B cells and CDgB− T cells. Bone marrow also has

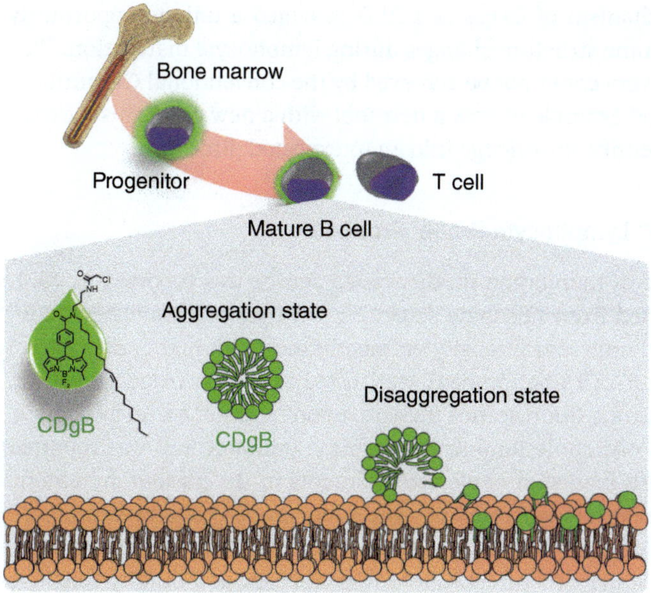

Figure 4.67 The mechanism of CDgB.

CDgB- and CDgB+ populations as expected. However, interestingly, the bone marrow had a new even brighter CDgB++ population! The three populations were separated and systematically analyzed with more than 20 antibodies using flow cytometry connected with mass spectrometry (CyTOF) technique [67]. CDgB- population is T and NK cells as expected. CDgB+ population is mature B cells and CDgB++ cells are immature B cells. Hematopoietic stem cells start to differentiate into a common lymphoid progenitor and divide into B-cell lineage and T-cell lineage. The younger cells may have a softer and more flexible membrane and has the highest uptake of CDgB generating CDgB++. When they differentiate into B-cell lineage, the membrane becomes slightly more rigid (CDgB+), but if they differentiate into T-cell lineage, the membrane becomes abruptly rigid (CDgB–). From this analysis, NK cells behave more similarly to T cells rather than B cells (Figure 4.68).

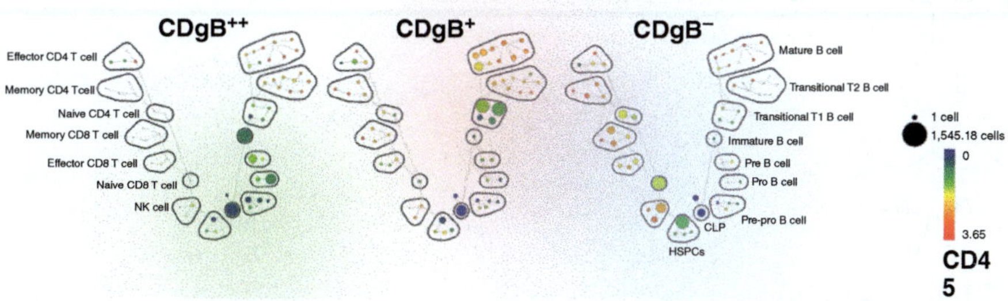

Figure 4.68 Differentiation of bone marrow cells. Source: Haw-Young Kwon et al. [67]/ American Chemical Society.

The totally new mechanism of CDgB as LOLD provided a unique opportunity of analyzing the membrane structure changes during lymphocyte maturation. This kind of insight or discovery could not be achieved by the conventional CD antibodies. This would be a good example of how a new tool with a new mechanism drives the advancement of scientific knowledge into an unexpected direction.

4.3.5 Activated CD8$^+$ Lymphocyte Probe: Probe41

An interesting example of membrane fluidity-based sensor was reported in 2021. The probe41 was selected from liposome screening with various cholesterol contents, showing higher fluorescent intensity on membranes with higher cholesterol contents. In the Jurkat T-cell line, probe41 was tested as a drug response sensor, and it demonstrated a strong fluorescence increase upon treatment with avasimibe. Avasimide is an Acyl-CoA: cholesterol acyltransferase inhibitor and was reported as a T-cell activator with increased cholesterol contents in the plasma membrane [68]. When CD8+ T cells were activated by avasimibe, probe41 showed enhanced fluorescence on the cell membrane [69]. Based on the fact that cholesterol increases the rigidity of the membrane structure, probe41 was proposed as a fluidity-sensitive probe (Figure 4.69).

While the application of probe41 is unique in T-cell activation, the mechanism of selectivity is conflicting with CDgB, which showed the preferable selectivity to more flexible membrane in the previous section. It is noted that CDgB and probe41 have structural similarity, possessing the same fluorophore of tetramethyl BODIPY with long hydrophobic carbon chain. So, it would be difficult for the two compounds to possess the same mechanism with opposite selectivity. With an extra amino acid and indole moiety, probe41 may have another binding target on the plasma membrane, either cholesterol itself or other components induced during the T-cell activation. To answer the riddle, a careful further investigation may be necessary.

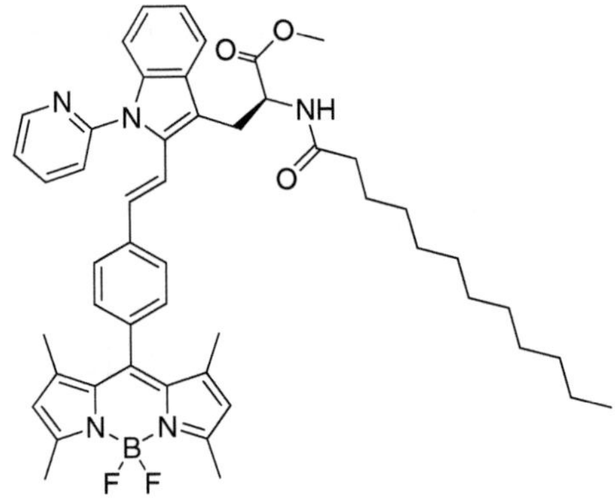

Probe41

Figure 4.69 Structure of Probe41 and CD8+ T-cell images.

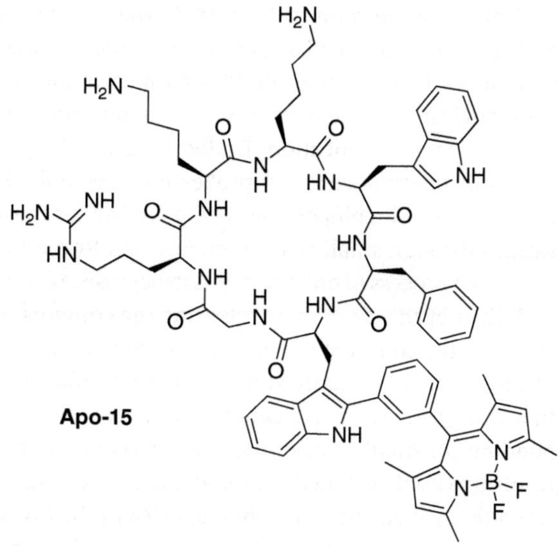

Figure 4.70 Structure of Apo-15.

4.3.6 Apoptotic Cell Probe: Apo-15

Cell membrane is composed of a double layer of phospholipids, and the inner and out leaflet composition is different. Inner leaflet is enriched with phosphatidylinositol and phosphatidylserine. When the cells go for apoptosis, the asymmetric order of the membrane is disturbed by phospholipid flip-flop. The phosphatidylserine exposed to the outer leaflet is the target of Annexin-V protein. Fluorescently labeled annexin V has been used as the standard probe for detecting apoptosis of cells. Positively charged cyclic peptide Apo-15 has been developed as a phosphatidylserine binder and applied to the apoptosis probe. The apoptotic neutrophil was selectively labeled by Apo-15, and the phagocytosis by macrophage was successfully imaged. Instead of any protein or carbohydrate targets, this is unique recognition of lipid components by a small-molecule probe for live-cell distinction, i.e. LOLD [70] (Figure 4.70).

4.4 Gating-Oriented Live-Cell Distinction (GOLD)

For the drug molecule's mechanism, a binding target may be necessary. Drug molecules should be regulators, by blocking the evil target's function or boosting friendly target's function. For the functional effect, direct or indirect physical contact and binding event should be involved in the action mechanism. Sensors may be in a similar situation. But, for successful probes, the existence of binding targets may not be always necessary. For example, the intracellular presence of the probe itself is good enough to distinguish the target cell, even though there is no specific binding partner of the probe in the cell. Unlike drugs, a probe's primary role is that of a reporter, not a regulator.

If there is a mechanism to increase the uptake and decrease the exit of the probe for the target cell, it will be possible to achieve the cell selectivity. Cells are compartmentalized by lipid bilayers. Plasma membrane separates the inner cellular space from environment, and intracellular membranes in organelles make further complex compartmentalization. Endocytosis on the plasma membrane is an apparent option of internalization of probes into the cells. Endocytosis could be either the uptake for fluids (pinocytosis: cell drinking) or macromolecules (phagocytosis: cell eating) through small vesicle (endosome) formation budding off into the cell. The reversible process of endocytosis is exocytosis, which fuses the endosome membrane with the plasma membrane releasing the contents into the extracellular space. Exocytosis is the common procedure to secrete high concentration of hormones or neurotransmitters originally stored in the granules in the cytoplasm. It is noteworthy that exocytosis does not generate any new small vesicles. The process to form a budding off small vesicle into the extracellular space is called ectocytosis and the resulting vesicle is called exosome. Therefore, endocytosis, exocytosis, and ectocytosis could be mega-scale mechanisms for probe transport.

For the molecular level of probe passage, transporter proteins on the membrane may play an important role. The material in and out through each membrane will decide the equilibrium for each molecule. If a molecule's concentration is higher in one side of a membrane than the other, a driving force will be generated by the concentration gradient. While a passive transportation is also possible, the process is often facilitated by transporter proteins, especially if the molecule is related to the biological function of the cell. Such transporters are called solute carriers (SLCs) and more than 400 members are known in humans and mice. SLCs on the plasma membrane usually facilitate the uptake of molecules into the cell, and this process does not require energy by utilizing the favorable concentration gradient.

In contrast, there are about 50 ATP-binding cassettes (ABC) transporters, usually for pumping out molecules from inside of the cell through the plasma membrane. As the name implies, this process is an ATP-dependent energy-consuming procedure, and the process can occur against the concentration gradient. In general, SLCs are usually responsible for intake of nutrients and ABCs are known to pump out toxic drug molecules from the cell. Therefore, overexpression of ABCs is often correlated to multidrug resistance mechanism. If a probe is substrate of a SLC and an ABC, the high expression of SLC will increase the probe accumulation in the cell, and the high ABC expression may have the opposite effect. The cell distinction mechanism based on the probe transportation could be defined as GOLD (Gating-Oriented Live-cell Distinction). GOLD is a unique mechanism as it does not require any binding target, unlike the conventional probe mechanisms such as POLD and COLD (Figure 4.71).

4.4.1 Cell Imaging Probes through Phagocytosis

Macrophages are one kind of immune cells and are representative phagocytes. Phagocytes engulf foreign particles, such as bacteria or dead cells, and sometimes digest the ingested materials in the intracellular acidic organelles, phagosomes.

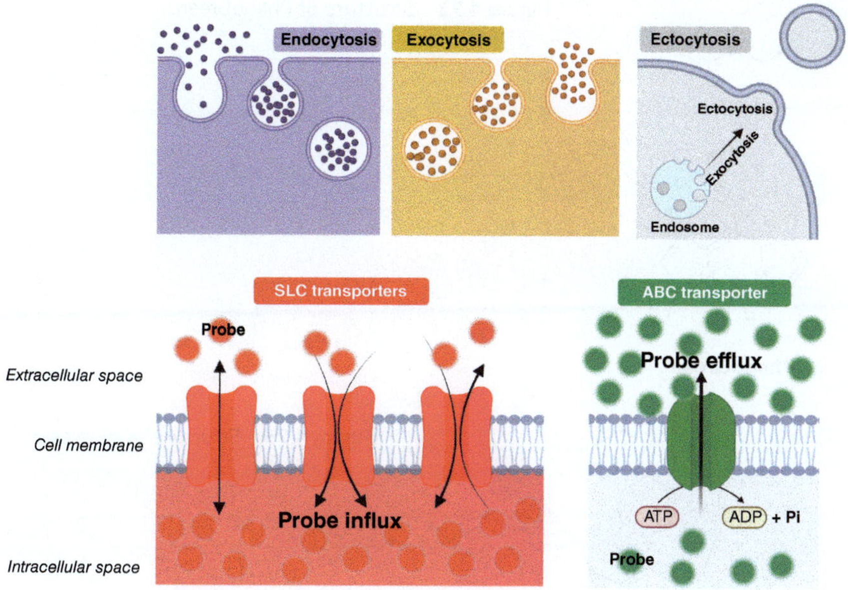

Figure 4.71 GOLD through endocytosis/exocytosis/ectocysosis and SLC/ABC function.

The precursor cell of macrophages, called as monocytes, circulates in the blood stream and leave the blood vessel near inflammation or infection area through extravasation. The monocytes mature into macrophages in the tissue and exhibit various immune responses. Therefore, sensing and imaging of macrophages could be the indication of inflammation in the tissue. Due to the phagocytic property, macrophages tend to engulf various materials including dyes, so in general they tend to stain stronger than any other cells around. Based on the higher internalization, various macrophage probes have been developed and demonstrated for in vitro and in vivo inflammation imaging.

PhagoGreen is one of the representative probes of this class for phagocytic macrophages. The amine group in PhagoGreen may quench the fluorescence of BODIPY by photoinduced electron transfer (PET) effect in neutral condition. After the cell entry, the localization of PhagoGreen in acidic organelle will protonate the amine group abolishing the PET effect and turn on the fluorescence of PhagoGreen, which makes the fluorescence signal outstanding in active phagocyte [71] (Figure 4.72).

There are various types of macrophages derived from the same monocytes, depending on the environmental differences. Osteoclasts are known to break down the bone tissue and the process is called bone resorption. Interestingly, osteoclasts are also monocyte-derived cells. To selectively image osteoclasts, a pH-sensitive BODIPY probe, pHocas-3, is developed. pHocas-3 is equipped with two bisphosphonate groups to bind the bone tissue, and pH sensitivity of the amino group turns on the fluorescence only at acidic condition where the active bone resorbing is carried out (Figure 4.73).

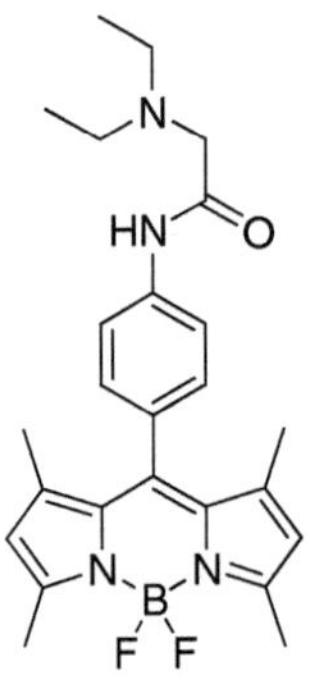

Figure 4.72 Structure of PhagoGreen.

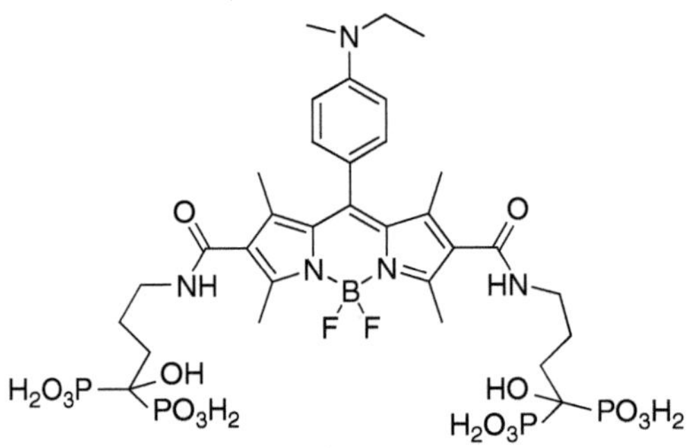

Figure 4.73 Structure of pHocas-3.

In metastatic lung tumor, the metastasis-associated macrophages (MAM) and their precursor cells are accumulated and promote the metastatic process via CCL2–CCR2 signaling [72]. CCL2 is a chemokine and CCR2 is its receptor. For the cell-type-specific internalization of the probe, this signaling machinery was hijacked by conjugating macrophage-activatable fluorophore (MAF) to mouse chemokine CCL2 (mCCL2). MAF is the same pH-sensitive BODIPY motif of PhagoGreen. The conjugated probe mCCL2-MAF may bind to MAM selectively and be internalized/activated by the acidic condition inside cells [73]. The activation of mCCL2-MAF is stepwise: two steps of internalization and activation by acidic pH as an AND gate. The AND gate probe may generate more specific signals with a lower background (Figure 4.74).

The nodal metastasis of cancer is crucial to prescribe suitable treatment for patients. macrophage-specific fluorescent probe (MFP) was developed to visualize sentinel lymph nodes during surgery, highlighting abnormalities related to inflammation and tumor infiltration with signal enhancement. MFP selectively visualized monocyte and macrophage cell populations in vitro, by live-cell imaging and flow cytometry, as well as in vivo, providing the intraoperative imaging platform and eliminating the need for invasive nodal dissections [74] (Figure 4.75).

In addition to macrophages, neutrophils are another important phagocyte in the immune system. Among white blood cells, neutrophils are actually the most

Figure 4.74 Structure of mCCL2-MAF.

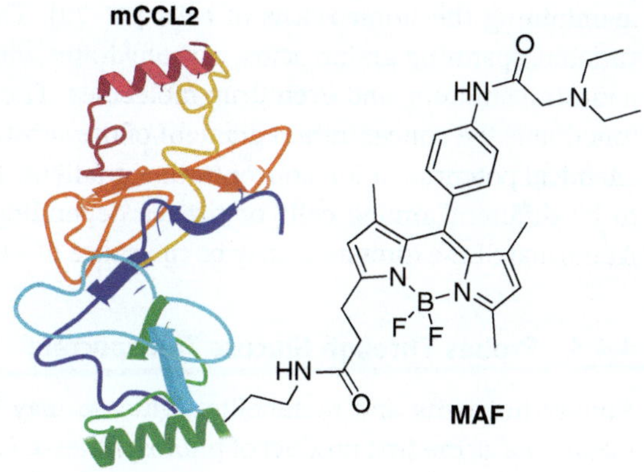

Figure 4.75 Structure of MFP.

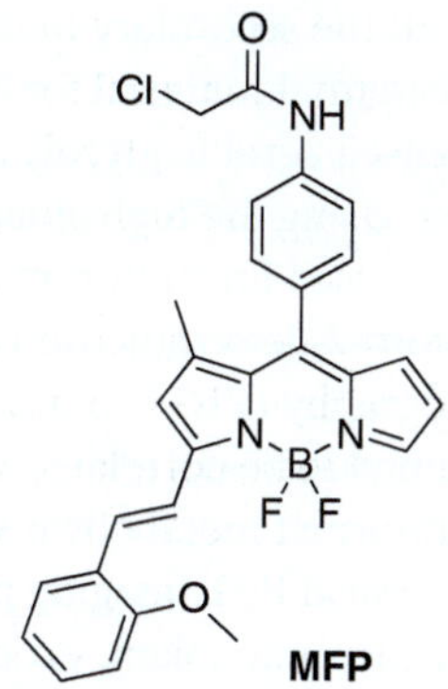

abundant cells making up 40–70% of all leukocytes. Unlike monocytes, neutrophils have a nucleus divided into 2–5 lobes. As the first line of immune defense, upon activation, neutrophils migrate into the wounded tissue, where they survive for 1–2 days. In contrast, the life span of macrophages is long, ranging from months to years [75]. The phagocytic property of neutrophils is used for autologous labeling of radioactive contrast agent for the whole-body imaging of inflammation using single-photon emission computed tomography (SPECT) (Section 9.4).

4.4.2 Probes Through SLC Transporters

More than 400 human SLCs have been discovered so far, and they are classified into 65 SLC superfamilies [76]. For example, the SLC1 family is a high-affinity glutamate and neutral amino acid transporter family with seven members (SLC1A1-7). SLC2 family is a carbohydrate transporter family and also called as GLUT transporters. SLC2A (or GLUT) has 1–14 subtype transporter proteins with several pseudogenes in the family. A related family is SLC5, sodium glucose cotransporter family, also known as SGLT with 12 members. Following these examples, SLCs are classified according to their known substrates and sequence homology.

SLC transporters are usually localized on the plasma membrane or intracellular organelles, and facilitate the membrane penetration of various small molecules, maintaining the homeostasis of cells [77–79]. The substrates of SLCs are highly variable, spanning amino acids, carbohydrates, vitamins, nucleic acids, metal ions, neurotransmitters, and even drug molecules. The driving force of SLC transporter function is the concentration gradient of the substrate, or indirectly utilizes electrochemical potential or ion concentration gradient. The expression of SLC is expected to be different among cells or tissues depending on their functions and needs. Abnormal SLC expressions may be the cause of various metabolic diseases.

4.4.3 Probes Through Glucose Transporters

Among nutrients and metabolites, glucose may be the most prominent primary metabolite as the first product of photosynthesis. Glucose is the common nutrient in almost all living things as the energy source and also the starting material to synthesize all the secondary metabolites such as amino acids and lipids. While glucose is the essential material for all cells, tumor cells demand much higher glucose due to increased aerobic glycolysis and aberrant material metabolism, called the Warburg effect. Using the high glucose uptake property of tumor cells, glucose derivatives are widely used for tumor imaging. For the clinical application, radioactive ^{18}F-labeled 2-fluoro-2-deoxyglucose (FDG) is almost the single dominant positron emission tomography (PET) contrast agent among many others. The high uptake of FDG is assumed to be correlated with the high glucose consumption for active metabolism or abnormal metabolism such as in inflammation or tumor progress. The details of the clinical PET imaging probe are discussed in Section 9.5 (Figure 4.76).

By the same token, various fluorescent derivatives of glucose were developed and used mainly for tumor imaging and energy consumption. 2-NBDG is the representative glucose derivative with the longest history and most number of examples [80]. By replacing the mediocre fluorophore, NBD with brighter and superior probes were developed and introduced. A Cy3-labeled glucose derivative at 1-a-position, **Cy3-Glc-α** demonstrated increased glucose uptake in tumor cells compared with normal cells [81]. A similar molecule design was applied to the two-photon probe AG2, and a deep tissue imaging of tumor was demonstrated [82]. A longer wavelength dye, Cy7-conjugated probe at 2-position of glucose, **CyNE 2-DG** also showed cancer selectivity with deep tissue imaging possibility [83]. Although the glucose derivatives proved to enter the cells through GLUTs by glucose competition assays, the specific GLUT target for each probe has not been fully elucidated (Figure 4.77).

The glucose transporter family, GLUT family, has 14 subtypes in humans, and is categorized into 3 classes based on sequence homology: Classes 1 (GLUT 1–4, 14), 2 (GLUT 5, 7, 9, 11), and 3 (GLUT 6, 8, 10, 12, 13). All GLUTs appear to possess 12 transmembrane domains, and one or more GLUT proteins are expressed in

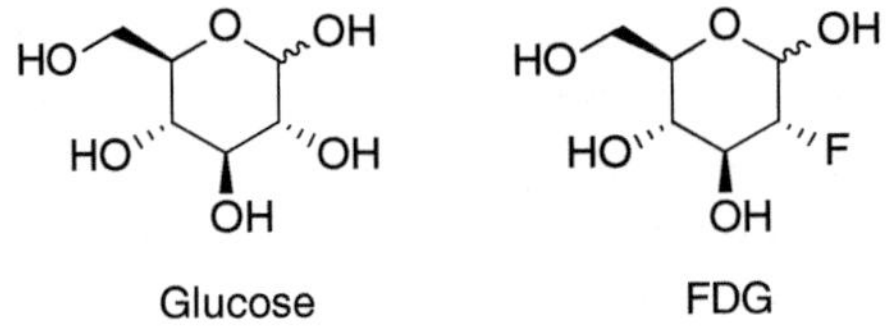

Figure 4.76 Structure of glucose and ^{18}F-FDG (the wavy bond represents the mixture of α and β anomers).

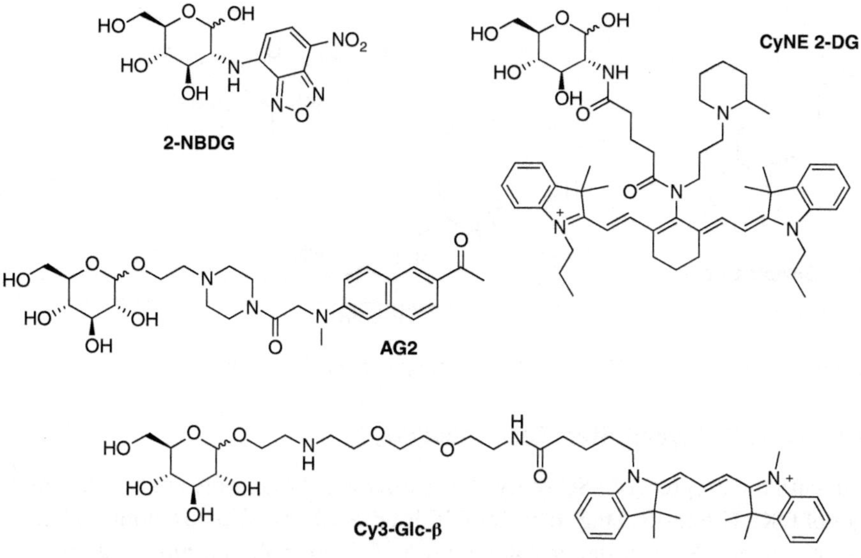

Figure 4.77 Structure of carbohydrate tumor probes.

practically every cell type of the body. Each GLUT protein has different kinetic and regulatory properties in a cell-type-specific manner. The most studied subtypes are GLUT1–4 with different cell distribution: GLUT1 (erythrocytes, brain, blood-brain barrier [BBB]), GLUT2 (liver, pancreatic β-cells, intestine, kidney, brain), GLUT3 (brain, testis), and GLUT4 (adipose, muscle). While the protein name (glucose transporter) implies that GLUT proteins may be selective facilitator only for glucose uptake, actually the substrate specificity is not very stringent tolerating various carbohydrates: GLUT1 (glucose, galactose, mannose, glucosamine), GLUT2 (glucose, galactose, mannose, glucosamine, fructose), GLUT3 (glucose, galactose, mannose, xylose), and GLUT4 (glucose, glucosamine) [84]. It is noted that GLUT2 is overexpressed in insulin-generating pancreatic β-cells, and GLUT4 is overexpressed in the main insulin-responding cells, adipose and muscle. Insulin is the hormone to activate glucose uptake by adipose or muscle. Pancreatic β-cells sense the glucose concentration and regulate insulin synthesis and secretion. When the blood glucose concentration is high, β-cells secrete insulin in response. Therefore, GLUT2 and GLUT4 have been studied widely in diabetes.

Streptozotocin (STZ) has been used to selectively destroy β-cells to induce a diabetic model animal. The selective toxicity of STZ only to β-cells without affecting other cells was quite a unique phenomenon. The structure of STZ is close to glucose and it was explicated that STZ is a selective substrate of GLUT2 and selectively killed GLUT2 overexpressed β-cells [85]. Fluorescent glucose derivative TP-β mimicking STZ structure was designed and demonstrated a selective staining of β-cells in pancreatic islets. Combined together with α-cell probe TP-α, TP-β could generate multicolor three-dimensional pancreatic islet images visualizing both α- and β-cells through two-photon microscopic technique [86]. This example shows that subtype-specific probe for each GLUT is achievable (Figure 4.78).

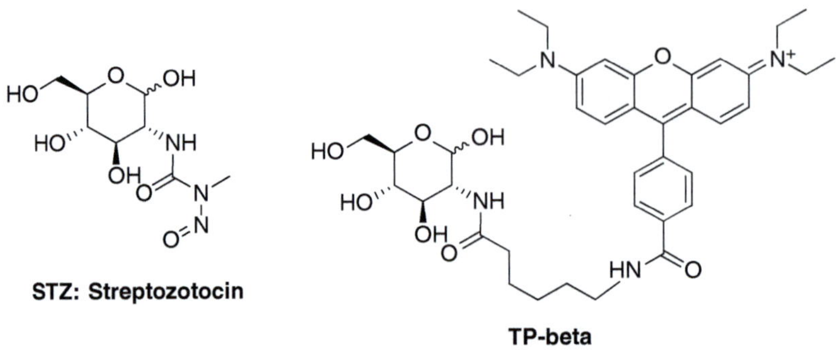

Figure 4.78 Structure of STZ and TP-β + 3D imaging of pancreatic islet.

4.4.4 Naïve Embryonic Stem Cell Probe: CDy9

Along with the progress of ESC study, it was discovered that there are two different stages of ESC. The earlier stage of naïve ESC is with a dome-shaped colony and later stage of primed ESC is a flatter colony. Both of them maintain pluripotency, but the molecular characteristics are slightly different and may have different profiles depending on the specific application [87]. Interestingly, mESC is more similar to naïve ESC, and hESC is closer to primed ESC (Figure 4.79).

In 2015, a new mESC probe CDy9 (**C**ompound of **D**esignation **y**ellow **9**) was developed. In comparison to previously developed ESC probe CDy1, CDy9 showed very restricted higher cell selectivity only to mESC without much tolerance [88].

In a further 2018 study, it was found that CDy9 even can distinguish two ESC stages, by selectively staining only naïve ESC, not primed ESC [89]. Following the biochemical characteristics, mESC was stained by CDy9, but not hESC. When mESC was primed, the staining disappeared. This shows that reprogramming of primed type hESC into naïve state could be monitored by CDy9 (Figure 4.80).

The target of CDy9 was identified by comparative analysis of overexpressed genes in naïve ESC. The confirmed target by gene knock-out was SLC13A5. This

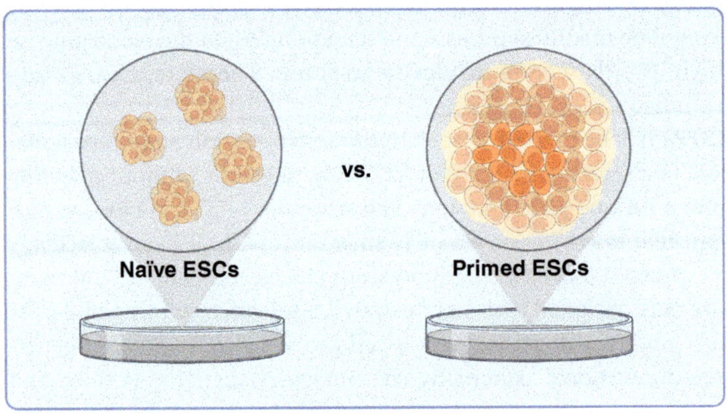

Figure 4.79 Naïve and primed ES colony shape.

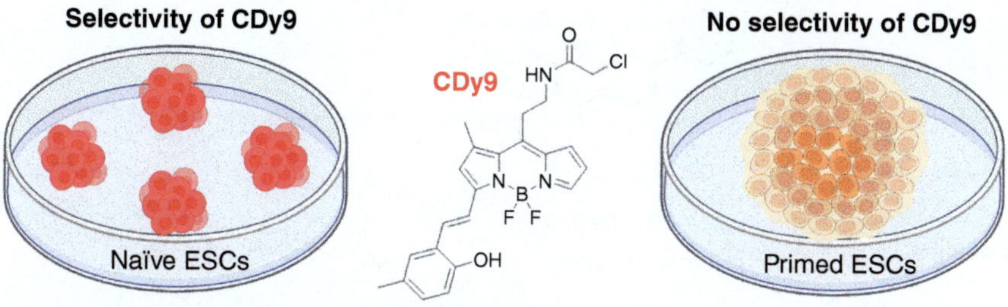

Figure 4.80 Structure of CDy9 and imaging of naïve vs. primed ESC.

transporter was originally known to be citric acid transporter, highly expressed in naïve ESC. The expression decreases over differentiation into primed ESC and generated the cell distinction clue through GOLD mechanism.

4.4.5 Neurotransmitter Mimetic Probes

Neurotransmitters are signaling molecules between neuronal cells, such as acetyl choline, glutamate, GABA, dopamine, epinephrine, and norepinephrine. Neurons have multiple dendrites to accept signals and once the signals exceed a threshold, they fire the electric signal through an axon. At the end of an axon, there is a synapse facing receiving cells synapse with a gap. In the presynaptic terminal, vesicles containing a high concentration of neurotransmitters wait for the electric signal of firing. Upon firing, vesicles undergo merging with the plasma membrane through an exocytosis to release neurotransmitters into the synapse gap. Neurotransmitters bind to receptors on the postsynaptic side and transfer the signal to the dendrite of receiving cells. In summary, the electric signal is converted into the chemical signal at the synapse. Neuronal cells are classified with the releasing neurotransmitters; if the neuron uses acetyl choline, it is a cholinergic neuron. Receiving neurons have receptors for various neurotransmitters (Figure 4.81).

Once the release of neurotransmitters is finished, majority of the signaling molecules are recovered by the original cells for recycling. For the reabsorption

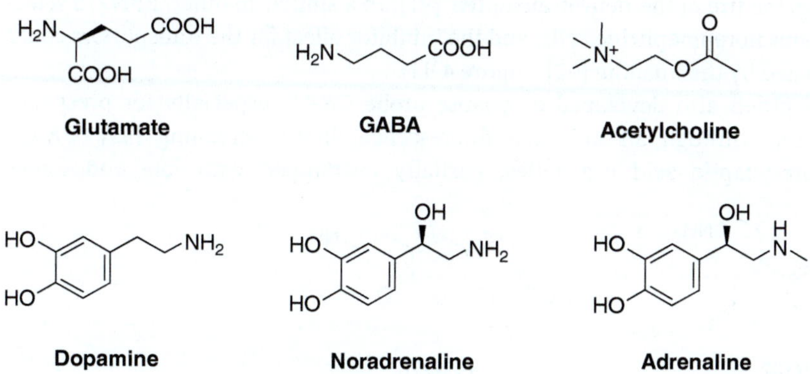

Figure 4.81 Structures of neurotransmitters.

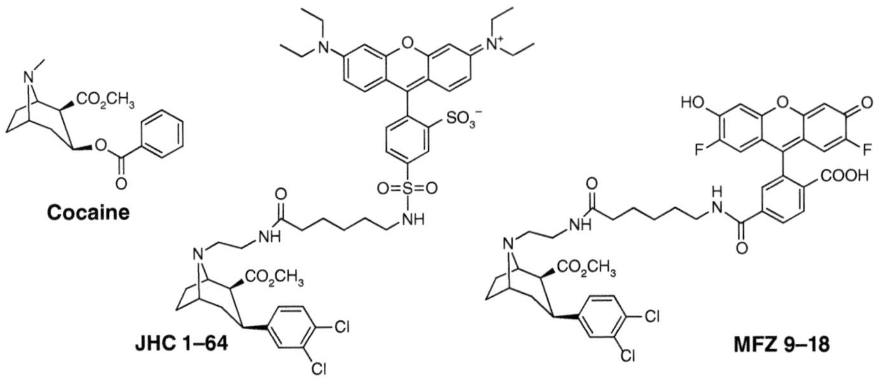

Figure 4.82 Structure of Cocaine, JHC 1–64, and MFZ 9–18.

of the neurotransmitters, SLC transporters take the role. For example, SLC6A1 for GABA, SLC6A2 (NET) for norepinephrine, and SLC6A3 (DAT) for dopamine. Cocaine is known to have a high affinity to DAT, and fluorescently labeled cocaine analogues, JHC 1–64 and MFZ 9–18, selectively stained dopaminergic neurons. Both JHC 1–64 and MFZ 9–18 showed specific labeling of DAT, and the binding was blocked by preincubation with an excess amount of dopamine. In addition, JHC 1–64 proved to have a prolonged dissociation rate, suggesting long-term DAT-tracking capability [90] (Figure 4.82).

Among neurotransmitters, dopamine and norepinephrine share catechol motif in the molecule. If there is a fluorescent mimetic probe for each neurotransmitter, the probe may selectively stain the firing cells through the reuptake SLC. Sames group in Columbia University has studied structurally similar derivatives of the neurotransmitters as "fluorescent false neurotransmitter" for selective neuronal cell imaging. FFN102 is a dopamine neuron probe and is absorbed through the DAT (SLC6A3) transporter, and FFN270 is a norepinephrine neuron probe and is absorbed through the NET (SLC6A2) transporter. Once they are in the cell, through the common transporter VMAT2 (SLC18A2: vesicle monoamine transporter 2), they are reloaded in the vesicle for the next firing. FFN102 not only stains dopamine neuron, but also visualizes the fire of the neurotransmitter [91]. In a similar manner, FFN270 selectively stains norepinephrine cells, and the inhibitor effect for the transporters could be evaluated by the imaging [92] (Figure 4.83).

Sames group also developed a synapse probe CX-G3 especially for presynaptic terminal through an unbiased fluorescence library screening [93]. CX-G3 stains presynaptic acid organelles, partially overlapped with late endosomes

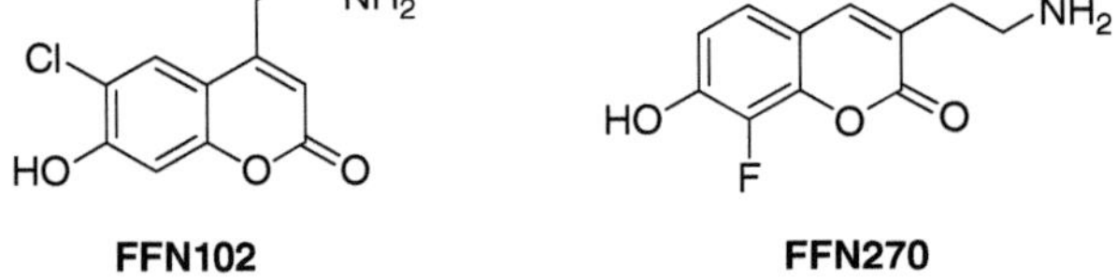

Figure 4.83 Structure of FFN102 and FFN270.

Figure 4.84 Structure of CX-G3.

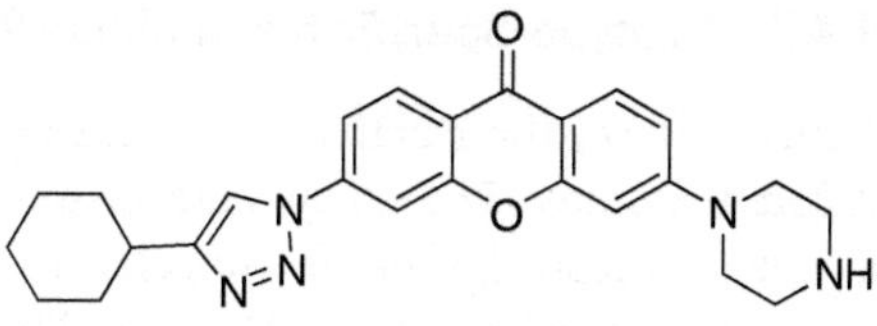

and lysosomes. While the exact molecular target was not elucidated, the uptake of CX-G3 seems through an SLC transporter. CX-G3 could penetrate BBB and exhibited the applicability to whole-body imaging (Figure 4.84).

4.4.6 Astrocyte Probe: SR101

Brain cells are mainly composed of neutrons, astrocytes, oligodendrocytes, and microglial cells. The first astrocyte probe, SR101 (sulforhodamine 101 or Texas Red), was reported in 2004 and demonstrated the live-cell imaging in brain cell culture and also in whole tissue of the cerebellum [94]. Although SR101 was widely used for astrocyte staining in hippocampus, the efficiency of astrocytes in ventrolateral medulla was not good. As SR101 is an organic anion, OATPs (Organic Anion Transporting Polypeptides) were proposed as candidates for SR101 uptake. The hypothesis was supported by reduced SR101 staining through rifampicin (OATPs inhibitor) treatment in astrocytes. It was also observed that neurons are also stained by SR101, but a fast extrusion was facilitated by ABC transporters. The same extrusion was not observed in astrocytes [95]. Further study identified the specific SLC target of SR101 as OATP1C1 (SLCO1C1) among multiple candidates of OATPs [96]. However, under certain conditions, SR101 was also reported to stain neurons and oligodendrocytes, raising questions about the selectivity mechanism. As oligodendrocytes do not express OATP1C1, the conditional staining of SR101 needs alternative explanation. By gap junction inhibitor, it was proposed that SR101-labling of oligodendrocytes requires uptake into astrocytes and then intercellular transfer through gap junction [97] (Figure 4.85).

Altogether, it seems the selective staining mechanism of SR101 is not a single one, but combined effect among coexisting competing cells [98]. While SR101 is the only astrocyte probe developed so far, due to its low specificity and unwanted excitatory side effect, it would be desirable to have more specific probes.

Figure 4.85 Structure of SR101 and selectivity mechanism.

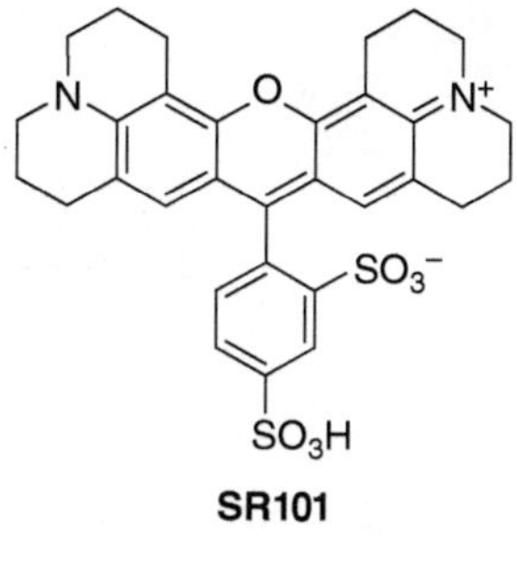

4.4.7 Subtype-Specific Macrophage Probes: CDg16, CDr17, CDg18

Depending on their activation status and direction, macrophages are defined by different names. The early state of macrophages differentiated from monocytes is called as resting state or M0 cells. The further activation or polarization of M0 branches either into M1 or M2 cells. The M1 activation is induced by lipopolysaccharide (LPS), tumor necrosis factor-alpha (TNF-α), and/or interferon alpha (INF-α), and the resulting M1 cells show usually aggressive immune response (pro-inflammatory). When M0 is activated by IL-4 (Interleukin 4) and/or IL-13, they polarize into M2 cells. M2 cells usually exhibit immune suppressive activity (anti-inflammatory). The M2 cells appearing near mature tumor tissue are especially called as tumor-associated macrophage (TAM). Interestingly, the phenotype of M1 and M2 seems totally opposite, and the discrimination of the two subtypes is important to understand their role in the biological system (Figure 4.86).

4.4.7.1 CDg16 for Activated Macrophage

Imaging macrophages could be used for inflammation site detection. However, if the probe stains all kinds of macrophages, the imaging site would be difficult to differentiate either an active inflammation site or a cancer tissue, and so on. For more specific imaging data, it would be ideal to develop M1- or M2-specific probe for each application. To achieve the goal, DOFL has been screened for M1 as the positive control and M0 as the negative control using mouse macrophage cell line, Raw 264.7 cells. Through an unbiased screening, M1-selective probe CDg16 (**C**ompound of **D**esignation **g**reen **16**) was discovered in 2019. CDg16 showed a consistent selectivity for not only M1 cells of the cell line, but also primary cells from mice, rats, and humans (Figure 4.87).

In whole-body imaging, CDg16 showed a strong fluorescent signal from LPS-induced inflammation model on paws and at virus-induced hepatitis in mice. Next challenging application was atherosclerosis. In atherosclerosis, lipids including cholesterol accumulate on the inner wall of arteries and inflammation increases.

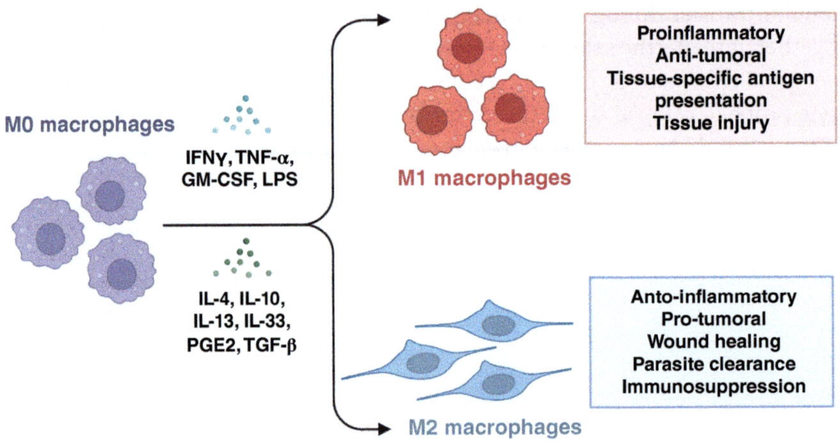

Figure 4.86 Polarization of macrophages.

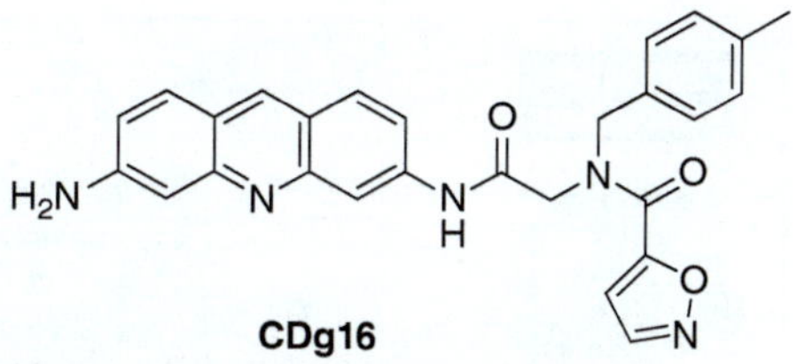

Figure 4.87 Structure of CDg16.

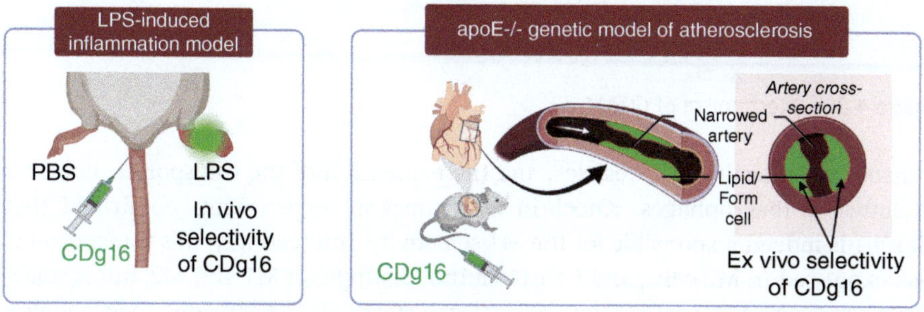

Figure 4.88 In vivo and ex vivo imaging of **CDg16.**

With the increased inflammation, activated macrophages with high lipid contents, called foam cells, were formed. The advanced inflammation site becomes vulnerable to rupture, but a better imaging contrast agent is not available yet. In mouse apoE−/− genetic model of atherosclerosis, **CDg16** demonstrated specific and strong imaging contrast in vivo and ex vivo (Figure 4.88).

CDg16 tends to stain acidic vesicles in M1 cells, but once the cell membrane is damaged, the CDg16 signal was washed out. These results imply that the selectivity mechanism of CDg16 may not be through a binding target, but may be a gating target such as a transporter. For the systematic screening of SLC transporters for **CDg16** target, Clustered Regularly Interspaced Short Palindromic Repeats (CRISPR) gene editing technique was employed. CRISPR was discovered in bacterial defense system against viruses. When viruses infect bacteria, most of the cases, viruses win and kill the bacteria. If bacteria survive the virus attack, they keep a small fragment of the virus gene as an immune memory. Bacteria have CAS9 (CRISPR associated protein 9) nuclease enzyme which cleaves any DNA which matches the saved immune sequence. When the same virus invades again, the virus gene is searched and cleaved by CAS9 enzyme. Applying the bacterial system into mammalian cells, a specific gene editing is possible. By introduction of Cas9 and guide RNA sequence into the cell, it is possible to cleave the matching DNA sequence in the genome. Therefore, the original CRISPR is a knockout tool. By mutations, dCas9 (deactivated Cas9) was developed, which can bind to the guide RNA matching sequence on DNA, but cannot cleave it. By attaching activation or suppressor domain to dCAS9, it is possible to enhance or suppress the gene expression of the bound site. The modified techniques are CRISPRa (activation) and CRISPRi (inhibition).

A CRISPRa library with 400 SLC transporters built in HeLa cells was used for the systematic screening of target of CDg16, and Slc18b1 was discovered. The polyamine

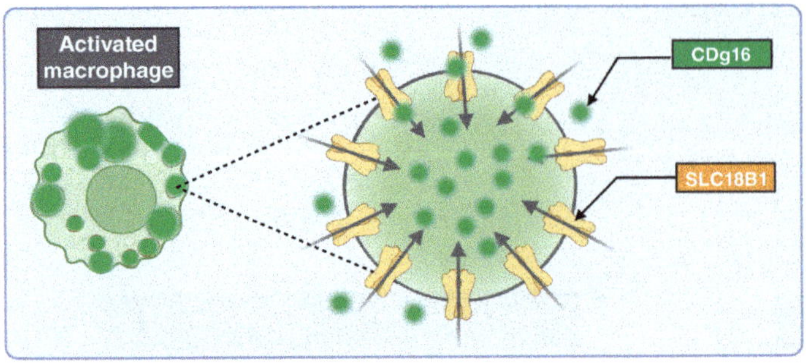

Figure 4.89 Mechanism of CDg16.

transporter is localized in vesicles, and the expression of the transporter increases in activated macrophages. Knock-in and knockout experiments confirmed that Slc18b1 is indeed responsible for the M1 selectivity. Interestingly, this transporter is also expressed in M2 cells, and CDg16 cannot distinguish M1 and M2, but is selective against naïve M0 cells. Therefore, CDg16 stains all the activated macrophages regardless of the polarization direction, either to aggressive M1 or suppressive M2 (Figure 4.89).

4.4.7.2 CDr17 for M1 Macrophage

For further discrimination between M1 and M2 macrophages, their preferred nutrients were targeted. Whereas M1 macrophages mainly use glucose transporters (GLUTs) to facilitate uptake of carbohydrates to sustain aerobic glycolysis, M2 macrophages prefer to take fatty acids by CD36 to feed TCA cycle and oxidative phosphorylation (OXPHOS) [99] (Figure 4.90).

Based on the hypothesis, fluorescently labeled carbohydrate library and fatty acid library were constructed and tested. From the carbohydrate library, M1-selective probe CDr17 (**C**ompound of **D**esignation **r**ed **17**) was discovered, which showed the selectivity over both M2 and M0 cells. GLUT1 was emerged as the candidate target

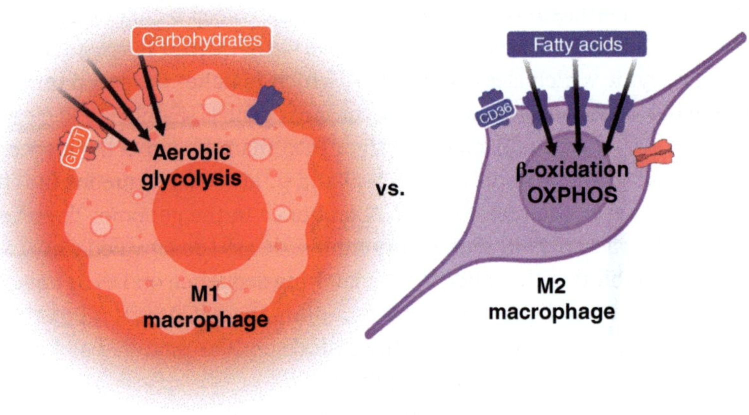

Figure 4.90 Metabolic difference between M1 and M2.

Figure 4.91 Structure of CDr17.

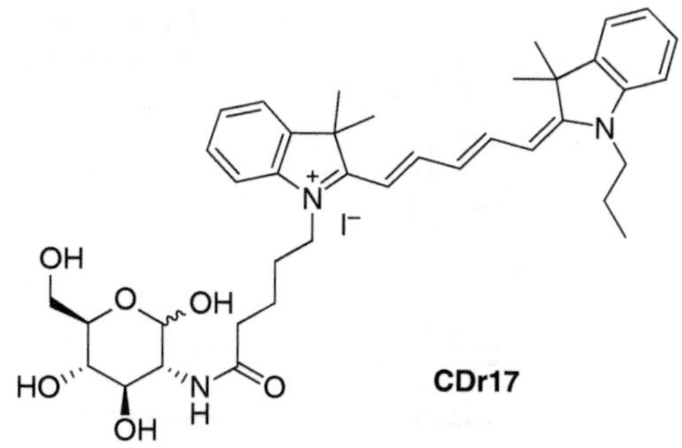

for CDr17 by mRNA expression analysis between M1 and M2. The biochemical and genetic studies validated GLUT1 as the relevant GOLD target for CDr17. It was also demonstrated that CDr17 can track M1 macrophages in vivo in a rheumatoid arthritis animal model. While there are probes for activated macrophages that have been reported and assumed for the selectivity, CDr17 is the first confirmed M1-selective probe over M2 macrophages (Figure 4.91).

4.4.7.3 CDg18 for M2 Macrophage

As a pair of CDr17, an M1 probe, CDg18 (**C**ompound of **D**esignation **g**reen **18**), was elucidated as an M2-selective probe from the fatty acid library. CDg18 stained lipid droplets selectively in M2 cells, not in M1 or M0 cells, even though they have lipid droplets. This observation implies that each subtype of macrophages may have different uptake preference for fatty acids through unique transporters. The detailed mechanism study showed that the selectivity of CDg18 is mainly by selective overexpression of SLC27A3 (long-chain fatty acid transport protein 3) in M2 cells, with a minor impact of CD36 transporter on the plasma membrane.

Due to the important role of tumor microenvironment maintenance, M2 cells attracted the attention as a promising antitumor treatment. The possible strategy would be selective killing of M2 cells or conversion of M2 into M1 cells, which are believed to be a more aggressive attacker against tumor cells. In this context, the reprogramming effectors are under hot study to control the interconversion between M1 and M2 cells. CDg18 is the first M2-selective probe reported so far and the color of CDg18 is compatible with CDr17 for bioimaging. Combining the two probes and the molecular effector, HS-1793, the real-time conversion of M2 into M1 was demonstrated with multicolor imaging (Figure 4.92).

4.4.8 B-Cell-Selective Probe Through GOLD Mechanism

After the discovery of CDgB (Section 4.3.4), a new B-cell-selective probe over T cell was developed. In comparison with the green color of CDgB, CDyB exhibits a yellow color emission, expanding the spectral window of B-cell probes. CDgB is a lipid-like molecule and the selectivity mechanism is revealed as a differential kinetics of lipid vesicle fusion to the flexible cell membrane in B cells faster than that in T cells.

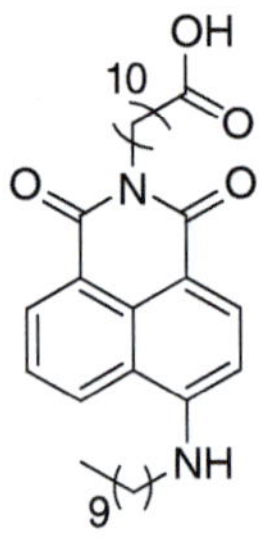

CDg18

Figure 4.92 Structure of CDg18 and the monitoring of M2 reprogramming into M1 cells.

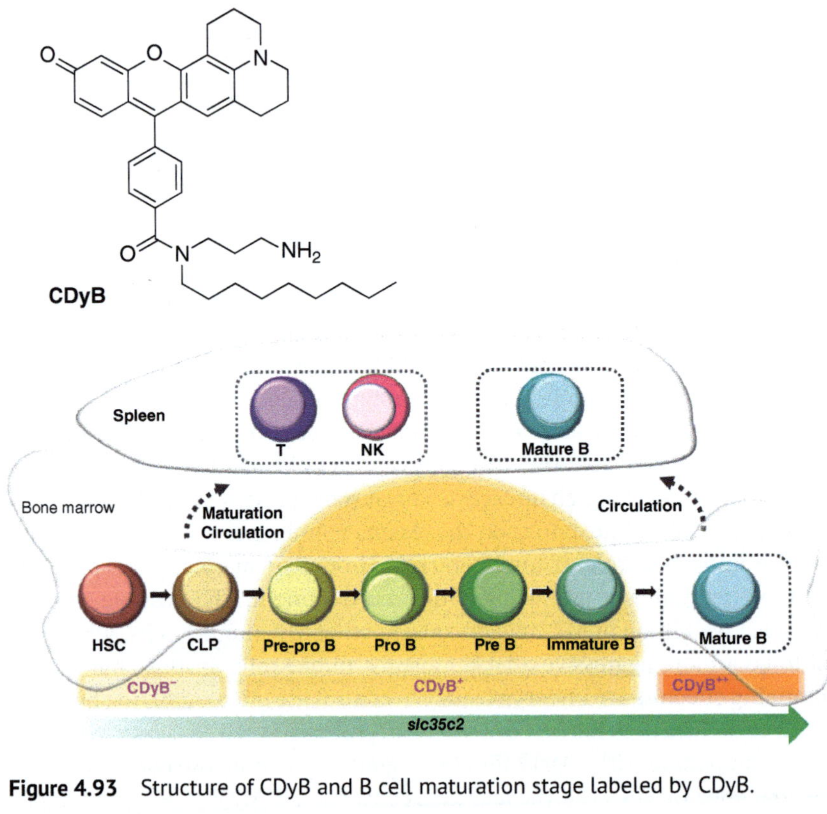

Figure 4.93 Structure of CDyB and B cell maturation stage labeled by CDyB.

The target of CDyB was elucidated as SLC35C2 transporter through a CRISPRi screening. SLC35C2 is exclusively expressed in B-cell ER and Golgi body membranes and transported CDyB into the lumen of organelles. Interestingly, CDgB preferred the younger B cells than the older ones during the maturation in the bone marrow, but CDyB showed the preference to the older B cells than the younger cells, reflecting the difference of the selectivity mechanism (Figure 4.93).

4.4.9 Bacteria Probes Through Transporters

Bacteria express their own specialized transporters that intake various nutrients from the environment. For instance, maltohexaose is a major source of glucose for bacteria and can be internalized into bacteria by the maltodextrin transporter, which is not expressed in mammalian cells. Utilizing the unique transporter system, maltohexaose-derived bacterial imaging agents, MDP-1 and MDP-2, were developed for the discrimination of bacteria from host cells. MDP-1 and MDP-2 enabled the visualization of both Gram-positive and Gram-negative bacteria and showed almost no signal in maltodextrin-transporter-deficient bacterial mutants and mammalian cells. In addition, MDP-2 allowed the sensitive and selective in vivo imaging of bacterial infection as low as 105 CFUs (colony-forming units) by intravenous injection in a mouse model [100]. Exploiting the same maltodextrin transporter, a maltotriose-conjugated Cy7 derivative, Cy7-1-maltotriose, was developed for fluorescence and photoacoustic imaging of bacterial infection in vivo [101] (Figure 4.94).

4.4.10 Probes Through ABC Transporters

While SLC transporters use concentration gradient as the driving force, ABC transporters use ATP as the energy source, and can pump out substrates even against the gradient. In humans, about 50 ABC transporters are known and many of them are related to drug resistance aspects. Among them, P-glycoprotein ABCB1, MRP1 (Multidrug Resistance Protein 1, also known as ABCC1), and BCRP (Breast Cancer Resistance Protein, also known as ABCG2) have been most extensively studied as the important players for multidrug resistance [102].

Among fluorescent probes, Hoechst 33342 and Rhodamine 123 were used for stem cell selection. As ABC transporters pump out the substrate, stem cells are collected from weak fluorescent populations, called "side population." Hematopoietic stem cells (HSE) were selected by this side population and ABCG2 was the responsible transporter protein [103] (Figure 4.95).

In a similar manner, ESC probe CDy1 was reported for isolation of drug-resistant multiple myeloma. The selected weakly stained cells were analyzed and found to overexpress ABCB1 transporter, implying that CDy1 is the substrate of drug-resistance pump ABCB1 [104].

In hESC study, KP-1 was selected as the selective hESC probe. From the mechanism study, KP-1 was identified as the substrate of ABCB1and ABCG2. Interestingly, hESC has low expression level of these transporter proteins compared to the feeder cells and show relatively stronger staining intensity. This is the first example of negative target for the stronger fluorescence intensity of target cell [105] (Figure 4.96).

In 2021, with a systematic ABC library using CRISPRa system, ABCB1 was rediscovered as the responsible transporter for ESC probe CDy1's selectivity [24]. In this case, ABCB1 expression level was higher in feeder cells than ESC, and the CDy1 signal is expected to be higher in ESC. As CDy1 has also a binding target protein in mESC, the mechanism of CDy1 for ESC selectivity is at least dual – POLD with ALDH2 as the binding target and GOLD with ABCB1 as the negative gating target. KP-1 also binds to ALDH2 in hESC, and it seems to share a similar complex mechanism with CDy1 (Figure 4.97).

In a separate study, CDg13 (**C**ompound of **D**esignation **g**reen **13**) was discovered as a novel NSC probe, showing the selected cells' preference to differentiate into

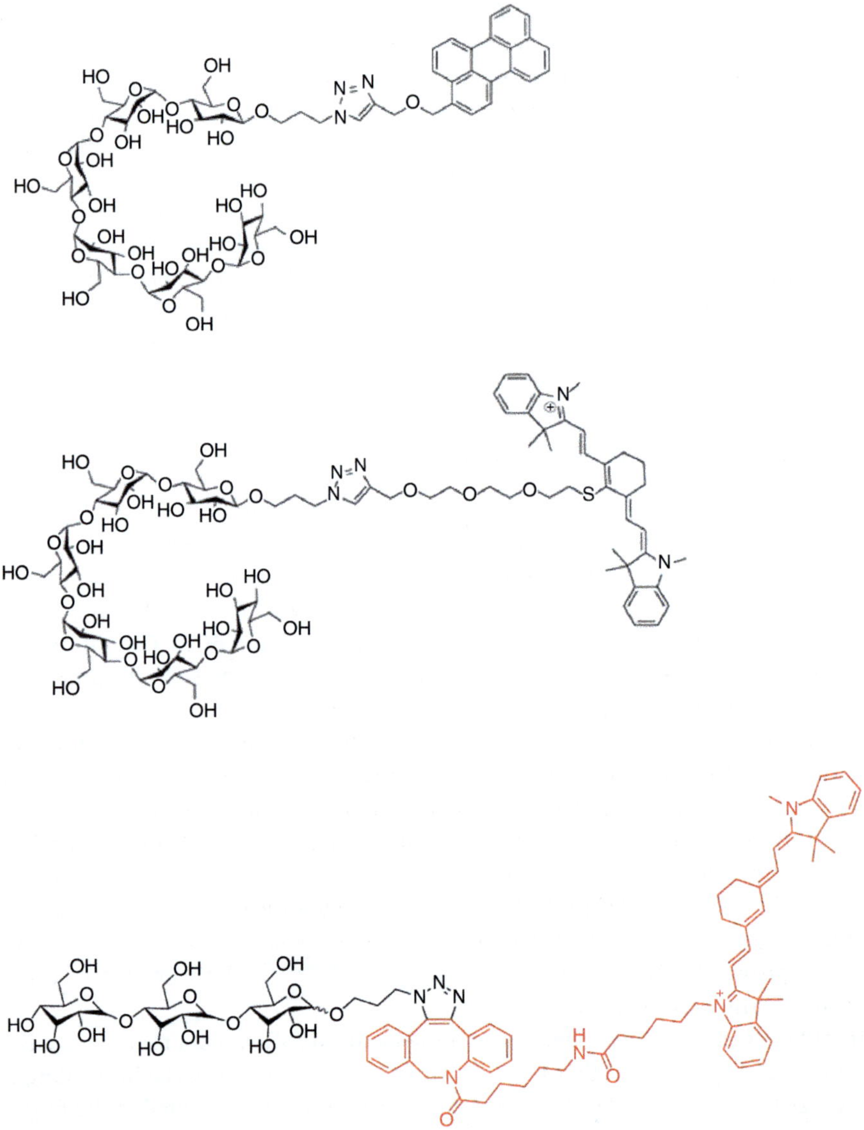

Figure 4.94 Structure of MDP-1, MDP-2, and Cy7-1-maltotriose.

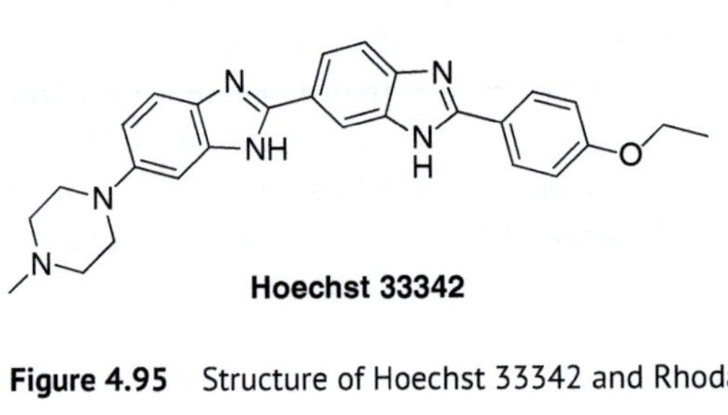

Hoechst 33342

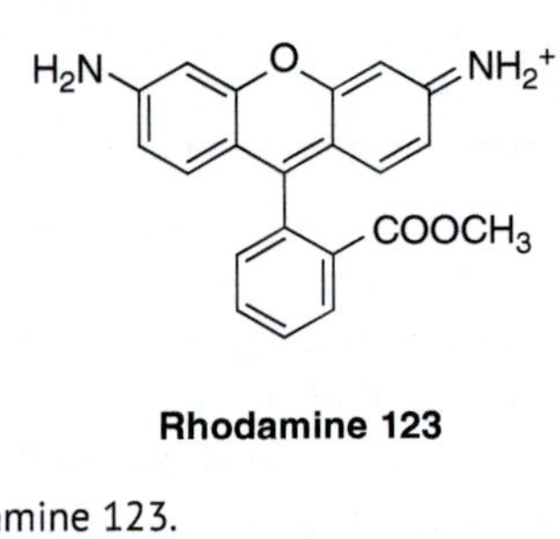

Rhodamine 123

Figure 4.95 Structure of Hoechst 33342 and Rhodamine 123.

Figure 4.96 Structure of KP-1.

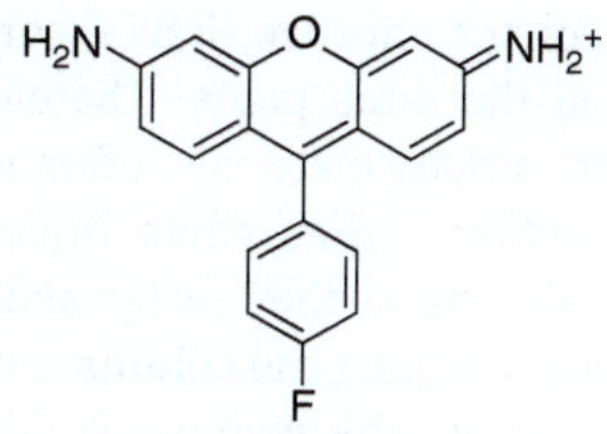

KP-1

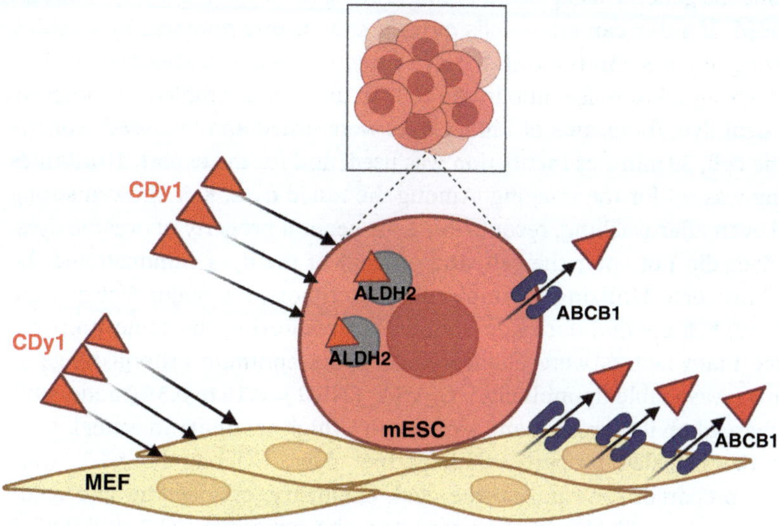

Figure 4.97 Dual mechanism of CDy1 of POLD and GOLD.

Figure 4.98 Structure of CDg13.

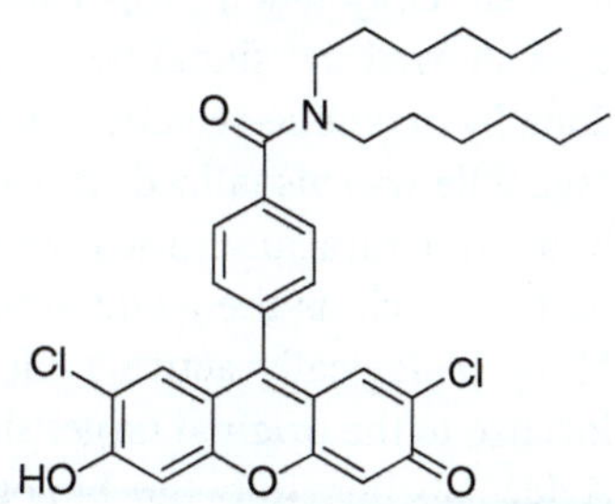

CDg13

neuronal lineages compared to previously reported CDr3 [26]. The cell selectivity mechanism was also identified as low expression of the negative target protein ABCG2 [106] (Figure 4.98).

4.4.11 Background-Free Tame Dye

Most of the organic dyes are composed of conjugated pi-electron system. The effective conjugation length is correlated with the wavelength of the absorption light.

For the effective light absorption, the conjugated pi-electrons need to be aligned on the same plane. Therefore, organic dyes tend to be hydrophobic and sticky to various materials, often showing strong background signal in the cell. Similar problem arises when organic chemists isolate the dyes from chromatographic columns, as they easily stuck to the column materials, lowering the recovery yield and clogging the columns. Introduction of charged motifs such as sulfonate group improves the general solubility problem of the dyes, but at the same time, decreases the cell permeability of the dye, hampering the biological application.

To overcome the general background problem of organic dyes, a tame dye concept was introduced. If a dye can enter cells quickly and cleanly removed by washing without leaving any trace in the cell, the dye could be used as a selective labeling agent with high signal-to-noise ratio in the cell. To uncover the molecular property of such an ideal dye, thousands of library dyes were tested and analyzed. For the entry into the cell, 30 minutes incubation was used, and for the export, 10 minutes after washing was set for the imaging. Among the tested dyes, 81% showed strong background even after washing, reconfirming the general property of organic dyes. 16% of the dyes did not enter the cell, and only 3% of the dyes demonstrated the ideal "tame" property. Utilizing 327 molecular descriptors, the major factors were searched through a chemoinformatic analysis for conferring the tame property. Finally, three main factors were elucidated with the optimum criteria range as SlogP = 1 to 4 (reasonable lipophilicity), Q_VSA_FNEG = 0.15 to 0.35 (moderately negatively charged on its surface), and logS = 2 to 6 (high solubility in water).

Based on the revealed property, click partner dyes, CO-1 and AzG-1, were designed and prepared; for catalyst-free click chemistry, cyclooctyne and azide pair was selected for cell imaging. As expected, the tame dye CO-1 and AzG-1 could enter the cells smoothly and washed out quickly without leaving a trait in the cell. Only when proper click partners were pre-labeled in the cells, the tame dyes showed the signal by the covalent click reactions. The pre-labeling could be done by organelle-selective molecule with click motifs, and by tame dye the target organelle was visualized. It was also demonstrated to label a protein incorporated by an unnatural amino acid with a click partner [107]. It is noteworthy that the click partner, such as the azide group, is much smaller than the fluorescent molecule. Many biologically active molecules may lose their property to enter the cell or localize to the original target site in the cell due to the bulky fluorescence labeling. Azide labeling into the biological molecules may minimize the disturbance of molecular property and the tame dye can be used for wider applications than the direct fluorescence tagging (Figure 4.99).

The original green tame dye was expanded to different colors to provide a palette of fluorescence imaging to blue, green, and yellow colors. Combined together with fluorescent proteins or probes with complementary colors, multicolor imaging was demonstrated. While most of the commercial click partner dyes are only applicable to cell surface target due to poor cell permeability, tame dyes could stain the intracellular target smoothly. As the functional application, azide-labeled carbohydrates were tested against different cancer cells. Different cells have their own preference for the carbohydrate uptakes and also incorporation into macromolecules through

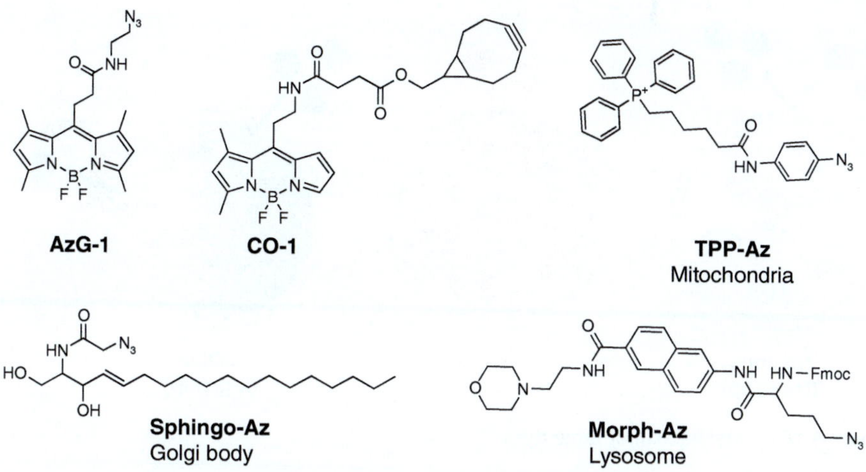

Figure 4.99 Structure of representative tame dyes and partners.

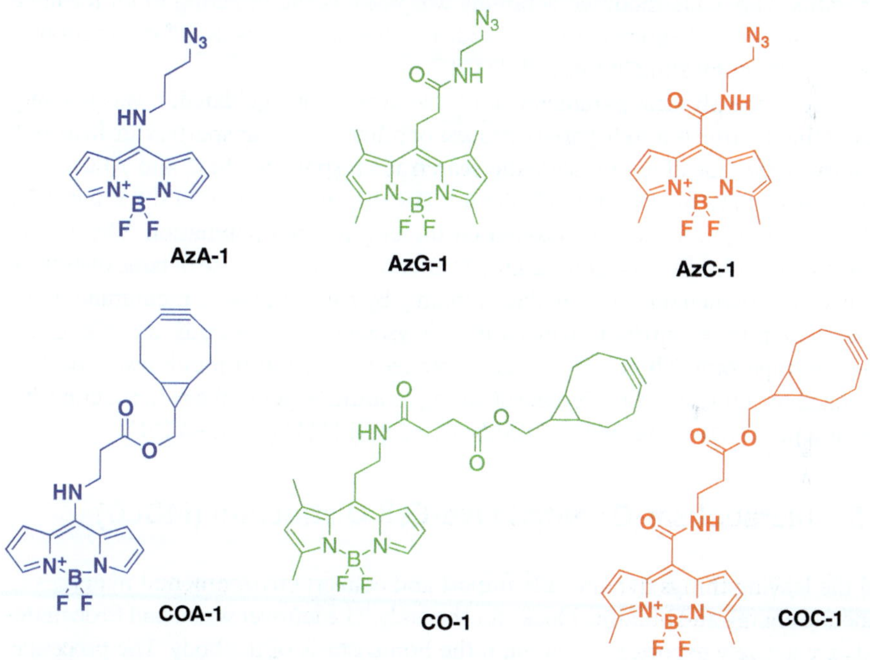

Figure 4.100 Tame dye palette.

selective glycosylation. The imaging of carbohydrates with tame dyes demonstrated differential carbohydrate incorporation according to cell types, casting a new possibility of the cell distinction [108] (Figure 4.100).

G4 (G-quadruplex) is a unique DNA or RNA secondary structure formed by G-rich domains. A macrocyclic hexaoxazole (6OTDs) is a known G4 ligand and for the intracellular imaging of G4, 6OTD was modified with the azide group and imaged by

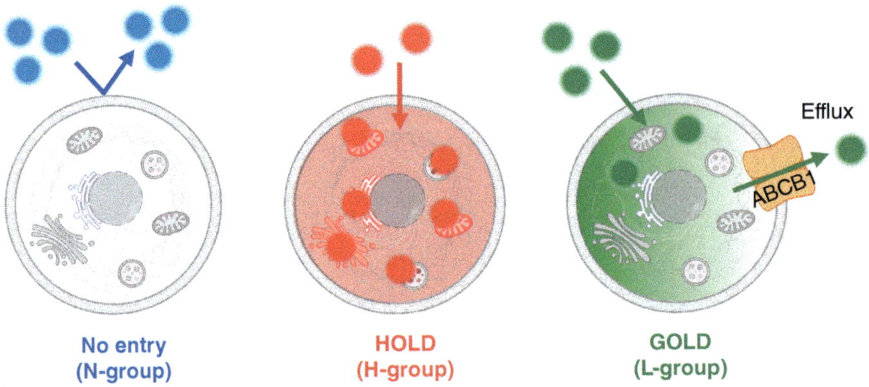

Figure 4.101 Mechanism of tame dyes.

CO-1 [109]. Another example of tame dye application is demonstrated for ceramidase activity measurement. Ceramide is a commonly used targeting moiety for Golgi apparatus. The azide-modified ceramide was used as the targeting motif for tame probes. The re-localization of the ceramide into lysosomes was used for acid ceramidase activity monitoring in live cells [110].

Although the physical parameters for tame dyes were elucidated, it was not very clear if the clearing is purely passive process or other active transporters are involved. To answer the question, a systematic search for responsible SLC and ABC transporters was done by CRISPR cell library. For the study, SLC CRISPRa and ABC CRISPRi library were searched to enrich the brighter cell population. The careful analysis with inhibitor and genetic target validation, the clearance of tame dyes from the cells was turned out to be mediated mainly by ABCB1 plasma membrane transporter with minor contribution by DIRC2 on lysosome. In this study, the role of SLC was not important. Therefore, the tame property occurred through fast exporting mechanism along with the absence of strong binding target in the cell. So, tame dye case is a pure GOLD mechanism, unlike CDy1 case [111] (Figure 4.101).

4.5 Metabolism-Oriented Live-Cell Distinction (MOLD)

All the leaving things dynamically import and convert environmental materials to generate energy and building blocks for the body. The leftover waste and toxic materials are actively exported to maintain the homeostasis of the body. The process is called metabolism and the involved materials are metabolites. The catalysts for the chemical conversion of metabolites are usually enzymes, and some examples are kinases, carbohydrate transferases, and proteases. Hijacking the unique metabolism of the target cell could be an excellent strategy to develop cell-type-specific probes.

4.5.1 Substrate for Proteases in Extracellular Matrix

Cells are isolated from the environment by a membrane. In multicellular organisms, outside of the cell membrane is not fully unfamiliar space, but often composed of a relevant biological network called extracellular matrix (ECM) . ECM is mainly composed of protein fibers and carbohydrate gels. ECM provides the structural rigidity of the tissue and connectivity of different cells. For the remodeling and controlling of ECM function, proteases play an important role, by cleaving the specific sequence of protein structure in ECM components. If one type of cell is surrounded by specific proteases in its ECM, the proteases could be sensible biomarkers for the distinction of the target cell, due to their close enough proximity.

The common design strategy for such probes is protease-specific peptides that are flanked by donor and acceptor fluorophores, which are fluorescence resonance energy transfer (FRET) pairs. In FRET, the donor's fluorescence emission is suppressed and the unused energy is transferred to the acceptor, increasing the emission. In general, the closer the donor and acceptor are, the higher efficiency of energy transfer is. For efficient FRET, the distance usually needs to be shorter than 5 nm range. Also, the overlap of the donor's emission spectrum and the acceptor's absorption spectrum is another important factor for efficient FRET. The more the overlap is, the higher the efficiency of FRET, increasing the detectable distance between the donor and the acceptor. When the donor and the acceptor are connected by a peptide, their distance is close enough to show high FRET. If protease cleaves the peptide in the middle, the donor and the acceptor separate and FRET efficiency drops. Commonly, donors have a shorter wavelength of excitation and emission than those of acceptors. For example, donors absorb blue light and emit green light. Acceptors may absorb green light and emit red light. When blue light is shed to the probe, the FRET suppresses green emission and increases red emission. After protease function, green emission will be increased, with decreased red emission (assuming the acceptor's absorption of blue light is not efficient). Typically, the emission ratio of donors and acceptors is used to measure the protease function (Figure 4.102).

Alternative option of two-color FRET is using quencher as an acceptor. In this design, the original donor's emission was suppressed by the quencher, and after release of the quencher, the emission becomes prominent. Based on

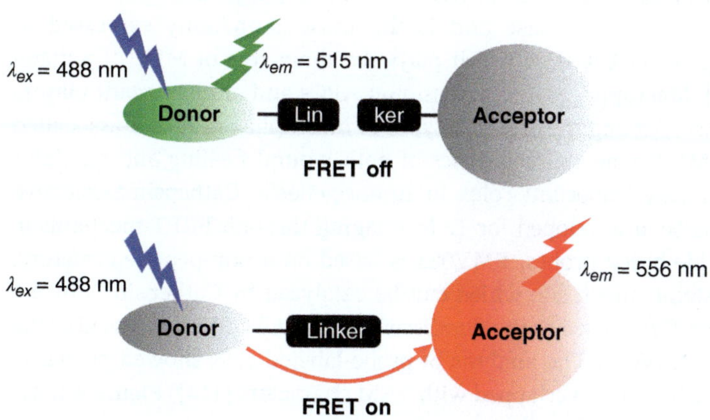

Figure 4.102 Design and mechanism of FRET-based probes.

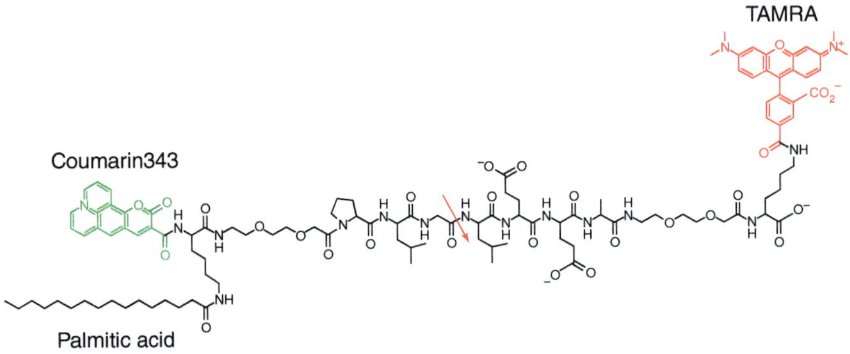

Figure 4.103 Structure of LaRee1 and ratiometric changes of fluorescence signal by MMP12.

this robust FRET mechanism, many protease substrates have been reported as cell-type-specific probes.

4.5.1.1 MMP12 Substrate for Activated Macrophage Probe

Matrix metalloproteinase 12 (MMP12) is mainly secreted by macrophages and has protease activity. FRET-based MMP12 probe LaRee1 was developed. In LaRee1 design, MMP-cleavable peptide was flanked by coumarin343 as the donor (ex/em 450/490 nm) and TAMRA as the acceptor (ex/em 545/575 nm). Notably, the donor side carries palmitic acid (C16 fatty acid) which can anchor the probe on the plasma membrane. In native macrophage, LaRee1 showed dominant TAMRA signal on the plasma membrane. Upon MMP12 addition to the cells, the signal shifted to coumarin343 emission, and interestingly the intracellular uptake of the donor side was observed after cleavage. LaRee1 demonstrated the detection of MMP12 of activated macrophages from a mouse model of pulmonary inflammation [112] (Figure 4.103).

4.5.1.2 Cathepsin S Substrate for Tumor-Associated Macrophage Probe

Cathepsin S is a cysteine protease and is the most abundantly expressed in antigen-presenting cells (APCs), where it plays the main role of MHC II antigen presentation [113]. Macrophages are professional APCs and are important players in immune response. Among the different types of macrophages, tumor-associated macrophages (TAM) display characteristics of both wound healing and regulator macrophages, and play important roles in tumorigenesis. Cathepsin-S-selective probe BMV083 has been developed for TAM imaging through FRET mechanism. Unlike other peptide-based probes, BMV083 is based on a non-peptide structure, but contains a cysteine attack site which can be catalyzed by Cathepsin S. In the design of FRET pair, Cy5 is used as the fluorescent motif and QSY21 was used as the quencher motif in BMV083. The analysis of probe-labeled cells showed primarily M2 macrophages, which are overlapped with TAM characters [114] (Figure 4.104).

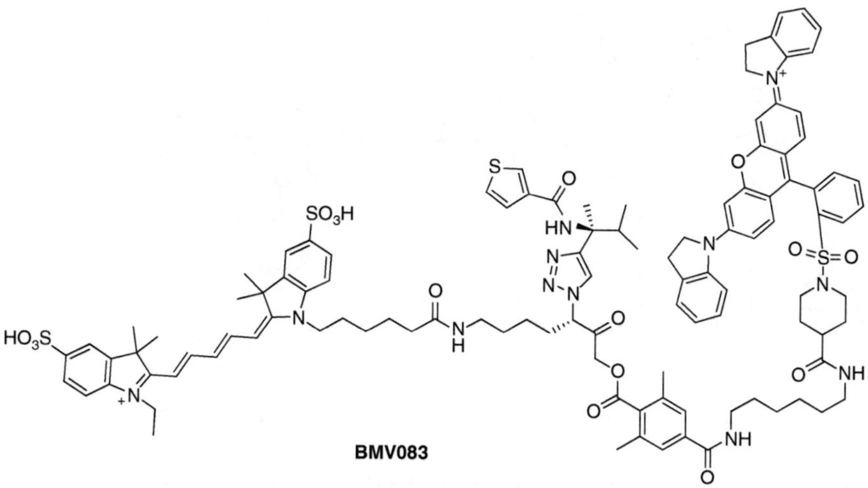

Figure 4.104 Structure of BMV083 and turn-on mechanism.

4.5.1.3 Elastase Substrate for Neutrophil Probe

Similar idea a of FRET probe design has been applied to elastase-targeting neutrophils. Neutrophil elastase is a serine protease mainly secreted by neutrophils, and has an important role in remodeling of ECM and defense against bacterial infections. Using an elastase-specific sequence, donor and acceptor fluorophores were introduced to generate the FRET probe, NEmo-2. Similar to the LaReel design, NEmo-2 also carries palmitic acid for plasma membrane anchoring. When tested in macrophages (that do not express elastase) with addition of elastase, a strong signal ratio change of donor vs. acceptor was observed. While the probe design was similar, in contrast to LaReel, no clear internalization of the donor fragment was observed. From intratracheal inflammation, NEmo-2 showed a strong signal ratio change, demonstrating the neutrophil elastase activity [115] (Figure 4.105).

By the same research group, the neutrophil elastase probe was modified by attaching additional DNA-binding motif to the FRET probe. Terminally stimulated neutrophils release large quantity of DNA to ECM and form sticky structure of NET. The sticky character of NET facilitates the physical capturing of bacteria and increase the antibacterial activity of neutrophils. In the design of the probe, famous DNA-binding probe Hoechst was introduced in the middle of FRET probe by click chemistry to give H-NE. The elevated protease activity was monitored by H-NE within ECM of sputum from cystic fibrosis patients. H-NE also showed neutrophil elastase activity at single-cell resolution within mouse lung slices [116] (Figure 4.106).

4.5.1.4 Granzyme Substrate for Natural Killer and Cytotoxic T Cell Probe

Granzymes are granule-associated serine proteases that play a pivotal role in the immune system. In humans, there are 5 granzymes (A, B, H, K, and M), and they are secreted by NK cells and cytotoxic T killer ($CD8^+$) cells to eliminate cancer cells or

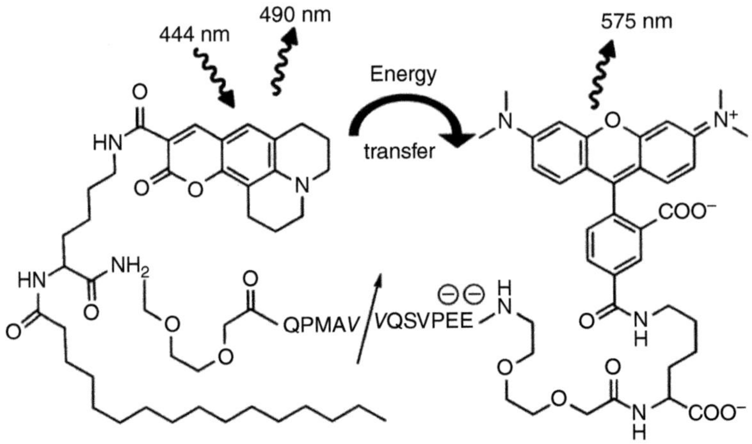

Figure 4.105 Structure of NEmo-2 and signal ratio change by neutrophil elastase.

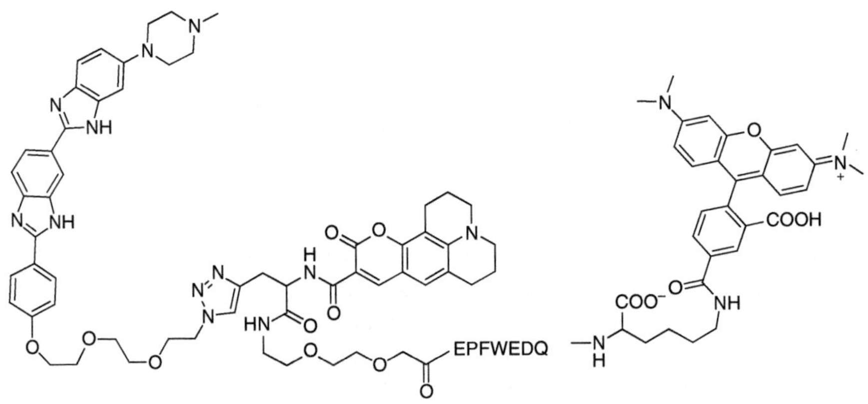

Figure 4.106 Structure of H-NE and imaging neutrophil.

infected cells with bacteria or viruses. Granzymes are stored in granules in cytotoxic lymphatic cells, and the granules also contain perforin. Perforin forms a pore on the target cell membrane, allowing granzymes to enter the cell cytoplasm. Granzymes induce target cell death by morphological changes, chromatin decondensation, and membrane blebbing [117]. Usually, antibodies cannot discriminate the active and inactive forms of granzymes. Designing small-molecule probes would be challenging to achieve subtype-specific granzymes due to their structural similarity. To achieve the selective recognition targeting granzyme A, a combinatorial scanning has been performed and it provided selective binding motif information. To selectively recognize only active form of granzyme A, a serine-reactive aryl phosphate was introduced for activity-based probe. Conjugated with Cy5, SK15.5 is the fluorescence-based irreversible probe for granzyme A. Using SK15.5, granzyme A activity could be monitored both in cell extract or live NK92 cell line [118]. NK92 is a NK-like cell line commonly used in the research (Figure 4.107).

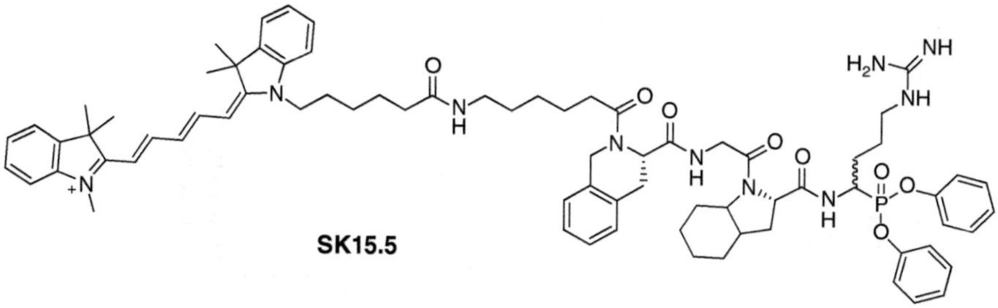

SK15.5

Figure 4.107 Structure of SK15.5.

The same group also report FRET-based granzyme B probe, qTJ71. qTJ71 is designed on the peptide substrate sequence for granzyme B, flanked by Cy3 and dizo quencher dye. Hydrolyzed by granzyme B, the release fluorophore of qTJ71 generated turn-on fluorescence signal in another NK cell line, YT cells. The effect of granzyme B inhibitor was also monitored in live-cell imaging [119] (Figure 4.108).

Instead of fluorescence reporters, chemiluminescent probe was also developed in 2021 for granzyme activity detection. The chemiluminescent probe 1 carries granzyme B cleavable site, and the chemiluminescent reporter is released and generated light. Unlike fluorescence, chemiluminescence does not require excitation light source and is useful for in vivo imaging. Using chemiluminescent probe 1, the anticancer activity of NK92 cells were demonstrated in a mouse model [120] (Figure 4.109).

Granzyme B substrate sequence was also utilized in a fluorogenic substrate to monitor T-cell function. In the design of $CyGbP_F$, the fluorogenic NIR dye was connected to cleavable peptide sequence and loaded on 2k size poly ethylene glycol (PEG). The hydrophilic PEG helps the passive targeting of the probe to the tumor site of living mice after systemic administration. Many solid tumors have leaky nature of vascular structure hyperpermeable to macromolecules, and this is the basis of passive targeting to the tumor site. In cell comparison study, $CyGbP_F$ increased the fluorescence only in CD8+ killer cells, but not in cancer cells or macrophages. By systemic injection study, $CyGbP_F$ showed good overlap with CD8+ antibodies [121] (Figure 4.110).

4.5.2 Microglia Probe: CDr10 and CDr20

Microglia are immune cells residing in the brain and their character is similar to macrophages. Actually, they share the same ancestor cells and they have migrated to the brain in the early stages of embryonic development. So, microglia can be considered as specialized brain macrophages. If macrophage is federal police, microglia would be a local sheriff in the brain. There are several more known tissue-specific macrophages: Kupffer cells in liver, alveolar macrophages in lungs, Langerhans cells in skin, and so on. [122].

4.5.2.1 CDr10a and b for Microglia Imaging among Brain Cells

In the brain, there are four major cell types: neutrons, astrocytes, oligodendrocytes, and microglia cells. While the other three cell types originate from common

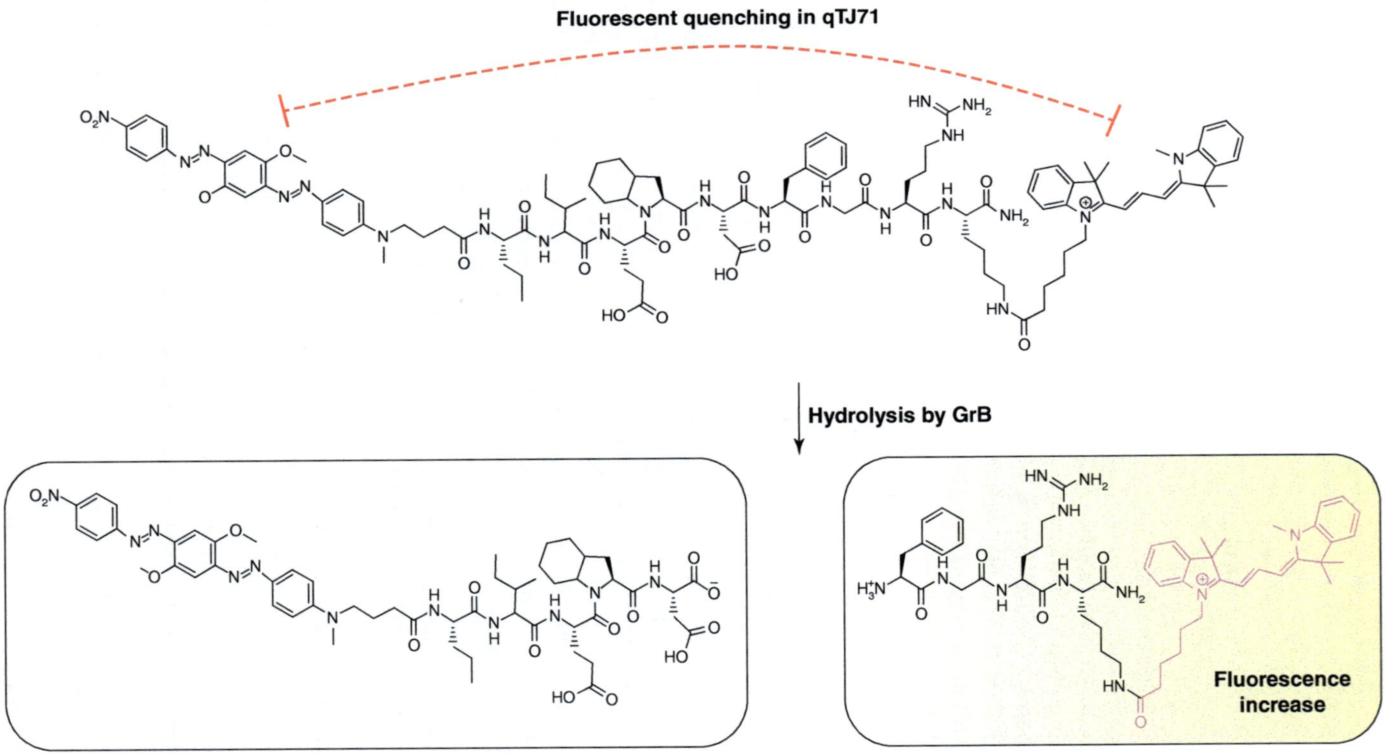

Figure 4.108 Structure of qTJ71.

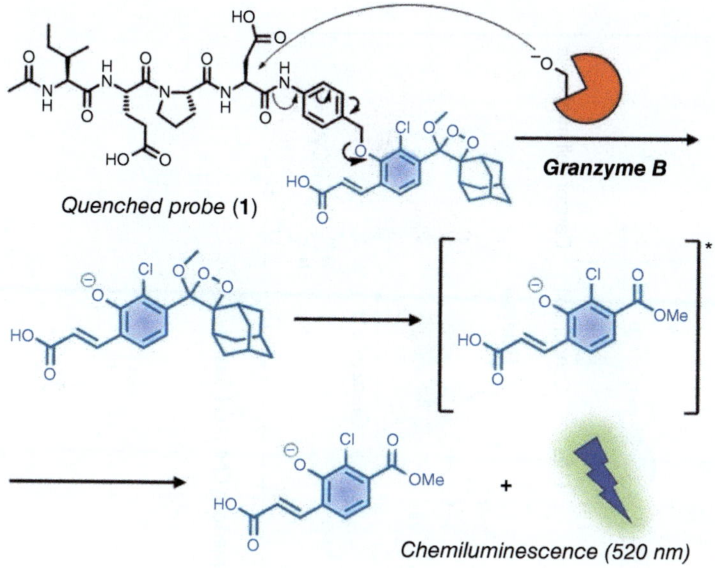

Figure 4.109 Structure of chemiluminescent probe 1. Source: Dr. Jamie I. Scott et al. [120]/ John Wiley & Sons/CC BY 4.0.

neuronal stem cells, only microglial cells are migrated cell lineages from outside of brain. In the context of cell selectivity among various cells residing in the brain, the primary brain cell culture would be a natural choice of the screening platform. Differential adhesion isolation method for brain cells allowed the growth of three cell populations among four types: neutrons, astrocytes, and microglial cells. Through an unbiased library screening of 5000 DOFL compounds, microglia-selective probes CDr10a and b (**C**ompound of **D**esignation **r**ed **10a** and **b**) were discovered [123] (Figure 4.111).

While CDr10a has acetyl group, CDr10b is equipped with CA group, a thiol-reactive functional group. While both of the probes selectively stained microglia over neurons or astrocytes, CDr10a signal is relatively vulnerable and easily washed out by media replacement. In contrast, CDr10b signal remained more resistant even to cell fixing condition with methanol. The result suggests that CDr10b forms a covalent bond with the target protein in microglia, and cytochrome c oxidase was identified as the fluorescently labeled protein from microglia cell extract. Considering the microglia selectivity of CDr10a even without CA motif, the relevance of the binding protein as the origin of the selectivity is yet to be elucidated. CDr10b tends to stain activated microglia with LPS compared to nonactivated counterparts. The stable signal and low toxicity of CDr10b allowed a video recording of microglia-attacking brain tumor cells in real time. The glioma cells express GFP and the vital microglia cells are labeled with red color of CDr10b.

In a separate study, CDr10b was demonstrated to suppress NF-kB activation and nitric oxide induction in activated macrophages induced by LPS [124]. Further study showed that CDr10b also suppresses the Toll-like receptor pathways in macrophages

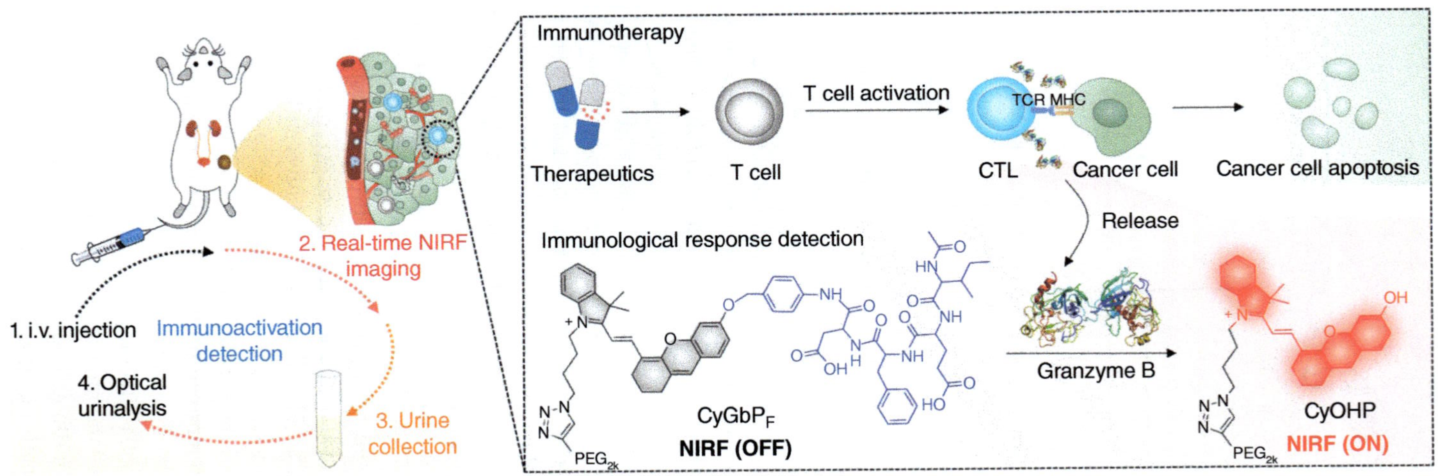

Figure 4.110 Structure of $CyGbP_F$ and the action mechanism. Source: Shasha He et al. [121]/American Chemical Society.

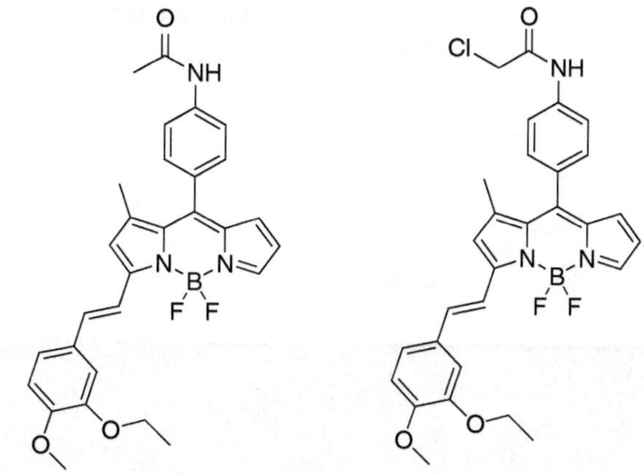

CDr10a CDr10b

Figure 4.111 Structure of CDr10a and CDr10b.

activated by TLR3 or TLR4 agonists [125], implying the functional connection of microglia and macrophages. Considering the fact that the functional effect can be induced only at high concentrations (20–100 μM range), it may be possible to use CDr10b as the imaging probe at low concentrations (<1 μM), and therapeutic application at high concentrations.

Interestingly, CDr10b could stain mid-intestinal structures in zebrafish embryo by simple immersion. Using laser-induced injury of the intestine, CDr10b could provide a useful readout of the integrity and repair of the epithelial cell barrier [126]. While useful, the connection of the tissue staining with microglia staining is not clear.

4.5.2.2 Microglia Probe CDr20 through Ugt1a7c

While CDr10 showed a good microglia selectivity in cultured brain cells, when the cell density reaches confluent condition, the cell discrimination became dull. In the same token, CDr10 did not work well on the brain slice culture for the selective staining. To improve the property of CDr10, a series of SAR study was performed in the confluent cultures of mouse primary glial populations containing microglia, astrocytes, neurons, and oligodendrocytes. This screening system is a more stringent condition to identify suitable microglia probes that might work in whole brain. The best compound from the campaign was disclosed as CDr20 (**C**ompound of **D**esignation **r**ed **20**). The new probe CDr20 showed prominent microglia selectivity in congested cell culture and also in whole brain slice, unlike CDr10 [127] (Figure 4.112).

For the cell selectivity, the hydroxyl group on CDr20 was indispensable. Interestingly, CDr20 did not emit a fluorescence signal in an aqueous environment, which enabled time-lapse imaging of microglia without removing CDr20 from the culture media. A high signal-to-noise ratio in fluorescent staining, CDr20 was a highly efficient tool for physical cell isolation by fluorescent-activated cell sorting. To test its ability to label microglia in an intact brain, a whole, live, day 13 mouse embryo was incubated with CDr20 to noninvasively label microglia in the live embryo (Figure 4.113).

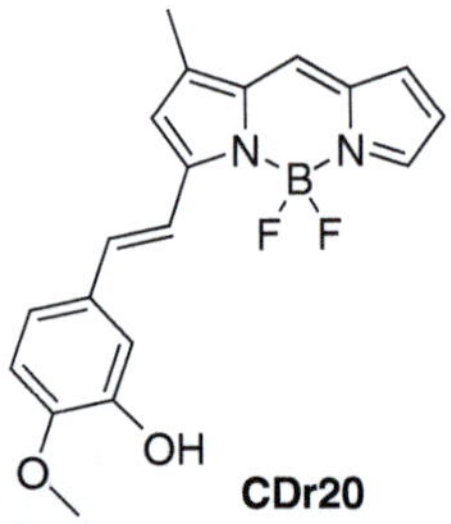

Figure 4.112 Structure of CDr20.

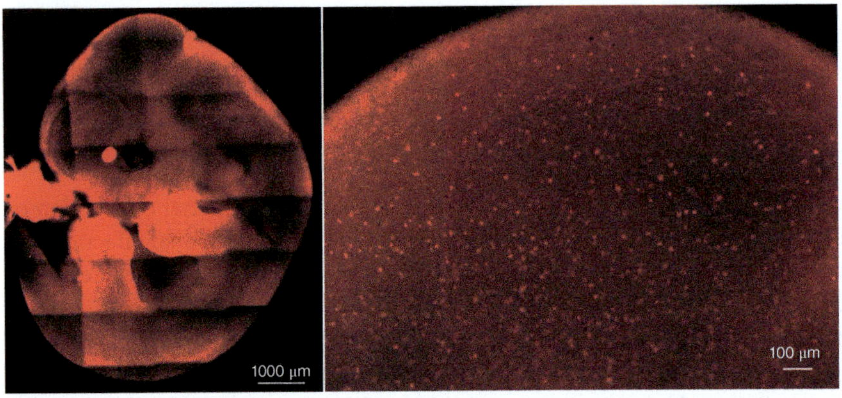

Figure 4.113 Microglia staining in whole live embryo. Source: Ref. [127], from John Wiley & Sons.

In Alzheimer's disease mouse model, intravenously injected CDr20 penetrated BBB and stained microglia in the live adult mouse brain. Live brain imaging was carried out by craniotomy above the primary somatosensory cortex with two-photon microscopy.

To identify the target protein of CDr20, a genome-scale CRISPR-Cas9 knockout screening in BV2 microglial cell line with the Brie library was conducted [128]. As a result, Ugtla7c, a UDP-glucuronosyltransferase emerged as a potential target of CDr20. Biochemical analysis confirmed that Ugtla7c transfers glucuronic acid to the hydroxyl group on the aromatic ring of CDr20. Quantum chemical calculations showed that CDr20 emission can be quenched through excited-state proton transfer (ESPT) by losing one proton at the hydroxyl group upon photoexcitation. The blocking of hydroxyl group with glucuronic acid was the physical origin of the fluorescence turn-on of CDr20. His is a unique example of Metabolism-Oriented Live-cell Discrimination (MOLD) (Figure 4.114).

4.5.3 Neutrophil Probe: NeutropG

Granulocytes are one of the major components of leukocytes along with monocytes and lymphocytes. As the name suggests, granulocytes contain plenty of granules in the cell, and they are classified into three types of cells, such as neutrophils, basophils, and eosinophils, according to the property of the granules (Section 2.2).

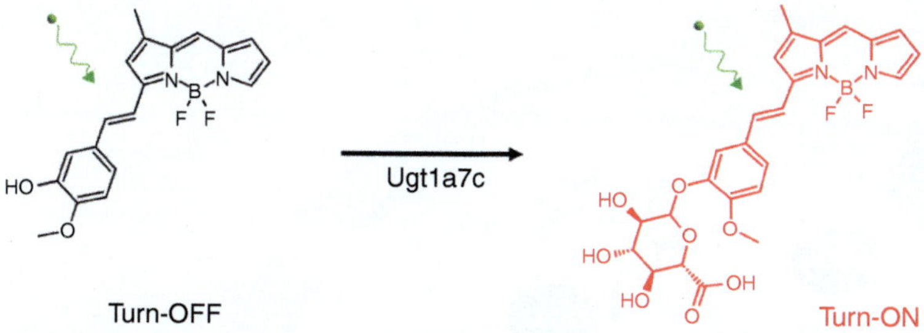

Figure 4.114 The mechanism of fluorescence turn-on of CDr20 by Ugt1a7c.

Figure 4.115 Structure of NeutropG.

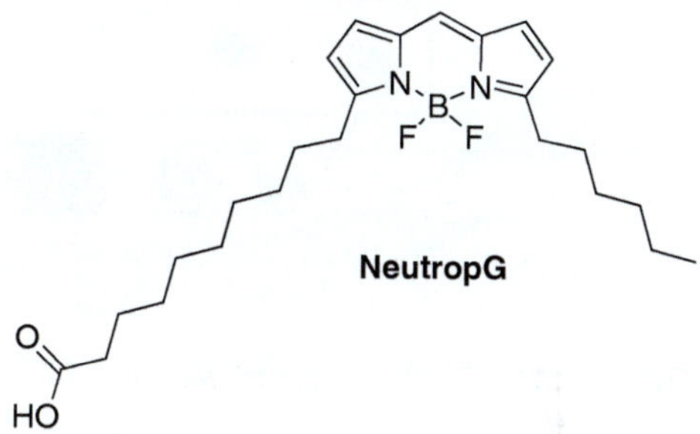

While color-generating staining method to distinguish the three cell types has been used for a long time, fluorescent probe for live granulocytes is relatively rare.

Neutrophils are the most abundant granulocytes, constituting 60–65% of leukocytes. Eosinophils comprise about 2–3%, and basophile occupies 0.5–1% of total leukocytes. For unbiased screening approach, ideally same numbers of cell preparation would be desirable. The biased population of granulocytes make it difficult to prepare for the even numbers of each cell type. As an alternative approach, the whole human blood was used for the screening platform using flow cytometry as the read out. From the phenotypic screening, a fluorescent fatty-acid-like probe was selected as the neutrophil-selective probe and dubbed NeutropG (**Neutrop**hil **G**reen) (Figure 4.115).

From fluorescence imaging, NeutropG stained lipid droplets in the neutrophils. Basophils and eosinophils also biosynthesize lipid droplets, but interestingly their lipid droplets are not stained by NeutropG. As NeutropG is a fatty-acid-like molecule, the fatty acid transporters were speculated as responsible biomarkers, but it was not the case. A careful analysis with inhibitors elucidated the enzymes in the biosynthetic pathway for lipid drops are exclusively expressed in neutrophils, but not in other granulocytes. The enzymes are acylCoA synthetase (ACSL) and acyl-CoA: diacylglycerol acyltransferase (DGAT), and their specific inhibitors Triascin C and Amidepsin D proved their essential role in the biosynthesis of lipid droplets in neutrophils (Figure 4.116).

The nontoxic property of NeutropG allowed the imaging of dynamic chasing and engulfing bacteria by neutrophil with its identity tag. So far, among the three granulocytes, live eosinophil probe is not developed yet. For basophils, a histamine

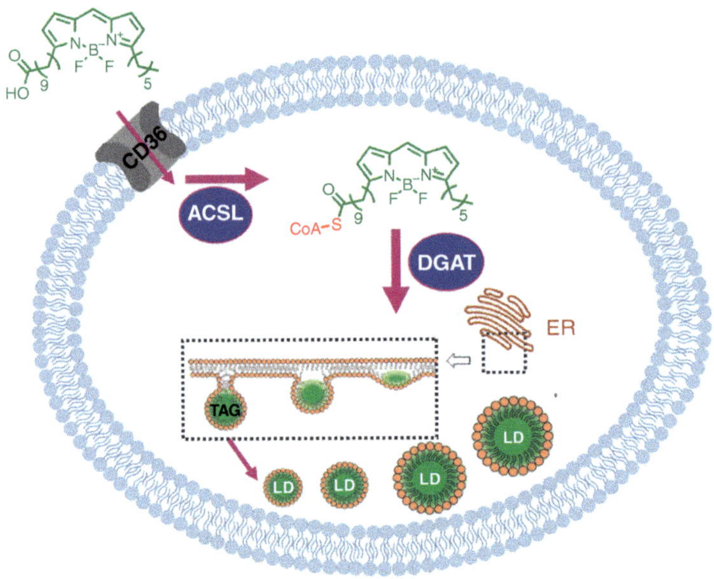

Figure 4.116 Biochemical pathways of lipid droplets in neutrophil.

sensor, histamine blue (Section 7.3.3), has demonstrated the live basophil staining, discriminating basophils from eosinophils and neutrophils [129].

References

1 Savage, D.C. (1977). *Annu. Rev. Microbiol.* 31: 107–133. https://doi.org/10.1146/annurev.mi.31.100177.000543.

2 Sender, R., Fuchs, S., and Milo, R. (2016). *PLoS Biol* 14: e1002533. https://doi.org/10.1371/journal.pbio.1002533.

3 Bernard, A.R. (1984). *Leucocyte Typing: Human Leucocyte Differentiation Antigens Detected by Monoclonal Antibodies*. Berlin: Springer-Verlag.

4 Clark, G., Stockinger, H., Balderas, R. et al. (2016). *Clin. Transl. Immunology* 5: e57. https://doi.org/10.1038/cti.2015.38.

5 Strebhardt, K. and Ullrich, A. (2008). *Nat. Rev. Cancer* 8: 473–480. https://doi.org/10.1038/nrc2394.

6 (a) International Human Genome Sequencing Consortium (2001). *Nature* 409: 860–921. https://doi.org/10.1038/35057062. (b) Venter, J.C., Adams, M.D., Myers, E.W. et al. (2001). *Science* 291: 1304–1351. https://doi.org/10.1126/science.1058040.

7 Collins, F.S., Morgan, M., and Patrinos, A. (2003). *Science* 300: 286–290. https://doi.org/10.1126/science.1084564.

8 Mann, M. and Jensen, O.N. (2003). *Nat. Biotech.* 21: 255–261. https://doi.org/10.1038/nbt0303-255.

9 Buccitelli, C. and Selbach, M. (2020). *Nat. Rev. Genet.* 21: 630–644. https://doi.org/10.1038/s41576-020-0258-4.
10 Slatko, B.E., Gardner, A.F., and Ausubel, F.M. (2018). *Curr. Protoc. Mol. Biol.* 122: e59. https://doi.org/10.1002/cpmb.59.
11 Van Dijk, E.L., Jaszczyszyn, Y., Naquin, D., and Thermes, C. (2018). *Trends Genet.* 34: 666–681. https://doi.org/10.1016/j.tig.2018.05.008.
12 Hwang, W.S., Ryu, Y.J., Park, J.H. et al. (2004). *Science* 303: 1669–1674. https://doi.org/10.1126/science.1094515.
13 Hwang, W.S., Roh, S.I., Lee, B.C. et al. (2005). *Science* 308: 1777–1783. https://doi.org/10.1126/science.1112286.
14 Kazutoshi Takahashi, Shinya Yamanaka Cell 2006, 126, 663–676 https://doi.org/10.1016/j.cell.2006.07.024
15 Zhou, H., Wu, S., Joo, J.Y. et al. (2009). *Cell Stem Cell* 4: 381–384. https://doi.org/10.1016/j.stem.2009.04.005.
16 Hou, P., Li, Y., Zhang, X. et al. (2013). *Science* 341: 651–654. https://doi.org/10.1126/science.1239278.
17 Im, C.-N., Kang, N.-Y., Ha, H.-H. et al. (2010). *Angew. Chem. Int. Ed.* 49: 7497–7500. https://doi.org/10.1002/anie.201002463.
18 Kang, N.Y., Yun, S.W., Ha, H.H. et al. (2011). *Nat. Protoc.* 6: 1044–1052. https://doi.org/10.1038/nprot.2011.350.
19 Vendrell, M., Park, S.-J., Chandran, Y. et al. (2012). *Stem Cell Res.* 9: 185–191. https://doi.org/10.1016/j.scr.2012.06.006.
20 Park, S.J., Yeo, H.C., Kang, N.Y. et al. (2014). *Stem Cell Res.* 12: 730–741. https://doi.org/10.1016/j.scr.2014.03.002.
21 Cho, S.-J., Kim, S.-Y., Park, S.-J. et al. (2016). *ACS Cent. Sci.* 2: 604–607. https://doi.org/10.1021/acscentsci.6b00099.
22 Vukovic, J., Bedin, A.-S., Bartlett, P.F., and Osborne, G.W. (2013). *Stem Cells Dev.* 22: 2341–2345. https://doi.org/10.1089/scd.2012.0660.
23 Kanatsu-Shinohara, M., Morimoto, H., and Shinohara, T. (2016). *Biol. Reprod.* 94: 13. https://doi.org/10.1095/biolreprod.115.135707.
24 Miyamoto, N., Go, Y.-H., Ciaramicoli, L.M. et al. (2021). *RSC Chem. Biol.* 2: 1590–1593. https://doi.org/10.1039/D1CB00147G.
25 Obokata, H., Wakayama, T., Sasai, Y. et al. (2014). *Nature* 505: 641–647. https://doi.org/10.1038/nature12968.
26 Yun, S.-W., Leong, C., Zhai, D. et al. (2012). *Proc. Natl. Acad. Sci. U.S.A.* 109: 10214–10217. https://doi.org/10.1073/pnas.1200817109.
27 Corti, S., Locatelli, F., Papadimitriou, D. et al. (2006). *Stem Cells* 24: 975–985. https://doi.org/10.1634/stemcells.2005-0217.
28 Leong, C., Zhai, D., Kim, B. et al. (2013). *Stem Cell Res.* 11: 1314–1322. https://doi.org/10.1016/j.scr.2013.09.002.
29 Chouw, A., Facicilia, G., Pratiwi, E.D. et al. (2020). *Cytotherapy* 22: S75–S76. https://doi.org/10.1016/j.jcyt.2020.03.120.
30 Kantawong, F., Saisuwan, C., Soeratanapant, P. et al. (2021). *Evid. Based Complement. Alternat. Med.* 2021: 8818618. https://doi.org/10.1155/2021/8818618.

31 Yun, S.-W., Leong, C., Bi, X. et al. (2014). *Chem. Commun.* 50: 7492–7494. https://doi.org/10.1039/C4CC02974G.

32 Zhang, W.C., Shyh-Chang, N., Yang, H. et al. (2012). *Cell* 148: 259–272. https://doi.org/10.1016/j.cell.2011.11.050.

33 Lee, Y.-A., Kim, J.-J., Lee, J. et al. (2018). *Angew. Chem. Int. Ed.* 57: 2851–2854. https://doi.org/10.1002/anie.201712920.

34 Kim, J.-J., Lee, Y.-A., Su, D. et al. (2019). *J. Am. Chem. Soc.* 141: 14673–14686. https://doi.org/10.1021/jacs.9b06068.

35 Wagner, B.K., Carrinski, H.A., Ahn, Y.-H. et al. (2008). *J. Am. Chem. Soc.* 130: 4208–4209. https://doi.org/10.1021/ja077656d.

36 Kim, Y.K., Lee, J.-S., Bi, X. et al. (2011). *Angew. Chem. Int. Ed.* 50: 2761–2763. https://doi.org/10.1002/anie.201007626.

37 Kim, Y.K., Lee, J.K., Lee, J.-S. et al. (2011). *Mol. Biosyst.* 7: 2375–2378. https://doi.org/10.1039/C0MB00074D.

38 Lee, J.-S., Kang, N.-y., Kim, Y.K. et al. (2009). *J. Am. Chem. Soc.* 131: 10077–10082. https://doi.org/10.1021/ja9011657.

39 Agrawalla, B.K., Chandran, Y., Phue, W.-H. et al. (2015). *J. Am. Chem. Soc.* 137: 5355–5362. https://doi.org/10.1021/ja5115776.

40 Bonnevie-Nielsen, V. and Lernmark, A. (1981). *Acta Biol. Med. Ger.* 40: 77–81. https://pubmed.ncbi.nlm.nih.gov/7020308.

41 Conget, J.I., Sarri, Y., González-Clemente, J.M. et al. (1994). *Pancreas* 9: 157–160. https://doi.org/10.1097/00006676-199403000-00003.

42 Lukowiak, B., Vandewalle, B., Riachy, R. et al. (2001). *J. Histochem. Cytochem.* 49: 519–528. https://doi.org/10.1177/002215540104900412.

43 Kang, N.-.Y., Lee, S.-.C., Park, S.-.J. et al. (2013). *Angew. Chem. Int. Ed.* 52: 8557–8560. https://doi.org/10.1002/anie.201302149.

44 Kang, N.-Y., Lee, J.Y., Lee, S.H. et al. (2020). *J. Am. Chem. Soc.* 142: 3430–3439. https://doi.org/10.1021/jacs.9b11173.

45 Lemaire, K., Ravier, M.A., Schraenen, A. et al. (2009). *Proc. Natl. Acad. Sci. U.S.A.* 106: 14872–14877. https://doi.org/10.1073/pnas.0906587106.

46 Kim, J.-Y., Sahu, S., Yau, Y.-H. et al. (2016). *J. Am. Chem. Soc.* 138: 402–407. https://doi.org/10.1021/jacs.5b11357.

47 Abdul Hamid, A.I., Nakusi, L., Givskov, M. et al. (2020). *BMC Microbiol.* 20: 22. https://doi.org/10.1186/s12866-019-1635-z.

48 Sharon, N. and Lis, H. (2004). *Glycobiology* 14: 53R–62R. https://doi.org/10 .1093/glycob/cwh122.

49 Krishnamoorthy, L., Bess, J.W. Jr., Preston, A.B. et al. (2009). *Nat. Chem. Biol.* 5: 244–250. https://doi.org/10.1038/nchembio.151.

50 Tateno, H., Toyota, M., Saito, S. et al. (2011). *J. Biol. Chem.* 286: 20345–20353. https://doi.org/10.1074/jbc.m111.231274.

51 Robertson, R.T., Levine, S.T., Haynes, S.M. et al. (2015). *Histochem. Cell Biol.* 143: 225–234. https://doi.org/10.1007/s00418-014-1301-3.

52 Tateno, H., Onuma, Y., and Ito, Y. (2014). *Methods Mol. Biol.* 1200: 313–318. https://doi.org/10.1007/978-1-4939-1292-6_26.

53 Lee, S.-C., Kang, N.-Y., Park, S.-J. et al. (2012). *Chem. Commun.* 48: 6681–6683. https://doi.org/10.1039/C2CC31662E.

54 Ghosh, K.K., Ha, H.-H., Kang, N.-Y. et al. (2011). *Chem. Commun.* 47: 7488–7490. https://doi.org/10.1039/C1CC11962A.

55 Kwon, H.-Y., Liu, X., Choi, E.G. et al. (2019). *Angew. Chem. Int. Ed.* 58: 8426–8431. https://doi.org/10.1002/anie.201902537.

56 Kwon, H.Y., Kim, J.Y., Lee, J.Y. et al. (2018). *Chem. Commun.* 54: 11865–11868. https://doi.org/10.1039/C8CC05544K.

57 Kwon, H.Y., Kim, J.Y., Liu, X. et al. (2019). *Biomater. Sci.* 7: 3594–3598. https://doi.org/10.1039/C9BM00152B.

58 Kim, S.J., Kim, J., Kim, B. et al. (2020). *Biochem. Biophys. Res. Commun.* 527: 646–653. https://doi.org/10.1016/j.bbrc.2020.04.153.

59 Beknke, O., Tranum-Jensen, J., and van Deurs, B. (1984). *Eur. J. Cell Biol.* 35: 189–199. PMID: 6519066.

60 Kim, J.Y.H., Lee, J.W., Lee, W.S. et al. (2012). *ACS Comb. Sci.* 14: 395–398. https://doi.org/10.1021/co300007a.

61 Kim, H., Jo, A., Ha, J. et al. (2016). *Chem. Commun.* 52: 7822–7825. https://doi.org/10.1039/C6CC02184K.

62 Er, J.C., Leong, C., Teoh, C.L. et al. (2015). *Angew. Chem. Int. Ed.* 54: 2442–2446. https://doi.org/10.1002/anie.201408614.

63 Nakamura, F., Maejima, H., Kawamura, M. et al. (2018). *Bioorg. Med. Chem. Lett.* 28: 2206–2209. https://doi.org/10.1016/j.bmcl.2018.04.067.

64 Li, J., Sun, K., Muroi, M. et al. (2019). *J. Cell. Mol. Med.* 23: 6283–6294. https://doi.org/10.1111/jcmm.14514.

65 Kang, K., Park, Y.S., Park, M. et al. (2016). *Nano Lett.* 16: 675–680. https://doi.org/10.1021/acs.nanolett.5b04458.

66 Seo, J., Lanara, C., Choi, J.Y. et al. (2021). *Adv. Healthc. Mater.* 10: e2000583. https://doi.org/10.1002/adhm.202000583.

67 Kwon, H.-Y., Das, R.K., Jung, G.T. et al. (2021). *J. Am. Chem. Soc.* 143: 5836–5844. https://doi.org/10.1021/jacs.1c00944.

68 Yang, W., Bai, Y., Xiong, Y. et al. (2016). *Nature* 531: 651–655. https://doi.org/10.1038/nature17412.

69 Kaplaneris, N., Son, J., Mendive-Tapia, L. et al. (2021). *Nat. Commun.* 12: 3389. https://doi.org/10.1038/s41467-021-23462-9.

70 Barth, N.D., Subiros-Funosas, R., Mendive-Tapia, L. et al. (2020). *Nat. Commun.* 11: 4027. https://doi.org/10.1038/s41467-020-17772-7.

71 Vázquez-Romero, A., Kielland, N., Arévalo, M.J. et al. (2013). *J. Am. Chem. Soc.* 135: 16018–16021. https://doi.org/10.1021/ja408093p.

72 Kitamura, T., Qian, B.-Z., Soong, D. et al. (2015). *J. Exp. Med.* 212: 1043–1059. https://doi.org/10.1084/jem.20141836.

73 Fernandez, A., Thompson, E.J., Pollard, J.W. et al. (2019). *Angew. Chem. Int. Ed.* 58: 16894–16898. https://doi.org/10.1002/anie.201910955.

74 Yoo, J.S., Lee, S.-C., Jow, Z.Y. et al. (2014). *Cancer Res.* 74: 44–55. https://doi.org/10.1158/0008-5472.CAN-13-2129.

75 van Furth, R. and Cohn, Z.A. (1968). *J Exp Med.* 128: 415–435. https://doi.org/10.1084/jem.128.3.415.

76 https://www.bioparadigms.org/slc/intro.htm

77 Schlessinger, A., Matsson, P., Shima, J.E. et al. (2010). *Protein Sci.* 19: 412–428. https://doi.org/10.1002/pro.320.

78 Schumann, T., König, J., Henke, C. et al. (2020). *Pharmacol. Rev.* 72: 343–379. https://doi.org/10.1124/pr.118.015735.

79 Höglund, P.J., Nordström, K.J.V., Schiöth, H.B., and Fredriksson, R. (2011). *Mol. Biol. Evol.* 28: 1531–1541. https://doi.org/10.1093/molbev/msq350.

80 Zou, C., Wang, Y., and Shen, Z. (2005). *J. Biochem. Biophys. Methods* 64: 207–215. https://doi.org/10.1016/j.jbbm.2005.08.001.

81 Park, J., Lee, H.Y., Cho, M.-H., and Park, S.B. (2007). *Angew. Chem. Int. Ed.* 46: 2018–2022. https://doi.org/10.1002/anie.200604364.

82 Tian, Y.S., Lee, H.Y., Lim, C.S. et al. (2009). *Angew. Chem. Int. Ed.* 48: 8027–8031. https://doi.org/10.1002/anie.200901175.

83 Vendrell, M., Samanta, A., Yun, S.-W., and Chang, Y.-T. (2011). *Org. Biomol. Chem.* 9: 4760–4762. https://doi.org/10.1039/C1OB05519D.

84 Mueckler, M. and Thorens, B. (2013). *Mol. Aspects Med.* 34: 121–138. https://doi.org/10.1016/j.mam.2012.07.001.

85 Schnedl, W.J., Ferber, S., Johnson, J.H., and Newgard, C.B. (1994). *Diabetes* 43: 1326–1333. https://doi.org/10.2337/diab.43.11.1326.

86 Agrawalla, B.K., Lee, H.W., Phue, W.-H. et al. (2017). *J. Am. Chem. Soc.* 139: 3480–3487. https://doi.org/10.1021/jacs.6b12122.

87 Weinberger, L., Ayyash, M., Novershtern, N., and Hanna, J.H. (2016). *Nat. Rev. Mol. Cell Biol.* 17: 155–169. https://doi.org/10.1038/nrm.2015.28.

88 Chandran, Y., Kang, N.-Y., Park, S.-J. et al. (2015). *Bioorg. Med. Chem. Lett.* 25: 4862–4865. https://doi.org/10.1016/j.bmcl.2015.06.037.

89 Cho, S.-J., Kim, K.-T., Kim, J.-S. et al. (2018). *Biomaterials* 180: 12–23. https://doi.org/10.1016/j.biomaterials.2018.07.007.

90 Eriksen, J., Rasmussen, S.G.F., Rasmussen, T.N. et al. (2009). *J. Neurosci.* 29: 6794–6808. https://doi.org/10.1523/JNEUROSCI.4177-08.2009.

91 Rodriguez, P.C., Pereira, D.B., Borgkvist, A. et al. (2013). *Proc. Natl. Acad. Sci. U.S.A.* 110: 870–875. https://doi.org/10.1073/pnas.1213569110.

92 Dunn, M., Henke, A., Clark, S. et al. (2018). *Nat. Commun.* 9: 2838. https://doi.org/10.1038/s41467-018-05075-x.

93 Dunn, M., Boltaev, U., Beskow, A. et al. (2018). *ACS Chem. Neurosci.* 9: 673–683. https://doi.org/10.1021/acschemneuro.7b00263.

94 Nimmerjahn, A., Kirchhoff, F., Kerr, J.N.D., and Helmchen, F. (2004). *Nat. Methods* 1: 31–37. https://doi.org/10.1038/nmeth706.

95 Schnell, C., Hagos, Y., and Hülsmann, S. (2012). *PLoS ONE* 7: e49398. https://doi.org/10.1371/journal.pone.0049398.

96 Schnell, C., Shahmoradi, A., Wichert, S.P. et al. (2015). *Brain Struct. Funct.* 220: 193–203. https://doi.org/10.1007/s00429-013-0645-0.

97 Hagos, L. and Hülsmann, S. (2016). *Neurosci. Lett.* 631: 13–18. https://doi.org/10.1016/j.neulet.2016.08.010.

98 Hülsmann, S., Hagos, L., Heuer, H., and Schnell, C. (2017). *Front. Cell. Neurosci.* 11: 44. https://doi.org/10.3389/fncel.2017.00044.

99 Jha, A.K., Huang, S.C.-C., Sergushichev, A. et al. (2015). *Immunity* 42: 419–430. https://doi.org/10.1016/j.immuni.2015.02.005.

100 Ning, X., Lee, S., Wang, Z. et al. (2011). *Nat. Mater.* 10: 602–607. https://doi.org/10.1038/nmat3074.

101 Zlitni, A., Gowrishankar, G., Steinberg, I. et al. (2020). *Nat. Commun.* 11: 1250. https://www.nature.com/articles/s41467-020-14985-8#citeas.

102 Young, H.C. and Yu, A.-M. (2014). *Curr. Pham. Des.* 20: 793–807. https://doi.org/10.2174/13816128200514021416521 2.

103 Zhou, S., Schuetz, J.D., Bunting, K.D. et al. (2001). *Nat. Med.* 7: 1028–1034. https://doi.org/10.1038/nm0901-1028.

104 Hawley, T.S., Riz, I., Yang, W. et al. (2013). *Am. J. Hematol.* 88: 265–272. https://doi.org/10.1002/ajh.23387.

105 Hirata, N., Nakagawa, M., Fujibayashi, Y. et al. (2014). *Cell Rep.* 6: 1165–1174. https://doi.org/10.1016/j.celrep.2014.02.006.

106 Kim, B., Feng, S., Yun, S.-W. et al. (2016). *ChemBioChem* 17: 2118–2122. https://doi.org/10.1002/cbic.201600490.

107 Alamudi, S.H., Satapathy, R., Kim, J. et al. (2016). *Nat. Commun.* 7: 11964. https://doi.org/10.1038/ncomms11964.

108 Alamudi, S.H., Su, D., Lee, K.J. et al. (2018). *Chem. Sci.* 9: 2376–2383. https://doi.org/10.1039/C7SC04716A.

109 Yasuda, M., Ma, Y., Okabe, S. et al. (2020). *Chem. Commun.* 56: 12905–12908. https://doi.org/10.1039/D0CC04957C.

110 Casasampere, M., Izquierdo, E., Casas, J. et al. (2020). *Chem. Sci.* 11: 13044–13051. https://doi.org/10.1039/D0SC03166F.

111 Ciaramicoli, L.M., Kim, H.S., Alamudi, S.H., and Chang, Y.-T. (2022). *Chem. Asian J.* 17: e202200229. https://doi.org/10.1002/asia.202200229.

112 Cobos-Correa, A., Trojanek, J.B., Diemer, S. et al. (2009). *Nat. Chem. Biol.* 5: 628–630. https://doi.org/10.1038/nchembio.196.

113 Zavašnik-Bergant, T. and Turk, B. (2006). *Tissue Antigens* 67: 349–355. https://doi.org/10.1111/j.1399-0039.2006.00585.x.

114 Verdoes, M., Edgington, L.E., Scheeren, F.A. et al. (2012). *Chem. Biol.* 19: 619–628. https://doi.org/10.1016/j.chembiol.2012.03.012.

115 Gehrig, S., Mall, M.A., and Schultz, C. (2012). *Angew. Chem. Int. Ed.* 51: 6258–6261. https://doi.org/10.1002/anie.201109226.

116 Guerra, M., Halls, V.S., Schatterny, J. et al. (2020). *J. Am. Chem. Soc.* 142: 20299–20305. https://doi.org/10.1021/jacs.0c08130.

117 Law, R.H.P., Lukoyanova, N., Voskoboinik, I. et al. (2010). *Nature* 468: 447–451. https://doi.org/10.1038/nature09518.

118 Kołt, S., Janiszewski, T., Kaiserman, D. et al. (2020). *J. Med. Chem.* 63: 3359–3369. https://doi.org/10.1021/acs.jmedchem.9b02042.

119 Janiszewski, T., Kołt, S., Kaiserman, D. et al. (2020). *J. Biol. Chem.* 295: 9567–9582. https://doi.org/10.1074/jbc.ra120.013204.

120 Scott, J.I., Gutkin, S., Green, O. et al. (2021). *Angew. Chem. Int. Ed.* 60: 5699–5703. https://doi.org/10.1002/anie.202011429.

121 He, S., Li, J., Lyu, Y. et al. (2020). *J. Am. Chem. Soc.* 142: 7075–7082. https://doi.org/10.1021/jacs.0c00659.

122 Epelman, S., Lavine, K.J., and Randolph, G.J. (2014). *Immunity* 41: 21–35. https://doi.org/10.1016/j.immuni.2014.06.013.

123 Leong, C., Lee, S.C., Ock, J. et al. (2014). *Chem. Commun.* 50: 1089–1091. https://doi.org/10.1039/C3CC45715J.

124 Gu, G.-J., Lim, S.-J., Ahn, S.-i. et al. (2014). *Eur. J. Pharmacol.* 742: 42–46. https://doi.org/10.1016/j.ejphar.2014.08.036.

125 Gu, G.-J., Ahn, S.-i., Kim, J.-S. et al. (2015). *Int. Immunopharmacol.* 28: 29–33. https://doi.org/10.1016/j.intimp.2015.05.017.

126 Sander, V., Patke, S., Lee, J.Y. et al. (2017). *Molecules* 22: 454. https://doi.org/10.3390/molecules22030454.

127 Kim, B., Fukuda, M., Lee, J.-Y. et al. (2019). *Angew. Chem. Int. Ed.* 58: 7972–7976. https://doi.org/10.1002/anie.201903058.

128 Doench, J.G., Fusi, N., Sullender, M. et al. (2016). *Nat. Biotechnol.* 34: 184–191. https://doi.org/10.1038/nbt.3437.

129 Gao, M., Lee, S.H., Park, S.H. et al. (2021). *Angew. Chem. Int. Ed.* 60: 23743–23749. https://doi.org/10.1002/anie.202108536.

5

Ex Vivo Tissue Imaging Probes

Tissue is composed of a group of related and connected cells. The cells in a tissue may not be identical, but they work together for a designated function. To understand the tissue structure and conditions, imaging techniques with cellular resolution and clear contrast would be necessary. In pathology and histology, H&E staining has been the gold standard tissue imaging method. In H&E staining, H (hematoxylin) stains acidic targets such as nucleic acid in blue and E (Eosin Y) stains basic targets in red color. Hematoxylin is oxidized into hematein, and through aluminum ion bridge makes colorful complex with acid motifs (Figure 5.1).

For H&E staining, the tissue sample needs prefixing (e.g. typically in 10% formaldehyde for one day) and imbedding with paraffin. Paraffin is a long-chain (C20-40) hydrocarbon and melts in the 47–64 °C range. The tissue is emerged in a warmed paraffin solution, and a solid block is formed by chilling. The resulting tissue block is called formaldehyde-fixed paraffin-embedded (FFPE) tissue. The FFPE block is sliced in micrometer thickness using a microtome and mounted on the glass slide. For the staining, the paraffin needs to be removed from the FFPE tissue by an organic solvent such as xylene and rehydrated by slow exchange of the solvents through ethanol, and eventually to water. As the first step of H&E protocol, hematoxylin (with aluminum salt) staining is performed for several minutes (typically 5–10 minutes), followed by water washing. Usually, the stain is applied excessively and the degree of blue color is adjusted by acid alcohol washing. For the second step, eosin Y is applied to the tissue sample either in water or an alcohol solution for several minutes (shorter time is required in alcohol solution). Rinse and dehydration by ethanol and xylene finalize the H&E slide preparation for microscopy. The optical microscope usually generates a strong blue color for nucleus and weak pink for the cytosol. Basic granules are stained in strong blue as in basophils, and acidic granules in strong red as in eosinophils. The end results of the color contrast highly depend on the solution composition and exposure/washing time. Usually, the stained tissue samples are subjected to the pathologists for reading or interpretation, which requires a long-term experience to extract out useful information. Overall, the procedures are laborious and time consuming with high possibility of image variation depending on the used condition. In addition, the biosamples are prefixed, limiting the real-time monitoring of biological processes (Figure 5.2).

Sensors and Probes for Bioimaging, First Edition. Young-Tae Chang and Nam-Young Kang.

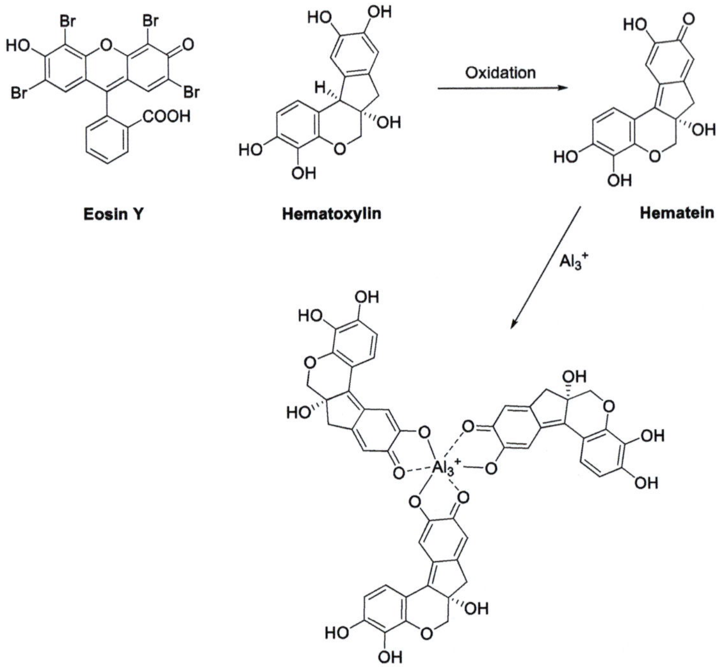

Figure 5.1 Structure of H&E and hematein.

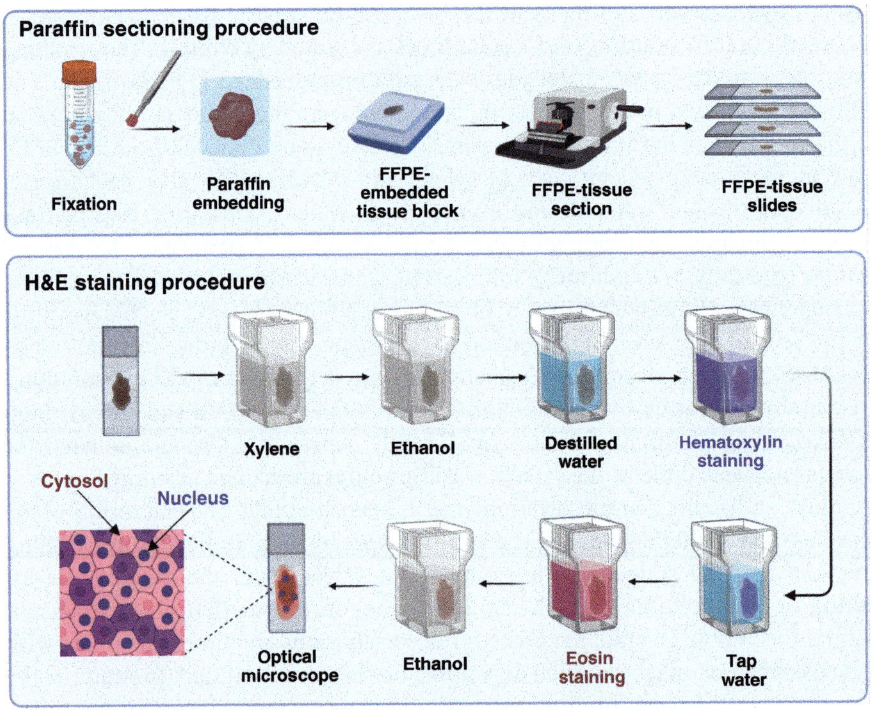

Figure 5.2 Procedure of paraffin sectioning and H&E staining.

5.1 Immunohistochemistry

For the molecular information above organelle-based H&E imaging, antibodies have been employed for tissue imaging, which is called IHC (ImmunoHistoChemistry: immuno means antibody, histo means tissue, and chemistry means binding here). Antibodies recognize specific proteins in the tissue and can provide the protein localization information through a proper visualization procedure. Similar to H&E staining procedure, HIC usually starts with deparaffinization of FFPE tissue and rehydration. Formaldehyde fixing induces heavy crosslinking in the tissue, and the epitope (antibody-binding site in antigen, i.e. target protein) could be blocked, so the unmasking antigen step is usually introduced. Heating in citrate buffer (pH 6) or Tris buffer (pH 9) is commonly used for antigen retrieval, while enzymatic treatment is another option instead of chemical treatment. To reduce the nonspecific binding of antibodies, bovine serum albumin (BSA) or milk proteins are often used as a blocking agent to cover the nonspecific hydrophobic surfaces of the tissue, such as exposed membranes. Then, primary antibody is applied to the tissue to induce specific binding with the target protein. There are several options for the visualization of the primary antibody. For the direct colorization, primary antibody could be covalently labeled with a coloring agent. Labeling primary antibody is often a big burden, especially if the antibody is a new entity and there is no available commercial source. Secondary antibody, which recognizes the Fc (fragment crystallizable: tail part of antibody) of primary antibody, is a common choice for the indirect method. Secondary antibody could be labeled with a coloring agent, or even further amplification using avidin–biotin complex (ABC) is also possible. Avidin or streptavidin is a tetrameric protein which has four binding sites to biotin with super high affinity, and they have been widely used in biological detection methods. Their binding affinity with biotin is so high as the K_d value is near to fM, and the binding is practically considered to the semi-covalent binding level. Utilizing the multivalent binding nature of avidin or streptavidin, further amplification could be achieved (Figure 5.3).

For the visualization, conventionally horseradish peroxidase (HRP) enzyme has been popularly used, by either binding to secondary antibody or biotin or avidin/streptavidin. With hydrogen peroxide, HRP typically oxidizes

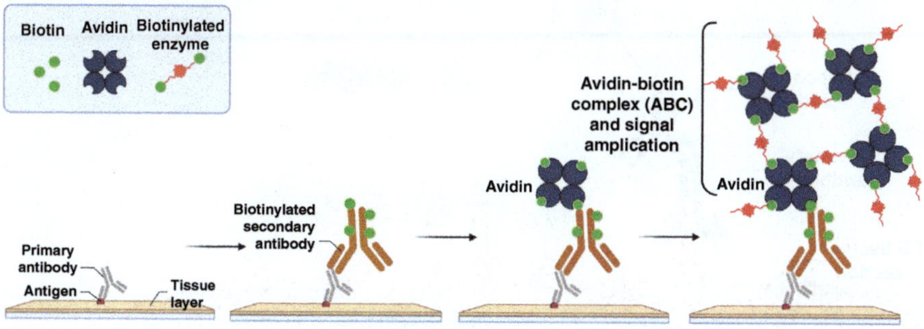

Figure 5.3 ABC and signal amplification.

3,3′-diaminobenzidine (DAB) into a dark brown polymer for visualization. The enzymatic reaction is catalytic, so another amplification could be achieved at this step. As the counterstaining of the brown color antibody location, hematoxylin is often used to visualize the nucleus of the cells in blue (Figure 5.4).

With the advancement of fluorescent labeling reagents, the IHC was also extended to the fluorescent version. Similar to HRP-DAB-based IHC, fluorescent labeling can be applied either to primary/secondary antibody or biotin/avidin/streptavidin, depending on the strategy. While the fluorescent labeling provides high sensitivity for subcellular imaging, if needed, HRP could also be used for further fluorescence amplification, so-called tyramide signal amplification (TSA) method. TSA incorporates HRP labeling and tyrosine-containing substrates. Tyrosine is converted into a reactive radical species catalyzed by HRP and can react with nearby tyrosine motifs in the target protein, forming covalent bonds and amplifying the signal. The substrate may contain fluorescent molecules or biotin for further amplification (Figure 5.5).

In fluorescent IHC, multicolor labeling is also possible. First option may be different color labeled primary antibody sets for multiple targets. In indirect labeling, secondary antibody could be chosen based on the species from which the primary antibody was generated. If the primary antibody was generated in mouse,

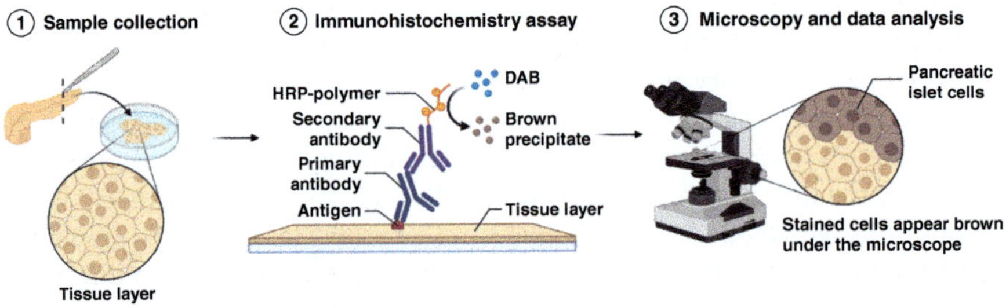

Figure 5.4 HRP-DAB reaction and example of IHC (wiki).

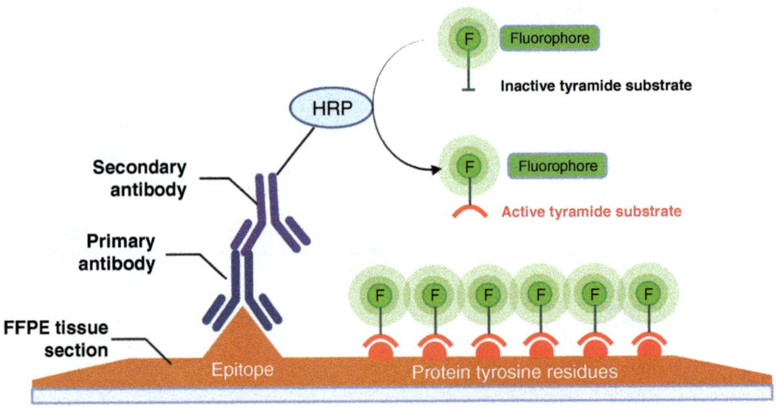

Figure 5.5 TSA amplification.

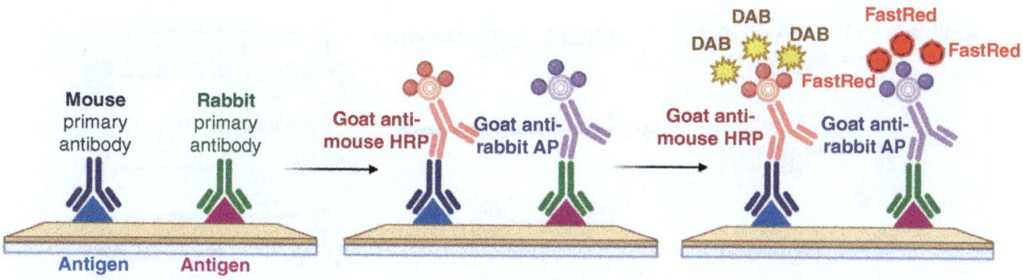

Figure 5.6 Multiplex IHC. AP, alkaline phosphatase.

anti-mouse Fc is the choice of the epitope for the second antibody recognition. By proper combination of primary antibody from different animals and matching secondary antibodies with different color labeling, multiplex (multicolor multitarget) imaging is possible in a single tissue sample. Similar to hematoxylin's role in conventional IHC, fluorescent IHC commonly uses Hoechst as the counterstain for blue nuclei (Figure 5.6).

If the fluorescence emission spectrum is sharp, many color multiplexing may be possible. But, most of the fluorescent dyes have broad excitation and emission spectra with significant overlap among different dyes. As a result, the maximum number of multiplexing dyes is practically limited to 4 or 5 at best. Therefore, if there is a need to test 10 antibodies at the same time, it is difficult to use a single tissue sample to test them all. One approach to overcome the limitation is recycling the same tissue by repeated antibody treatment and removal procedure. First of all, the antibodies are divided into multiple sets with comparable color tagging. After the imaging with one set is finished, the used antibodies are removed by chemical and thermal procedures, and the tissue is reused for the next set of antibodies. Each of the antibody removal steps potentially damages the tissue, so a careful sample handling is important to not lose the target proteins in this method.

To overcome the limit of usable antibody number, an interesting antibody labeling method with oligonucleotides was introduced. Instead of fluorescence labeling, the primary antibodies are labeled with DNA oligomers with different sequences. After antibody treatment to the tissue altogether, the fluorescently labeled complementary DNA oligomers are used as separate sets for the imaging. The affinity of cDNA oligomer to the target sequence can be easily adjusted by the length of the complementary oligomer. After one set imaging is done, the dissociation of the cDNA oligomers could be achieved by much milder chemical or thermal treatment than the removal of antibodies (Figure 5.7).

Another interesting solution is labeling antibodies with heavy metal ions and detecting them by inductively coupled plasma mass spectrometry (ICP-MS). Most heavy metal ions have unique masses with integer number difference, and more than 100 different masses can be distinguished among stable isotopes. This technique is not direct optical imaging method, but pseudo-images can be reconstructed to provide spatial information, enhancing the multiplexing number dramatically. If this method is applied into flow cytometry, it is called mass cytometry or CyTOF (cytometry with time of flight detector in mass spectrometry) (Figure 5.8).

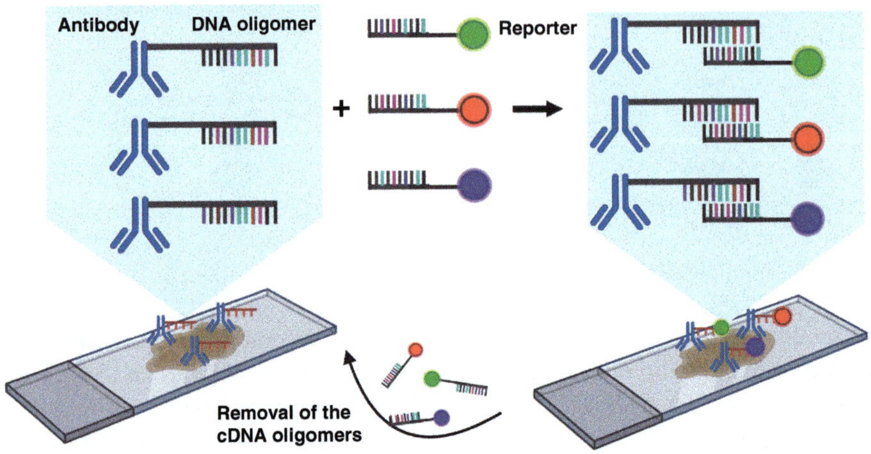

Figure 5.7 Multitags by DNA oligomer.

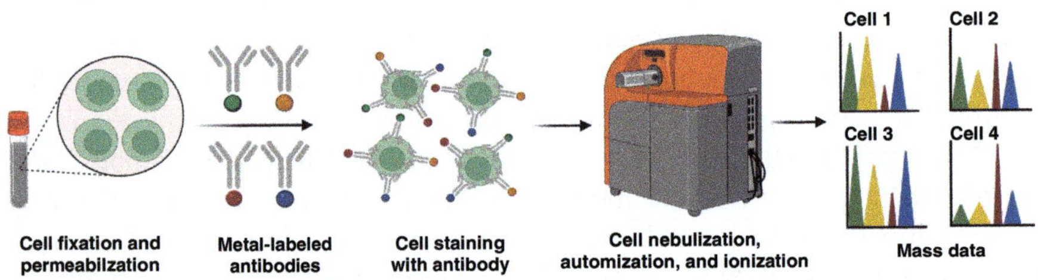

Figure 5.8 Cytometry with time of flight.

If the tissue sample is a three-dimensional (3D) chunk, rather than a 2D slice, the penetration of light and antibody through the deep tissue could be the major hurdle. For light penetration, a clearing technique may be required to make the tissue optically transparent. The opaque property of the tissue (such as brain) is usually due to the heavy scattering caused by lipid molecules. Therefore, the clearing techniques are mainly focused on the removal of lipid molecules without losing the integrity of the tissue structure. Various versions of solutions have been introduced [1], and CLARITY is one of the most popular procedures. Using CLARITY as the clearing technique, a whole brain could be imaged with multiple fluorescent antibody showing the intact structure of neuronal connections in the brain [2].

Another interesting approach for tissue imaging was introduced as expansion microscopy (ExM). To increase the spatial resolution of imaging, the tissue is physically expanded by applying polymer network to the tissue sample. Monomers are applied to the proteins in the sample and the polymerization makes the network among the proteins. Even after digestion of the proteins, still part of the proteins remains on the network representing the relative positions. Through gelation and expansion, the tissue can be swelled to enlarge the distance of each protein fragment. If the epitope of the proteins survives the fragmenting digestion, the visualization of proteins by antibodies will generate much higher spatial resolution images than those of conventional microscopy. This is a physical method of super-resolution imaging, overcoming the Abbe's limitation of diffraction [3]. Maintaining the epitopes during the digestion and even isotropic expansion of the tissue through swelling would be important for the high-quality and undistorted image generation (Figure 5.9).

Combined together with the extension of tissue, instead of antibodies, a carbohydrate and amine staining protocol with fluorescent reagent was introduced to generate information-rich tissue imaging. The vicinal diol motif of carbohydrates is treated with $NaIO_4$ to induce oxidative cleavage exposing aldehyde groups, which are subsequently labeled with fluorescent hydrazine. Amine groups mainly from proteins are labeled with amine-reactive NHS ester and the nucleus is

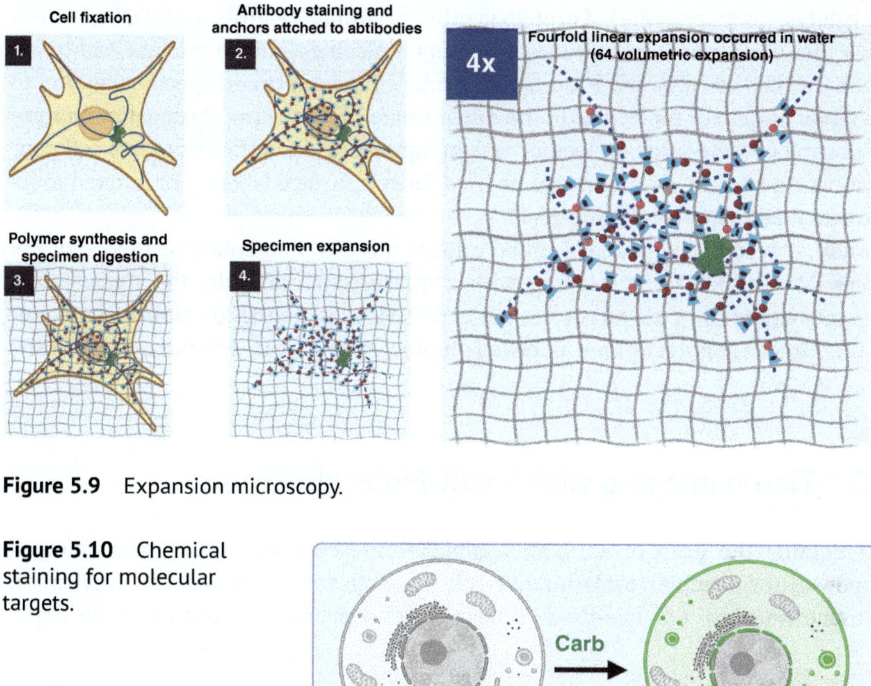

Figure 5.9 Expansion microscopy.

Figure 5.10 Chemical staining for molecular targets.

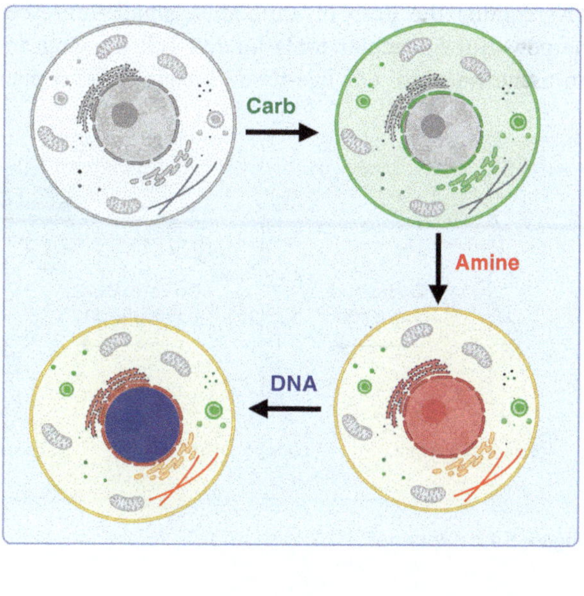

counterstained with Hoechst. The three staining reagents cover the main biological molecules, namely, carbohydrates, proteins, and nucleic acids. Especially with the enhanced spatial resolution by extension, highly information-rich tissue images could be obtained even without using any antibodies. This technique is an advanced fluorescence version of conventional H&E staining of tissues [4] (Figure 5.10).

5.2 Tissue Imaging with Nucleic Acid Probes

Antibodies are made of proteins. Our body is basically composed of an information system (nucleic acids) and a machinery system (proteins). However, some nucleic acids are known to have functional roles and rRNA may be the best example: rRNA is the catalyst (as an enzyme) of protein synthesis. Similar to antibodies, DNA or RNA molecules can be evolved into target-binding molecules. Such a nucleic acid-based probe is called an aptamer. Fluorescently labeled aptamers can specifically bind to the target protein or molecules in the cell or tissue. With the highly negative charges (of phosphate) and large molecular weight, aptamers also suffer from live cell penetration. As a result, aptamers can be used for cell surface labeling or applied to fix permeabilized samples (Figure 5.11).

Another type of nucleic acid probe utilizes the complementary sequence of the target DNA or RNA. By labeling with a fluorescent molecule, the nucleic acid oligomer can bind to the target site in the cell, visualizing the location and amount of the target. This technique is called fluorescence in situ hybridization (FISH) (Figure 5.12).

5.3 Tissue Imaging with Small-Molecule Probes

Overcoming the limit of antibodies, small fluorescent molecules have emerged as powerful molecular tools for live-cell imaging and expanded the application to tissue imaging. For live-tissue imaging, the conventional formaldehyde fixing

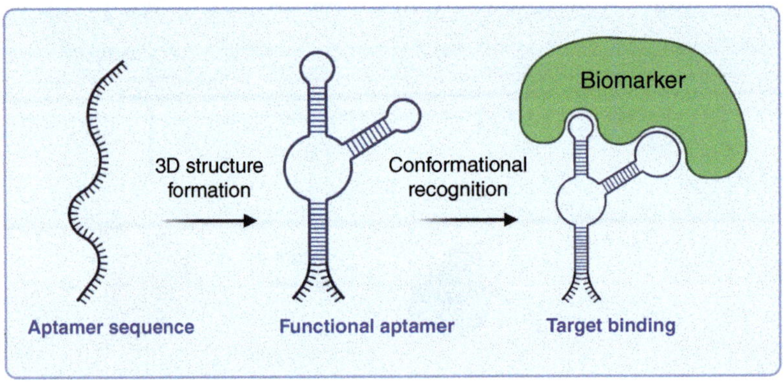

Figure 5.11 Aptamer.

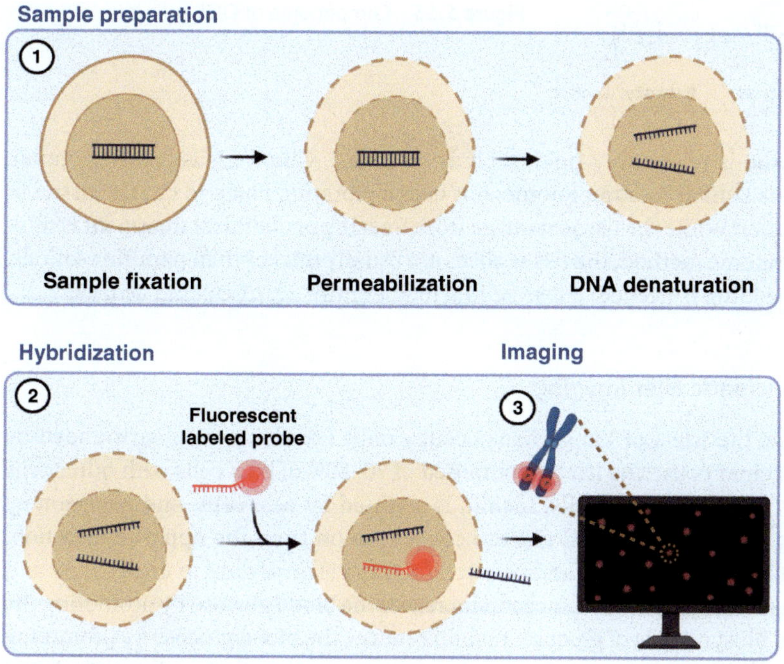

Figure 5.12 Fluorescence in situ hybridization.

in paraffin sections is not a compatible procedure. An alternative tissue sample preparation is cryosection or frozen section. Lowering the temperature hardens the tissue to some extent, but most of the tissues are still too soft for microtome cutting even at a low temperature. Deep freezing may induce crystallization of water and damage the tissue too much. Instead of paraffin, optimal cutting temperature compound (OCT) is used for soft tissue embedding in cryosection. OCT is a hydrophilic gel-like material, consisting of polyethylene glycol and polyvinyl alcohol. Tissue samples are embedded in OCT and frozen to −20 to −30 °C on a cryostat, which is practically a microtome in a freezer. The cryosection slices are typically thinner (~1 μm) than paraffin sections (5–10 μm). The quality of tissue sections is usually lower than paraffin sections, but the processing time is much shorter (10 minutes vs. multiple hours). That is why cryosection method is commonly used for real-time cancer tissue examination (currently through H&E staining) during the surgery. Defining the cancer boundary is an important guideline for surgeons to determine the range of surgical removal. The tissue damage from the rapid freezing may be unavoidable, but the hydrophilic polymers in OCT help the viability of the cells in the tissue. This is similar to using the hydrophilic organic solvent dimethyl sulfoxide (DMSO) or glycerol in a frozen storage of cells or tissues in liquid nitrogen for better viability. Before the imaging probe is applied, the OCT needs to be rinsed with water or buffer, and the staining could be done in a hydrated condition. The remaining paraffin or OCT could be the source of autofluorescence background (Figure 5.13).

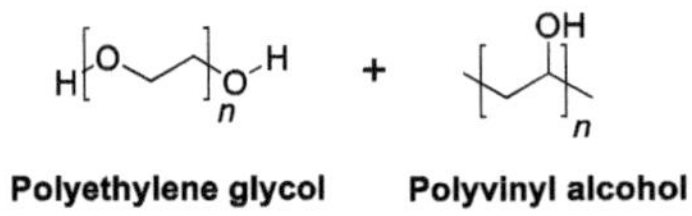

Figure 5.13 Components of OCT.

If the tissue is reasonably rigid, such as the brain, vibratome is another option. Vibratome is similar to a microtome, but uses a vibrating blade to cut the tissue in water or buffer. While the tissue damage from freezing or chemical treatment is minimal in vibratome method, the tissue slices are usually thicker than paraffin sections. When embedding is needed, a low-gelling temperature agarose could be used.

5.3.1 Pancreatic Islet Imaging

In pancreas, the islets of Langerhans occupy only 1% of the cells, surrounded by dominant acinar cells. The islet is composed of 70–80% of beta cells with other cells including 10–25% of alpha cells. Insulin is secreted by beta cells, and glucagon by alpha cells regulates the blood glucose concentration from the opposite direction, so the balance between alpha and beta cell function is important to understand and regulate diabetes symptoms. Glucagon increases the blood glucose by promoting the release and biosynthesis of glucose. Insulin reduces the blood glucose by promoting the uptake of glucose into muscles or other tissues. As diabetes is a disease induced by high blood glucose concentration, insulin has been used for treatment especially for type I diabetes, the where beta cells are destructed. Therefore, type I diabetes is insulin-dependent and is also known as juvenile diabetes. Currently, the diagnosis of diabetes is carried out by measuring blood glucose concentration and insulin response for glucose concentration control. The delated insulin response is called as insulin resistance and considered as an important diabetes symptom. Type II diabetes is a chronic disease with insulin resistance, and the beta cell status changes as the disease progresses. In early stages of diabetes, the beta cells produces more insulin to overcome the insulin resistance, but as the disease progresses the beta cell number and function decline. Therefore, the status of alpha and beta cells could be an essential information for the diagnosis and prognosis.

For discriminating between alpha and beta cells in the pancreatic tissue, insulin and glucagon antibodies would be an obvious choice as probes. With fluorescently labeled antibodies, the status of each cell and its relative compositions have been widely studied. However, insulin and glucagon are mainly stored inside cells, and the antibodies can be applied only after fixation/permeabilization of the tissue. Live tissue imaging requires cell-permeable probes such as small-molecule probes.

Newport Green is a zinc sensor and has been a popular imaging probe for pancreatic islets by targeting high zinc concentration in the beta cells. Combined with propidium iodide for dead cell detection, Newport Green could detect healthy beta cells in the islet [5]. During the islet separations for transplantation to diabetes patients, Newport Green has been used for confirming the functionally active, healthy islet collection (Figure 5.14).

In 2013, a new generation probe PiY (**P**ancreatic **i**slet **Y**ellow) was developed as a beta-cell-selective probe against not only alpha cells, but also against acinar cells [6]. The selectivity especially to acinar cell is important for the islet contrast considering the dominant population of acinar cells in pancreas. With the confirmation of beta-cell selectivity (Section 4.1.5.2), PiY has been applied to high-quality islet imaging in pancreatic tissue. Interestingly, the islet image of PiY was in much better contrast when intravenously injected tissue was compared to the direct in vitro treatment of the sectioned tissue. This may be due to the fact that the pancreatic islet is well connected to the blood vessel system; the alpha and beta cells need to sense the glucose concentration and also need to secrete the hormone into the blood stream. The traditional probe Newport Green does not work by the injection method. The high contrast of PiY staining facilitates the picking and isolation of pancreatic islets. With blood vessel staining with dextran, the 3D images of islets could be constructed with PiY (Figure 5.15).

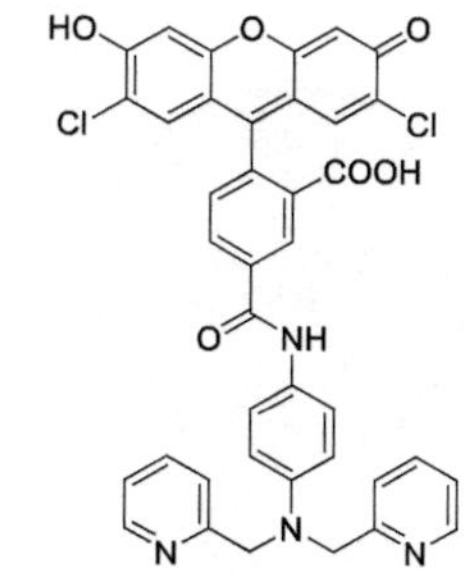

Newport Green

Figure 5.14 Structure of Newport Green and beta-cell images with propidium iodide.

Among two types of diabetes, type I model can be constructed by artificial beta-cell destruction. Streptozotocin (STZ) is a glucose-analogue-alkylating natural product, and particularly toxic to beta cells. Therefore, STZ-treated mice with broken beta cells are considered a type I diabetes model. In STZ-treated mouse pancreas, the diminished islet number and fluorescence signal of PiY were clearly demonstrated, reflecting the status of beta cells in the islet (Figure 5.16).

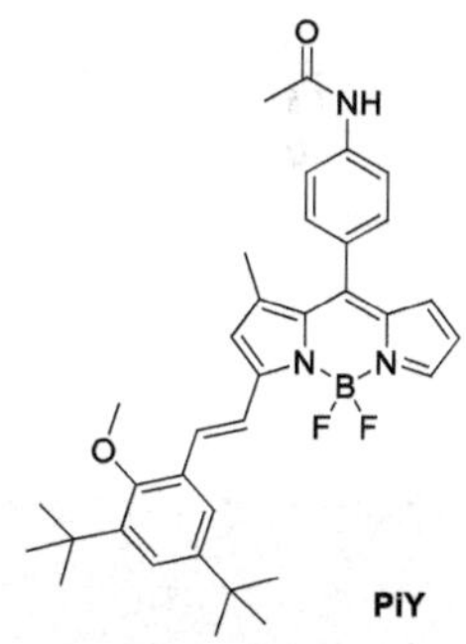

PiY

Figure 5.15 Structure of PiY and pancreatic tissue images with islet staining + 3D islet images.

In 2015, two photon alpha (TP-α) was developed as an alpha-cell-selective probe by unbiased phenotypic screening [7] and applied to pancreatic islet imaging. In mouse pancreatic islets, the minor alpha cells are localized on the surface area of the islet. Utilizing the two-photon property of TP-α, the alpha cells in mouse islets were imaged in a 3D mode using two-photon fluorescence tomography (Figure 5.17).

In 2017, as a pair of TP-α, a glucose derivative TP-β was developed as a beta-cell-selective probe. As a glucose analogue, TP-β, was expected to enter the cells through glucose transporters. As expected, TP-β uptake into beta cells was decreased in a dose-dependent manner with increasing glucose concentration. As pancreatic beta cells express high levels of Glut2, the target of TP-β is presumably Glut2 as the gating target, rather than the binding target. The blue color of TP-α and the red color of TP-β were expected to generate multicolor images of both alpha and beta cells. With the cocktail of TP-α and TP-β, the 3D images of both alpha and beta cells were beautifully achieved (Figure 5.18).

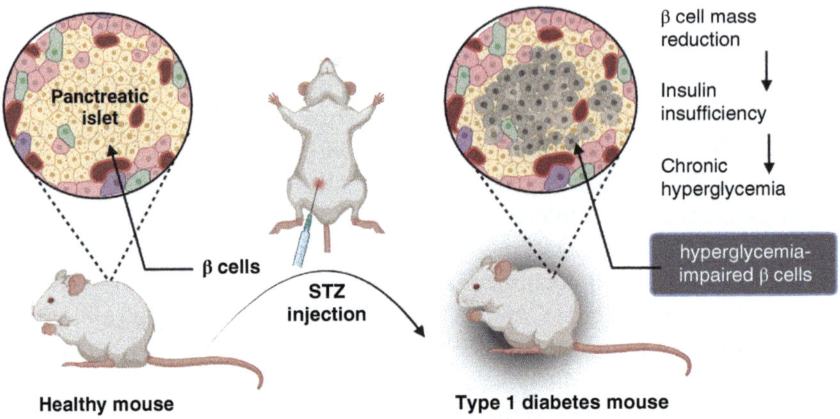

Figure 5.16 Images of STZ-treated pancreatic islet.

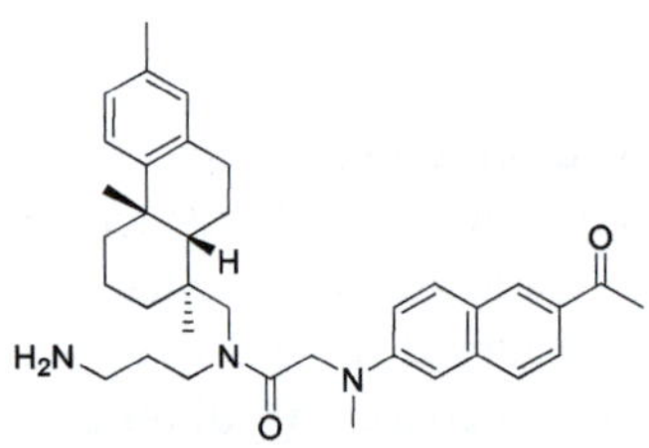

TP-α

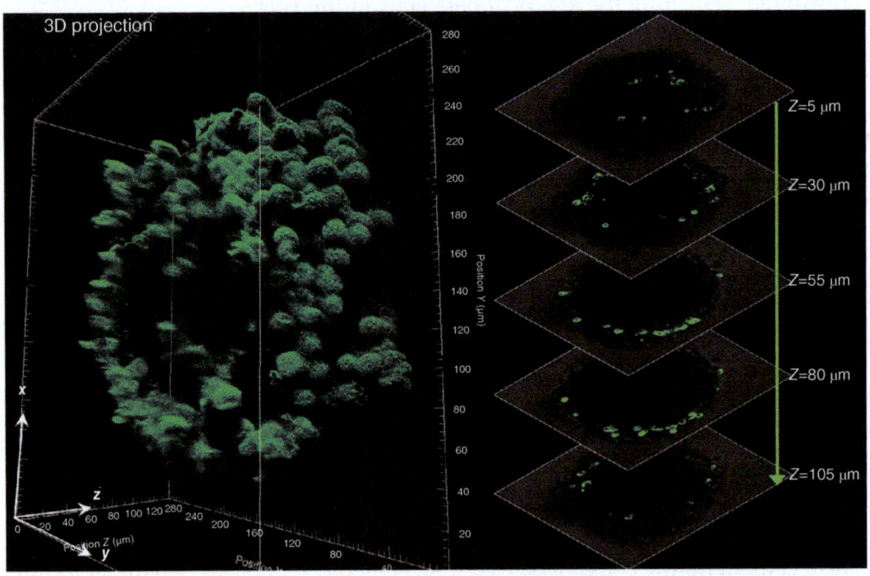

Figure 5.17 Structure of TP-α and 3D imaging of alpha cells of pancreatic islet.

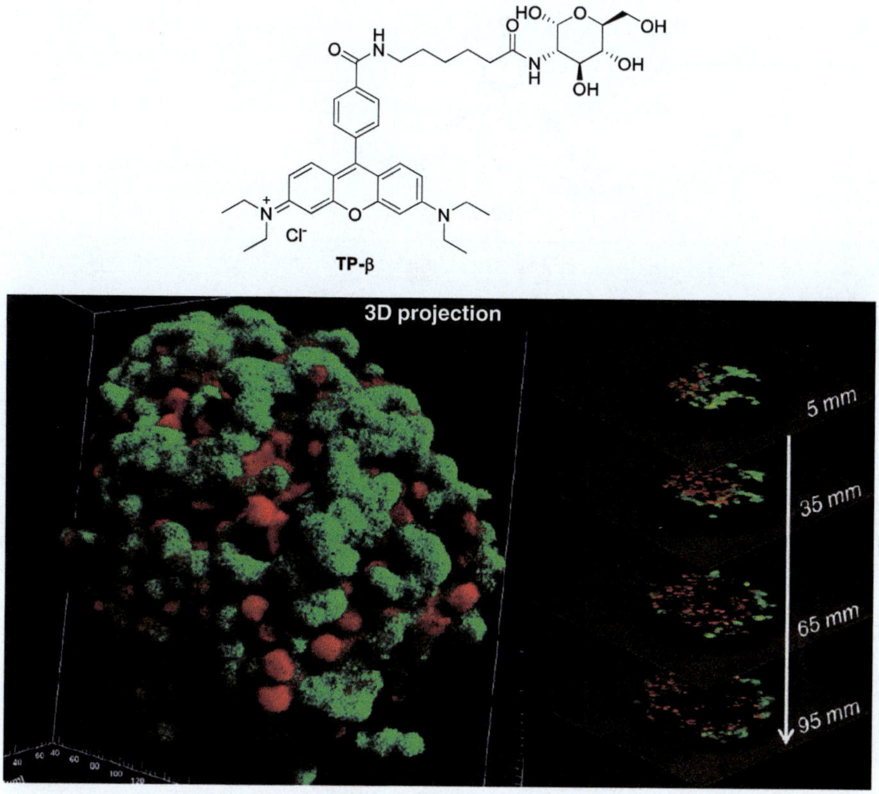

Figure 5.18 Structure of TP-β and alpha/beta cell staining image of pancreatic islet.

In 2020, an improved beta-cell probe, PiF, was introduced and compared with PiY for tissue staining performance [8]. PiF was superior to PiY in tissue section imaging, by shortening the process time from days to hours, accelerating the whole process. PiF also works well with tail vein injection, and STZ-treated type I diabetes mice were successfully distinguished from healthy mice. The facilitated tissue images allowed the processing of hundreds of pancreatic sections, generating statistically meaningful data (Figure 5.19).

A surrogate of insulin treatment is the transplantation of pancreatic islets to diabetes patients, but monitoring the survival of the transplanted islets is always an issue. The sensitive signal of PiF also allowed the detection of transplanted islets in the liver through fluorescence imaging. This is the first example of monitoring the transplanted islets in the body (Figure 5.20).

The color matching of PiF with TP-α also allowed the dual-color imaging of alpha and beta cells in the pancreatic islets, reflecting the relative position and distribution of the cells (Figure 5.21).

5.3.2 Neuronal Tissue Imaging

Brain is the biggest neuronal organ in the body, wired with a compact neuronal network. The sensory neurons deliver the signal from the peripheral region to the

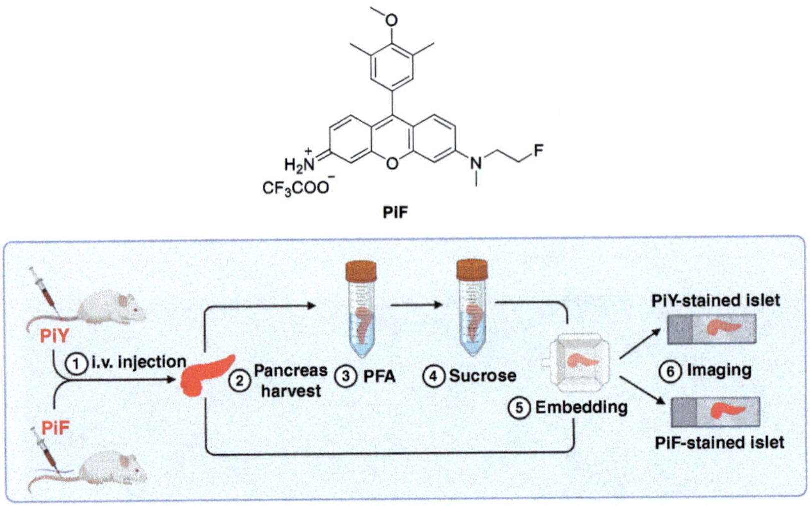

Figure 5.19 Structure of PiF and tissue processing scheme comparison.

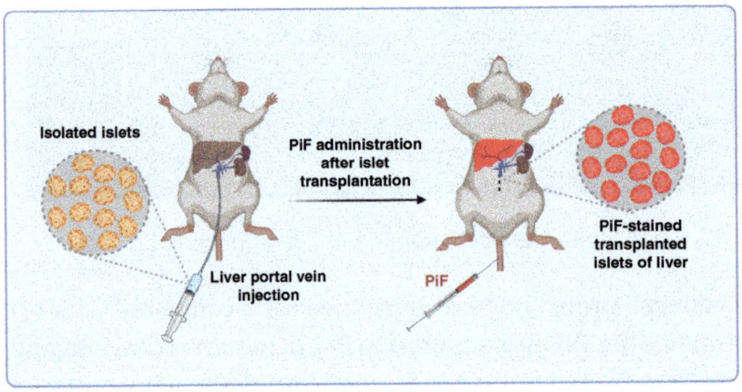

Figure 5.20 Transplanted islet detection through fluorescence tissue images.

brain, and the motor neurons deliver the signal from the brain to the muscles. Each neuronal cell receives input signals through multiple dendrites, and fire the output signal through the single axon. To understand the function of the brain, figuring out the connection structure between neurons will be the first step. For a long time, brain is believed to be composed of about 10% of neuronal cells for the neural network and 90% of supporting glial cells [9]. A recent study with an elegant cell counting technique found that the ratio of neurons to glia cells is about 1 : 1 [10]. According to the study, the number of neurons in an adult human is estimated to be 86 billion. The high density and entangled neuronal network in the brain make it so difficult to map the connections. Human Connectome Project is one of the collaborative efforts to build the complete neuronal connection model (www.humanconnectomeproject.org). One representative approach is making thin slices of the brain tissue and accumulate electron microscope images to build the full 3D neural network. Although electron microscopy generates ultra-high-resolution images of the

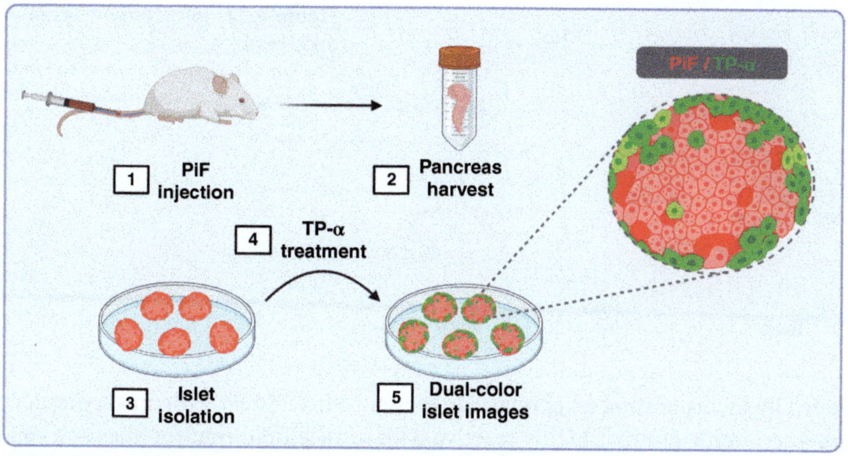

Figure 5.21 Dual-color pancreatic islet images of PiF and TP-α.

neuronal structures, the tissue needs special treatment with a metal coating, and the imaging requires vacuum conditions, which are far from the live tissue state.

For the optical visualization of the neuronal network, hydrophobic plasma-membrane dyes such as DiI has been popularly used. When these dyes are applied as a chunk in one position of the brain area, they are incorporated into the cell membrane nearby. Then they travel along the membrane structure, and the long axon and dendrite structures deliver the dyes to the remote site of the cells as far as they are connected by the same membrane structure. While this method could visualize the neuronal network to some extent, it takes a long time (several days) for the staining, and the dyes are not selective to the neuronal cells. They only show the membrane structure of the cells connected to the injection site. Even with the staining, the high density of the neuronal cells makes it difficult to distinguish one cell from another, and the optical limit of resolution makes the situation even more challenging. One trick to make the individual neuron distinct from other neurons is the brainbow technique, where genetic fluorescent reporter genes are shuffled to generate a random color combination [11].

In 2015, the first universal neuron-selective probe NeuO was developed and demonstrated to distinguish neuronal cells from other glial cells in a cell culture (Section 4.3.3). When NeuO was applied to a brain section, the principal cell layer of dentate gyrus (DG), the stratum granulosum, was sparsely labeled, and the principal cell layer of CA1, the stratum pyramidale, was densely stained. Through two-photon and confocal microscopic imaging, NeuO clearly revealed the neuronal structures from whole-brain staining of the neonatal brain.

Retina is part of the central nervous system and has various interneurons and sensory neurons such as photoreceptor cells. When NeuO was tested in retina sections, it clearly showed the neuronal structures such as the outer segment layer of photoreceptor cells and some bipolar cells. However, the limitation of NeuO is that it is easily washed out by repeated washings or cell permeabilization conditions. To increase the retention of the stain, a functionalized derivative of NeuO was

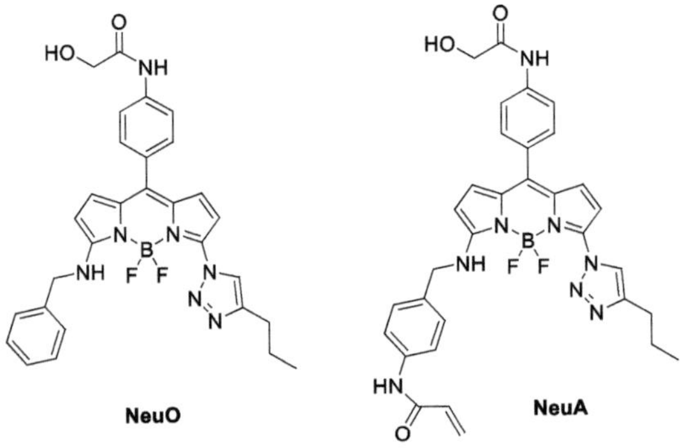

Figure 5.22 Structure of NeuO and NeuA.

developed by incorporating an acrylamide group: NeuA (NeuO with an acrylamide group). Acrylamide group is highly reactive with intracellular nucleophiles and can form covalent bond to macromolecules in the cell, thereby enabling the probe to be retained during cell washing or permeabilization. NeuO and NeuA showed almost same staining pattern for neurons with a stronger and sustained staining of NeuA compared to NeuO. For confirming the neuronal selectivity, the NeuA-stained retinas were examined with double staining immunohistochemistry (IHC) of anti-Recoverin antibody that stains photoreceptor cells and type 2 and 8 bipolar cells. NeuA and Recoverin signals were well overlapped in the outer segment layer (OSL) and the inner plexiform layer (IPL) (Figure 5.22).

5.4 Organoid as Model of Tissue and Organ

Organs are body's recognizable structures and are usually made of several tissues. Primary tissues or organs are sometimes problematic to obtain for imaging research and are usually difficult to maintain for a long-term period. If animals should be sacrificed for the tissue extraction, ethical issues will be involved. Especially for clinical tissue samples, the ethical and even legal concerns are more severe. As the surrogate of animal or primary tissue, organoids could be a good option. The suffix -oid means "resembling," so an organoid may be defined as an artificial organ produced in vitro. One type or mixtures of cells are cultured in vitro to form self-organized 3D structures, mimicking the function of naturally occurring tissues or organs. The cell sources can be pluripotent stem cells, adult stem cells, or somatic cells. To achieve a good organoid, a careful orchestration of spatiotemporal culture control is necessary. The proper signaling control by growth factors and optimum supportive materials such as extracellular matrix (ECM). Several tissue-mimicking organoids are reported for intestinal, hepatic, renal, and cortical tissues [12]. Not only for the primary normal organs, primary tumor cells can be used for tumoroid formation, which may mimic the tumor microenvironment [13].

In comparison to the conventional 2D cell culture, 3D organoids may better represent the natural characteristics and procedures, such as stem cell differentiation, cell–cell interaction, and cell movement. Once established, the storage and long-term maintenance of organoids should be much easier than primary tissues. Therefore, in general, organoids are amenable to live imaging techniques. If cells of human origin are used for organoid formation, interspecies complexity issues will be minimized.

The organoids could also be used in "organ-on-a chip" study for multiorgan interaction. As an interesting example, the female reproductive system was recreated by assembling the related organs. The EVATAR system has different female organs connected with tubes mimicking the blood stream. The female hormone production was regulated by the cooperative function of each organ and EVATAR successfully reproduced the female menstrual cycle [14]. As a modularized system, the organ on a chip may replace the current single organ donation and be used as an alternative transplantation option for patients.

5.4.1 Blood Vessel 3D Model

Blood vessels are localized in all the organs for oxygen and nutrient delivery to each cell of the body. Considering the importance, the visualization of the blood vessel would provide essential information on the tissue structure from the functional aspect. Angiogenesis (new blood vessel generation) is also a biomarker for cancer growth and could be a good target for blood vessel probes. One approach to blood vessel imaging was using fluorescently labeled dextran (carbohydrate polymer) as a blood flow probe, but different sizes of dextran may cause issues such as clogging capillaries or leakage into tissues. The inner surface of blood vessels is layered with endothelial cells and the outer surface is surrounded by smooth muscle and fibroblasts. Therefore, endothelial-cell-selective probe is expected to be a promising candidate for blood vessel imaging. However, except antibody probes, there was no other small-molecule probe for endothelial cells for a long time.

For an ideal blood vessel probe, a whole-body imaging-based screening would be appropriate for the in vivo selectivity against all other tissues in the body and toxicity issues. Practically, however, it would be difficult to use an animal for every probe candidate, especially when the candidate number is high as a library format. Conventional 2D cell culture of endothelial cells with a negative control such as fibroblasts would be an alternative option, but relevance of the cell culture to real blood vessel status is a big question. In such a situation, a 3D organoid-mimicking blood vessel would be much more sophisticated and closer to the real blood vessel environment than a simple 2D cell culture.

In 2021, a fluorescent probe CyA-B2 was developed through a screening of 3D blood capillary model [15]. The model was built by culturing endothelial cells and fibroblast cells in collagen microfibers, which was prepared by fibrillation and sonication of collagen I sponge. The collagen microfibers provide the ECM in the complex tissue of endothelial-cell-derived capillaries and supporting fibroblasts. The endothelial cells in this blood capillary model expressed significantly higher specific endothelial membrane proteins than those in the 2D culture. Through a screening with Cy5 range of library with 670 nm emission, the first blood-vessel-specific small-molecule probe CyA-B2 was elucidated (Figure 5.23).

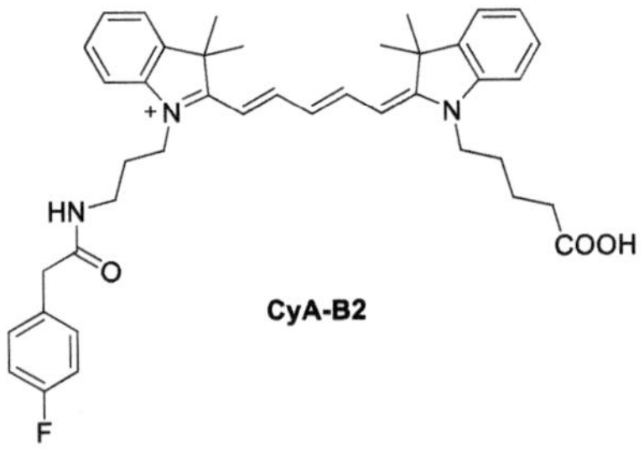

Figure 5.23 Structure of CyA-B2.

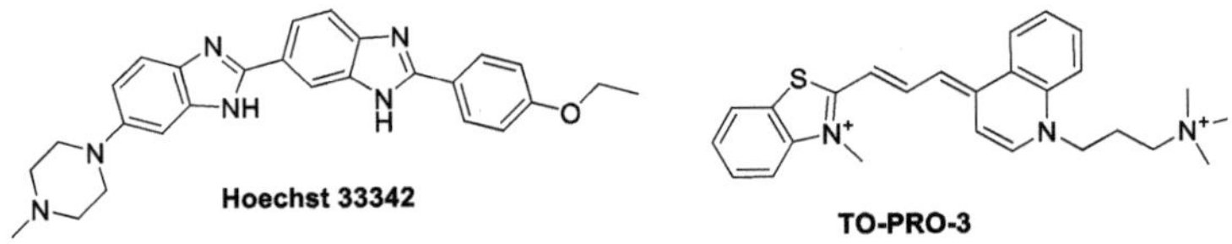

Figure 5.24 Structure of Hoechst 33342 and TO-PRO-3.

The performance of CyA-B2 was confirmed by ex vivo imaging of mouse aorta. By injection of CyA-B2 into excised aorta, the endothelial-cell-selective staining was confirmed by opening and mounting the aorta to the imaging microscope. The selective imaging of the blood vessel of CyA-B2 was also demonstrated under in vivo labeling by intravenous injection into mice.

5.4.2 Tumor Organoid for Drug Screening

Tumor organoids generated may serve as a high-fidelity tumor model in cancer biology, compared to conventional 2D cell culture. Especially, the patient-derived tumor organoid would be essential for the personalized precision medicine therapeutics. However, constructing scalable and reproducible tumor organoids for massive screening is in general challenging.

As a successful demonstration, Matrigel domes were used to serve as ECM for tumor organoid formation. For high-throughput screening, a fast automatic focusing would be necessary and an inverted microscopic configuration would be ideal to detect the contacted cells from the plate bottom. For high-content fluorescence imaging, Hoechst 33342 for live cells and TO-PRO-3 for dead cells, together with caspase-3/7-activated fluorescent reagents for apoptotic cells, were employed to demonstrate the therapeutic reagent effect on tumor organoids [16] (Figure 5.24).

References

1 Ueda, H.R., Ertürk, A., Chung, K. et al. (2020). *Nat. Rev. Neurosci.* 21: 61–79. https://doi.org/10.1038/s41583-019-0250-1.

2 Chung, K., Wallace, J., Kim, S.-Y. et al. (2013). *Nature* 497: 332–337. https://doi.org/10.1038/nature12107.

3 Chen, F., Tillberg, P.W., and Boyden, E.S. (2015). *Science* 347: 543–548. https://doi.org/10.1126/science.1260088.

4 Mao, C., Lee, M.Y., Jhan, J.-R. et al. (2020). *Sci. Adv.* 6: eaba4542. https://doi.org/10.1126/sciadv.aba4542.

5 Lukowiak, B., Vandewalle, B., Riachy, R. et al. (2001). *J. Histochem. Cytochem.* 49: 519–528. https://doi.org/10.1177/002215540104900412.

6 Kang, N.-Y., Lee, S.-C., Park, S.-J. et al. (2013). *Angew. Chem., Int. Ed.* 52: 8557–8560. https://doi.org/10.1002/anie.201302149.

7 Agrawalla, B.K., Chandran, Y., Phue, W.-H. et al. (2015). *J. Am. Chem. Soc.* 137: 5355–5362. https://doi.org/10.1021/ja5115776.

8 Kang, N.-Y., Lee, J.Y., Lee, S.H. et al. (2020). *J. Am. Chem. Soc.* 142: 3430–3439. https://doi.org/10.1021/jacs.9b11173.

9 Ullian, E.M., Sapperstein, S.K., Christopherson, K.S., and Barres, B.A. (2001). *Science* 291: 657–660. https://doi.org/10.1126/science.291.5504.657.

10 Azevedo, F.A.C., Carvalho, L.R.B., Grinberg, L.T. et al. (2009). *J. Comp. Neurol.* 513: 532–541. https://doi.org/10.1002/cne.21974.

11 Weissman, T.A. and Pan, Y.A. (2015). *Genetics* 199: 293–306. https://doi.org/10.1534/genetics.114.172510.

12 Keshara, R., Kim, Y.H., and Grapin-Botton, A. (2022). *Annu. Rev. Cell Dev. Biol.* 38: 447–466. https://doi.org/10.1146/annurev-cellbio-120320-035146.

13 Srivastava, V., Huycke, T.R., Phong, K.T., and Gartner, Z.J. (2020). *Curr. Opin. Cell Biol.* 66: 51–58. https://doi.org/10.1016/j.ceb.2020.05.003.

14 Xiao, S., Coppeta, J.R., Rogers, H.B. et al. (2017). *Nat. Commun.* 8: 14584. https://doi.org/10.1038/ncomms14584.

15 Abdul Sisak, M.A., Louis, F., Aoki, I. et al. (2021). *Small Methods* 5: 2100338. https://doi.org/10.1002/smtd.202100338.

16 Larsen, B.M., Cancino, A., Shaxted, J.M., and Salahudeen, A.A. (2022). *STAR Protoc.* 3: 101407. https://doi.org/10.1016/j.xpro.2022.101407.

6

In Vivo Whole-Body Imaging Probes

When chemical probes are used in a live whole body, there are several extra considerations for the imaging procedure. Toxicity of the probe would be the basic issue, similar to live-cell or tissue imaging. For securing enough concentration of the probe, the proper delivery of the probe into the target site would be one of the most important points. For cells or tissues, a simple incubation or immersion in the probe-containing medium would suffice in most of the cases. In a whole body system, there are several delivery methods, such as eye drops, skin creams, oral uptake, or injection. Injections could be done intravenously for whole-body distribution or locally as in subcutaneous (under the skin), intramuscular (into the muscle), and intraperitoneal (into the body cavity in the belly) injections. In addition to the binding affinity of the probe to the target, the pharmacokinetic property of the probe is also important, similar to drug cases. Pharmacokinetics is about how the body treats external material such as probes or drugs, and usually absorption, distribution, metabolism, and excretion (ADME) are the main factors. The delivery method should be designed carefully to make the probe available at the site of targets with the highest signal-to-background ratio at the optimum time point. As a result, the proper dose, delivery method, and imaging time window should be carefully designed and optimized.

Another important issue is the target site and matching relevant modes of imaging. If the target site is under the skin, the tissue penetration of excitation and emission light should be considered. The main source of light loss is the absorption and scattering in the tissue. In general, the light scattering is stronger for shorter wavelengths of light, so a near-infrared (NIR: typically, 780–2500 nm) light source and matching probe could help, but still only up to several hundred micrometers depth at best.

Photoacoustic imaging is a surrogate technique to fluorescence imaging for better tissue penetration. In photoacoustic imaging, the input is the same NIR light as an excitation source for the probe. When the probe is excited, some of the absorbed energy is converted to heat energy, and the thermal expansion and contraction caused by the following cooling procedure generate sound waves in the tissue. The sound could be collected and converted into the location information of the probe. As the sound wave has much better tissue penetration property than light, the output signal intensity could be much higher in photoacoustic imaging compared

Sensors and Probes for Bioimaging, First Edition. Young-Tae Chang and Nam-Young Kang.

to fluorescence emission. Combined with multipoint detection and tomographic imaging process, photoacoustic technique also can provide three-dimensional (3D) images for the probe distribution.

If the target site is deep in the body, the physical access of the optical imaging instrument should be secured. If the target is on the gastrointestinal tract, topologically speaking, the area could be considered outside of the body, same like the skin surface. Therefore, a catheter (tube with an optical camera installed) is an option to access to the target site. Depending on the position of the target in the gastrointestinal tract, the catheter could be inserted either from the mouth or rectum side. Bladder is also an accessible organ to catheter. If the target site is deeper than light penetration from the topological surface, invasive instrument insertion may be necessary for optical imaging.

6.1 ElaNIR for Elastin Imaging in Mouse

Elastin is an important component protein of the extracellular matrix. As the name implies, elastin is responsible for the elasticity in the connective tissue and skin. Elastin also wraps blood vessel walls providing the elastic property to cope with the heart's blood pumping. As we age, the elastin amount decreases, causing wrinkles and sagging skin. Also, the artery vessels become more vulnerable to hypertension. Visualizing elastin in vivo would be an ideal way to monitor the functional degree of aging. One possible approach for elastin-selective imaging probe development may be structure-based design. Unfortunately, elastin is a complex protein with polymeric properties and difficult to work with for the modeling. Also, elastin is located under the skin, so the penetration of light could be a hurdle for visualizing the elastin-bound probe. To overcome the technical issues, an NIR-range fluorescent library was developed for deep-tissue penetration of light and directly screened on the elastin-rich tissue, avoiding the design difficulty. Cross section of a rabbit's aorta was employed for the ex vivo imaging screening. With the unique lamella structure of elastin, the fluorescence tissue imaging guided the screening and eventually provided a selective elastin probe. With the NIR range of fluorescence property, the probe was dubbed ElaNIR (Elastin NIR) [1].

In the tissue section, ElaNIR clearly visualizes the lamella structure of elastin with low background in other areas. The high selectivity of ElaNIR was postulated due to the unique biphenyl group in ElaNIR, as the absence of the group diminished elastin selectivity in the library. The low background of ElaNIR may be due to the zwitterionic (containing both positive and negative charges) motifs. Usually hydrophobic motif of small molecules tends to nonspecifically bind to biological macromolecules or membrane structures. The zwitterionic compounds possess high hydrophilic properties and tend to give a low nonspecific tissue background [2] (Figure 6.1).

In addition to the ex vivo tissue slice, ElaNIR was also tested in live animals through intravenous injection in mice. Through the skin fluorescence, it was clearly demonstrated that young mouse (1 month old) is much brighter than old mouse

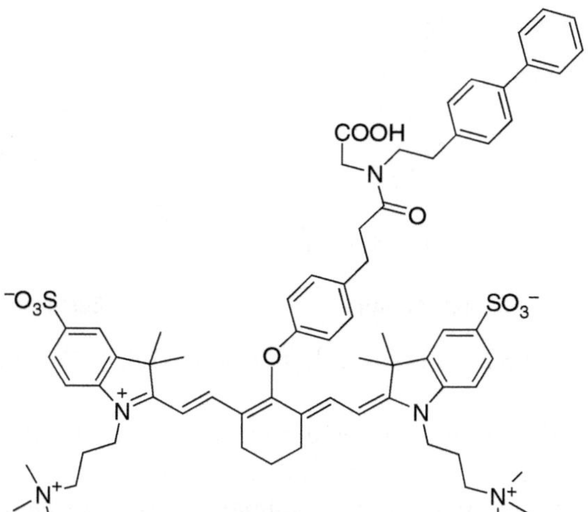

Figure 6.1 ElaNIR structure and staining pattern.

(10 month old). The fluorescence signal of ElaNIR penetrates the skin and reports the relative elastin amount under the skin. While the fluorescence images are only for the skin surface, i.e. 2D, photoacoustic imaging provided 3D tomographic images of the elastin under the skin.

6.2 Probes for Exposed Neuron in Zebrafish Embryo

Zebrafish (*Danio rerio*) embryo is a great whole-body imaging model for vertebrates. First of all, zebrafish embryo is optically transparent with a few skin spots. As a vertebrate, the embryo contains a complex central nervous system (CNS) of brain and spinal cord, which is quite similar to that of humans. The fish model also has internal organs and muscle structure. Even blood system including immune cells are comparable to other mammals. The basic organ structure is visible within a one-day-old embryo with a big yolk, and the general fish shape is visible in three to four days. The size of the zebrafish embryo is small, in several millimeters size, and it is relatively easy to maintain them in large stocks. With the full genome sequencing and an efficient knockdown system using morpholino oligonucleotides made zebrafish the standard genetic model among fish. As the model is a water-born fish, incubation with a probe or drug is straightforward. With these advantages, zebrafish have been popularly used as an advanced vertebrate animal model for chemical genetics and drug development.

The small size of the zebrafish embryo and easy chemical treatment allow medium-to-high throughput screening in 96-well plate format. Taking advantage of the optical transparency, DOFL has been applied to zebrafish embryos for phenotypic fluorescence-based screening, and two neuron-selective probes were reported in 2010 [3]. According to the color, the probes were dubbed ZeN (Zebrafish Neurotracer)-Green and Red. As the screening was performed by simple incubation of probes in the fish-growing media, without any injection method, ZeN probes

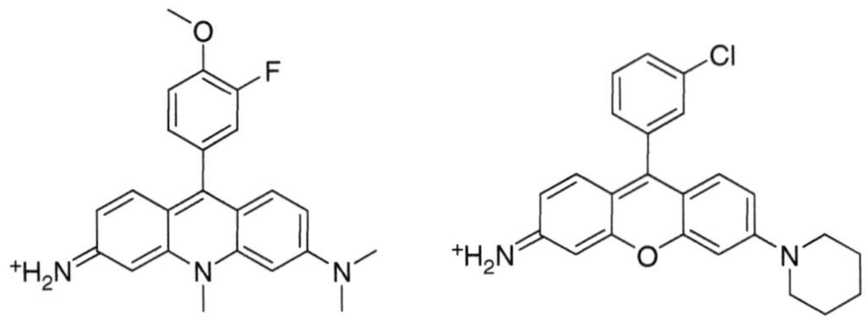

Figure 6.2 Structure of ZeN-Green and Zen-Red/neuronal organs stained by ZeN probes.

mainly stained exposed neuronal organs of zebrafish. The stained neurons are located in nose, ears, and neuromasts along the side of the body as the motion sensors. As an aquatic animal, the sensory neuron cells of zebrafish may be in direct contact with the environment for chemical/physical sensing, and the probes must reach the exposed neuron without any physical barrier. Other areas of the body are covered with skin, which may prevent easy access to external materials (Figure 6.2).

The ZeN probes share the rosamine motif, which is known to preferably bind to mitochondria of the cells. Assuming that the mitochondrial membrane potential may be important to the binding affinity of ZeN probes, they were expected to be used as a tool for toxic environmental surveillance. Indeed, various toxins such as lead (environmental heavy metal ion contaminant) or gentamicin (aminoglycoside as a known neuro toxin) treated zebrafish showed decreased ZeN staining, indicating that the probe staining is correlated with the healthy status of the neuronal cells. Compared to the neuronal-cell-based toxicity test, this whole-body-based system would provide a much more specific and sophisticated neurotoxicity screening platform. While ZeN probes could be used for selective healthy neuron staining, the application was limited only to the exposed neurons.

6.3 NeuO for Whole-Body Neuron Imaging in Zebrafish

In 2015, NeuO was developed through a systematic high-throughput cell-based screening as a neuron-selective probe and demonstrated its potential as a live-cell and tissue imaging probe [4]. When NeuO was added into the medium with zebrafish embryos, similar to ZeN probs, the exposed neurons such as the nose and motion sensors on the neuromasts were mainly stained. The internal neurons seemed not accessible by simple incubation in the medium as in the case of ZeN probes (Figure 6.3).

Through injection of NeuO into the brain ventricle space, the detailed structure of CNS (brain and spinal cord) and internal neurons in the eye/ear were elucidated through tomographic imaging, envisaging the potential of NeuO for rapid and nontoxic neuronal labeling probes [5].

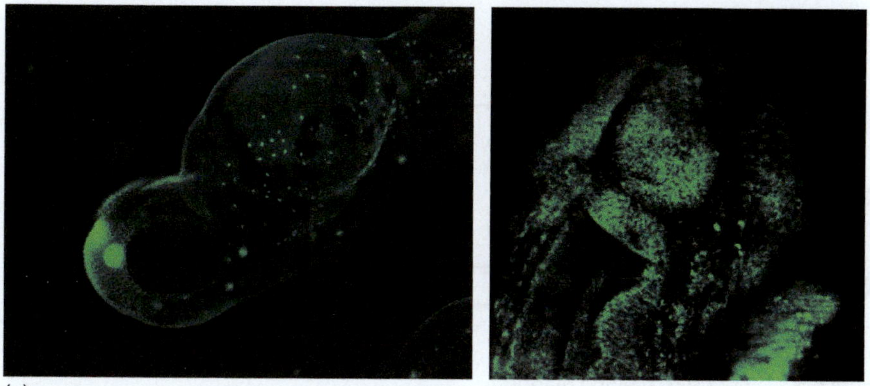

Figure 6.3 (a) Image of zebrafish with NeuO. (b) 3D image of fish with NeuO injection.

6.4 LipidGreen for Fatty Tissue Imaging in Zebrafish

Sometimes, unexpected fluorescence property is observed during various organic synthesis and it often leads to new fluorophore discovery. Korean Chemical Bank (https://chembank.org) collected more than half a million drug-like molecules for a nation-wide screening network to elucidate novel drug candidates. Among the molecules, unexpectedly fluorescent compounds were collected and characterized in cells and zebrafish embryos. One of the compounds demonstrated a lipid droplet staining pattern in 3T3-L1 adipocyte cells and reported with the name of LipidGreen in 2011 [6]. When incubated with zebrafish embryos, LipidGreen stained accumulated fat tissue in the body, and the selectivity was much higher with a lower background than the common lipid droplet probe Nile Red. During the early embryogenesis, LipidGreen was mainly deposited in the yolk as expected, and moved to the pancreatic area (15–20 days post fertilization: dpf) and near eye area (30 dpf). At 50 dpf, a new staining emerged around the pectoral fins, skull, and heart, and after 80 dpf, the fat deposits were distributed near the intestine and liver. LipidGreen seems to report the fatty tissue distribution in zebrafish body by simple incubation (Figure 6.4).

Interestingly, when the zebrafish was starved, the fat staining was reduced in size from 5 days and completely depleted after 14 days. With refeeding, several fat deposits were detected from three days and fully restored by eight days. The entire process was monitored by LipidGreen in real time, and LipidGreen performed as a whole-body fat probe. This fatty tissue monitoring system could be used for the screening of inhibitors for lipid droplet biosynthesis such as PF04620110. Such drugs could be diet-drug candidates, and the zebrafish platform is a sophisticated diet-drug screening system

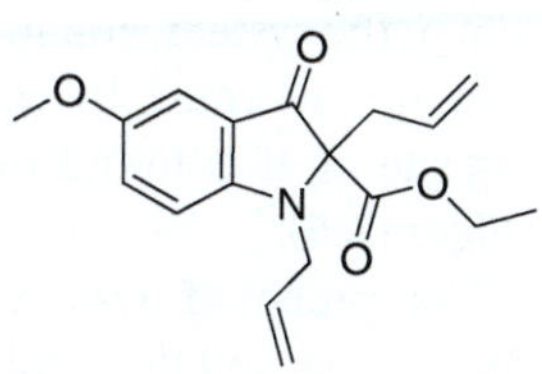

LipidGreen

Figure 6.4 Structure of LipidGreen.

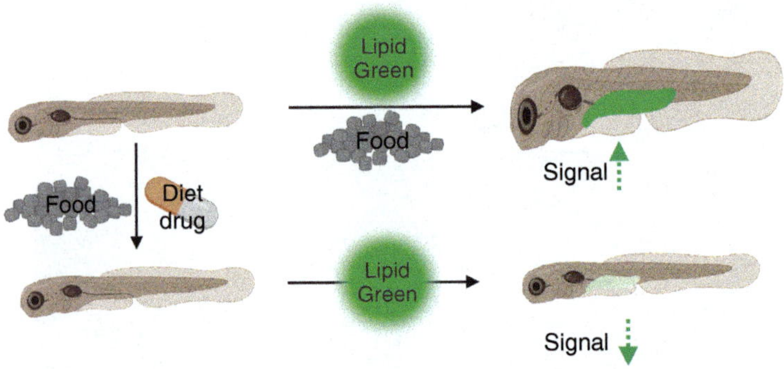

Figure 6.5 Diet drug screening platform.

using vertebrates. Because each embryo does not require LipidGreen injection, the screening system is a high-throughput format (Figure 6.5).

6.5 Blood Vessel Imaging in Zebrafish

Blood vessels are important for the material supply in the body, and imaging probes for the structure and status of blood vessels would provide important information of the body. New blood vessel formation (angiogenesis) is usually considered the biomarker of cancer growth, and monitoring or screening of angiogenesis would be a useful drug-screening platform. Similar to LipidGreen discovery, from whole-body screening of zebrafish embryos, a blood-vessel-staining fluorescent probe (4f) was discovered and reported in 2019 [7] (Figure 6.6).

In the beginning, probe 4f was expected to bind to endothelial cells, which compose the inner wall of the blood vessels. However, the analysis of the probe indicated that 4f resides in the bloodstream and flow together with other blood cells. Later, it turned out that 4f binds to the serum albumin and increases its fluorescence, i.e. 4f is an albumin sensor. Interestingly, through a simple incubation in the media, probe 4f somehow enters the zebrafish bloodstream and binds to albumin, turning on the fluorescence intensity. The definition of albumin is the most abundant protein in the system, and in this case, it is the most abundant protein in the blood, that is serum albumin. By competition assay, using human serum albumin, the binding site of 4f is found to be the ibuprofen site, not the warfarin or digitoxin site (Figure 6.6).

The probe 4f was tested in zebrafish with anti-VEGF inhibitor (PTK787/ZK222584), and the inhibition of angiogenesis was clearly visualized through blood vessel imaging. This example demonstrates that probe 4f could be used for the drug screening for angiogenesis using whole-body imaging of zebrafish. Compared to the cell culture system, the systemic zebrafish screening platform could provide much richer information, such as toxicity of the candidate drug molecules and other biological effects in the whole body.

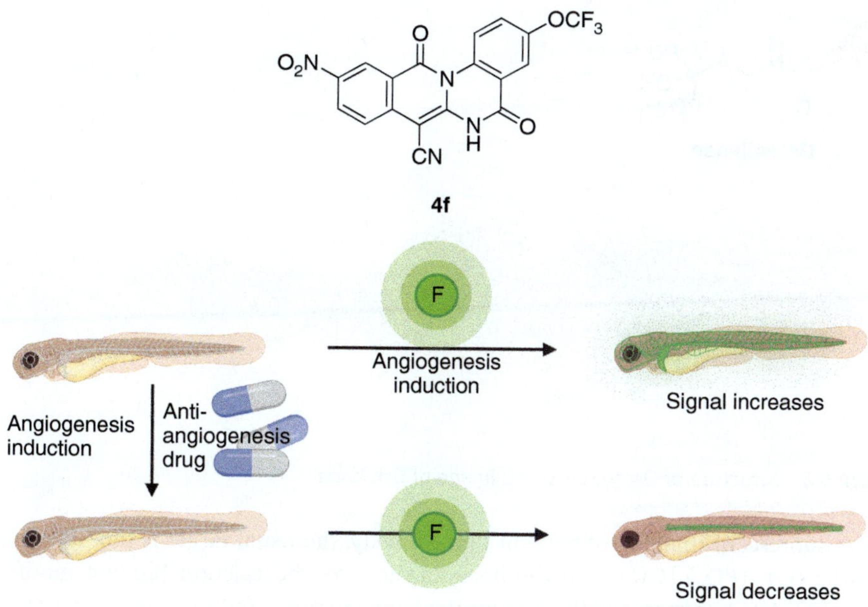

Figure 6.6 Structure of 4f and anti-angiogenesis drug screening platform.

6.6 Probes for Bone Imaging

The scaffold of the body is bone, and imaging bone would be central for information on body shape and structure. X-rays cannot penetrate bone and have been used for the imaging of bone. 3D imaging of bone structure is also available with tomographic techniques, commonly called as computed tomography (CT). While the terminology CT can be applied to different imaging techniques such as magnetic resonance imaging (MRI) or positron emission tomography (PET), when you hear a simple "CT," most likely it means X-ray CT. Usually X-ray CT provides anatomical information and is overlapped together with molecular probes (such as fluorescent probes) as a reference image.

For optical bone imaging, alizarin red S has been used by reflecting the calcium binding property. The dye was also used for calcium deposits in cell culture. Alizarin red S could show the bone structure in a better contrast, but the sampling procedure requires treating the body sample in an organic solvent such as isopropanol. Therefore, it is not comparable to live in vivo imaging (Figure 6.7).

Figure 6.7 Structure of alizarin red S.

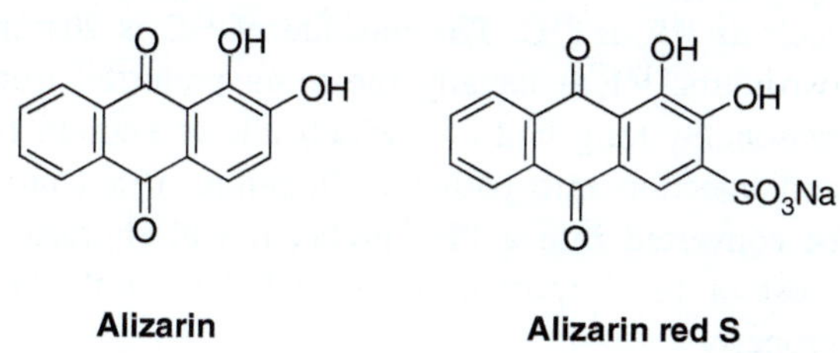

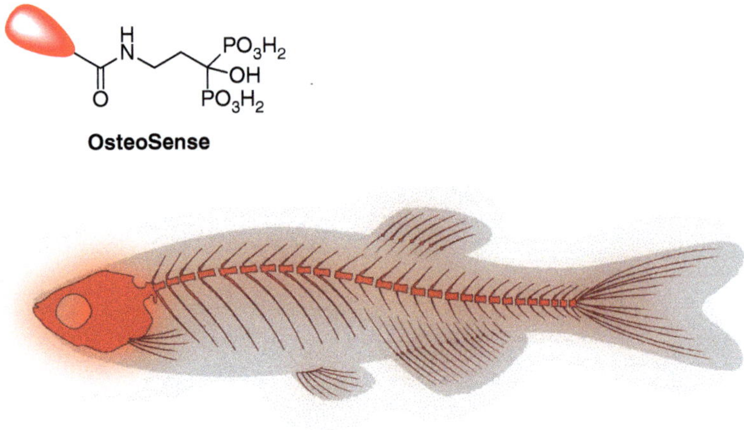

Figure 6.8 Structure of OsteoSense and image of fish bone.

For fluorescent imaging of bone in the live body, the usual target is the hydroxyapatite ($Ca_{10}(PO_4)_6(OH)_2$) on the bone surface. As the calcium binding motif, iminodiacetate has been widely used for the bone targeting and various fluorescent motifs were conjugated. One of the popular commercial bone probes is IVISense Osteo (OsteoSense 680, 750, and 800), which contains diphosphonate as the targeting motif. Although PerkinElmer sells three color bone probes (680, 750, 800 nm), the exact chemical structure has not been exposed yet. In the literature, an NIR diphosphonate probe Pam78 was reported for osteoblastic activity imaging [8] (Figure 6.8).

Due to the strong negative charges, both iminodiacetate and diphosphonate probes need to be administered through tail vein injection or intraperitoneal injection in mice. And, usually 24 hours of incubation is recommended to allow enough time for the probes to reach the bone surface.

6.7 Probes for Pancreatic Islet Imaging

Although fluorescent probes are useful for cell-based or ex vivo tissue imaging, whole-body in vivo imaging is difficult due to the penetration of input and output light for the probe activation and emission, especially for the deep organs of the human body such as the pancreas. For clinical application, PET contrast agent may be an ideal option, but it requires positron-emitting radioactive nuclei, such as ^{18}F or ^{11}C. The half-life of ^{11}C is 20 minutes, and that of ^{18}F is about two hours. ^{18}F is usually the more preferred isotope for PET probes due to its reasonably long half-life, which allows enough time for the chemical synthesis and injection into patients. Therefore, if a characterized fluorescent probe can be converted into a PET probe, it will shorten the development procedure, as most of the important biological data could be obtained by the fluorescence property.

In diabetes patients, the status of beta cells that produce insulin is important information for the diagnosis in addition to the currently available data on blood glucose concentration and insulin resistance. Therefore, the bioimaging probe for pancreatic islets, especially the main population of beta cells, would be an ideal tool for noninvasive diagnostic protocol. As beta-cell probe PiY stains pancreatic islets well through intravenous injection [9], the probe had a good potential to be developed into a PET probe by introducing ^{18}F. However, the introduction of ^{18}F into an already existing molecular probe (PiY in this case) carries always an extra burden to find the F attachment position and a method, requiring a thorough structure–activity relationship (SAR) study and chemical scheme development. In addition, when injected, PiY showed a stronger background in the liver than in the pancreas in the pharmacokinetic analysis. Liver is close to pancreas and may generate difficulties to isolate the PET signal, especially in small-animal models. Also, considering the possibility of islet transplantation into the liver, the liver localization of probe limits the possible applicability as a tracing probe.

The PiF (**P**ancreatic **i**slet **F**luorinated probe) development plan was designed to overcome the known problems of PiY. The key idea is to build a fluorescent probe that minimizes the chemical modification into a PET probe. If the fluorescent probe contains the stable isotope ^{19}F, by replacing the ^{19}F with ^{18}F, the fluorescent probe could be converted to a PET-active probe without any SAR study, as both of them are chemically and biologically almost identical. Based on this concept, insulin sensor molecules are collected as the candidate for beta-cell detection, and the chemical structures were shuffled to generate the next candidates containing ^{19}F. Among the candidates, the best-performing candidate in vivo with a low background in liver and fast excretion was selected and dubbed PiF. All the fluorescence-based characterization of PiF was satisfactory for in vitro and in vivo imaging of pancreatic islets, specifically beta cells targeting insulin. By simple isotope replacement, ^{18}F-PiF was successfully demonstrated for real-time uptake and secretion of the probe in live mouse imaging. This is the first example of an already built-in F-incorporating fluorescent probe being converted into a clinically relevant PET probe by simple atomic replacement [10] (Figure 6.9).

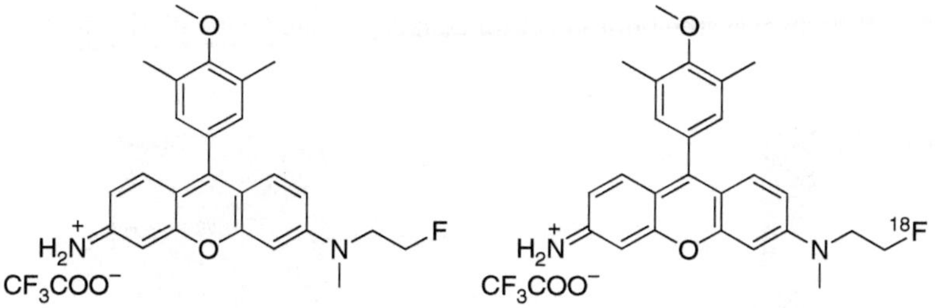

PiF **^{18}F-PiF**

Figure 6.9 Structure of ^{18}F-PiF.

6.8 Probes for Eye Imaging

Eye is a visual organ, and visual means light sensing. Eyes detect light and convert it to electrochemical signals in neurons and transfer the information to the brain. The images are reconstructed in the brain and the result is vision or visual perception. There is a phrase that eyes are the window to the soul. Well, at least, eyes are evolved to sense light and are thus fully "transparent" to visible light. In a sense, visible light was also defied by "sensible wavelength light" by "human eyes." Therefore, unlike all other tissues, there are no physical hurdles for optical imaging of eyes, as light can penetrate through the entire path from the cornea to the pupil and lens, and finally to the retina.

The structure of the eye ball is very similar to that of a camera. The light enters through the pupil and the amount of light is controlled by the pupil size, similar to the aperture in a camera. The input light is condensed through the lens and projected into the retina, which corresponds to the image sensor or film of the camera (Figure 6.10).

The exposed surface of the eye is covered by the cornea, a transparent protective tissue for the eye. Cornea is composed of collagen-rich stroma in between live epithelial and endothelial cell layers. As transparency is important, the healthy cornea does not have blood vessels, but nutrients and oxygen are supplied by tears. Cornea refracts light and contributes to the eyesight. The Lasik (laser-assisted in situ keratomileusis) surgery involves reshaping the cornea structure to adjust the refraction of light through the cornea. As the live tissue, the transplanted cornea could experience tissue rejection by the receiver's immune system.

Retina is the innermost light-sensitive screen in the eye. The collected light through the cornea and lens creates a focused 2D image on the retina. Similar to the camera sensor or photosensitive film, the retina has photoreceptor cells on the surface. Human retina has two types of photoreceptor cells, namely rods and cones. Rod cells mainly sense dim light with high sensitivity and generates black-and-white vision. Cone cells are responsible for color perception by responding to different wavelengths of light. In humans, there are three types of cone cells, namely blue, green, and red. That is how we have 3D color system. Light stimulation on the photoreceptor cells generates cascade of chemical and electrical signal conversion, resulting in nerve impulses which are transferred to the brain for imaging process. Interestingly, in human eyes, the photoreceptor cells are located beneath the neuron

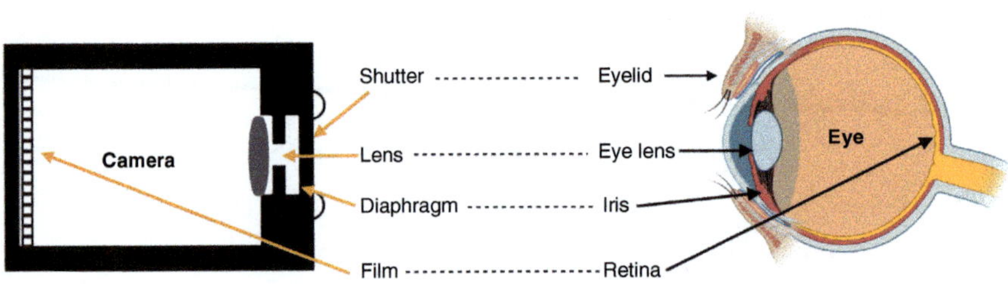

Figure 6.10 Structure of eye and camera.

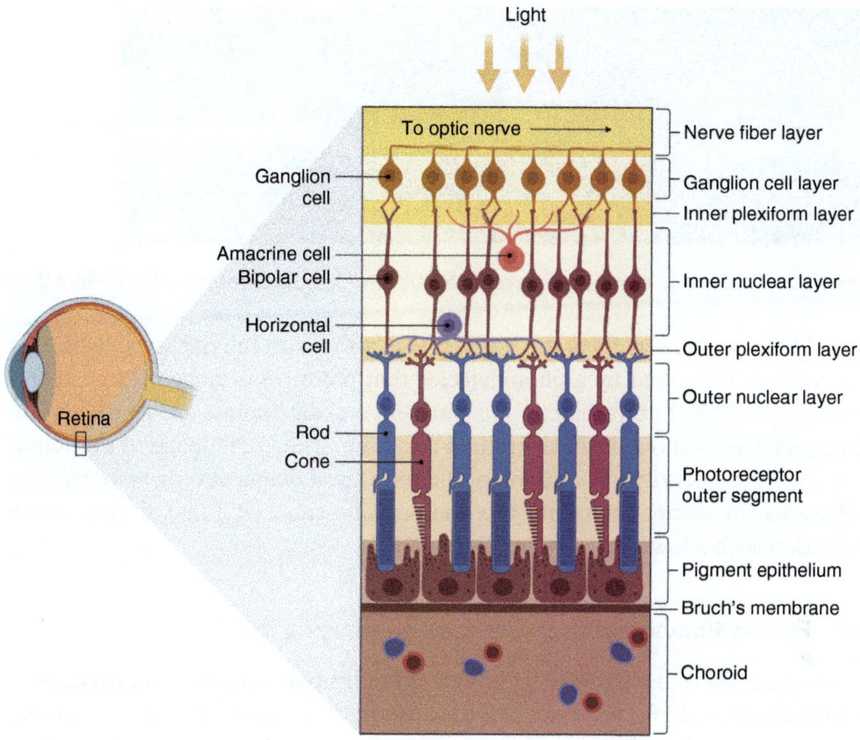

Figure 6.11 Structure of retina.

fiber layer. It is similar to the situation that the electric wiring is on top of the screen, rather than behind it. So, the neuronal cells should be optically transparent for clear vision. This may not be a necessary structure for eye construction, but just an option from the evolutionary pathway, as a reversed structure was also found in other animals, such as the cephalopod family. Because of the reversed structure, the human optic nerves are collected at one site and cross the retinal screen at least once. This spot does not have photoreceptor cells due to the wiring cells, and it is called a blind spot. As expected, the cephalopod retina does not have a blind spot, which may be a more ideal design from an engineering aspect. Behind the photoreceptor cells are lined with pigmented epithelial cells, which are supportive tissues and also prevent light reflection for clean vision (Figure 6.11).

6.8.1 Optical Coherence Tomography for Retina

The layered structure of retina can be imaged by optical coherence tomography (OCT) technique. OCT is a probe-free imaging technique by light coherence to achieve micrometer-resolution 3D images of tissues. OCT uses the term coherence, but actually it is an interferometry of a signal (probe) light with a reference light. For deeper light penetration, usually NIR light is employed in OCT imaging of retina. After irradiating the probe light to the retina, the echo time delay of light

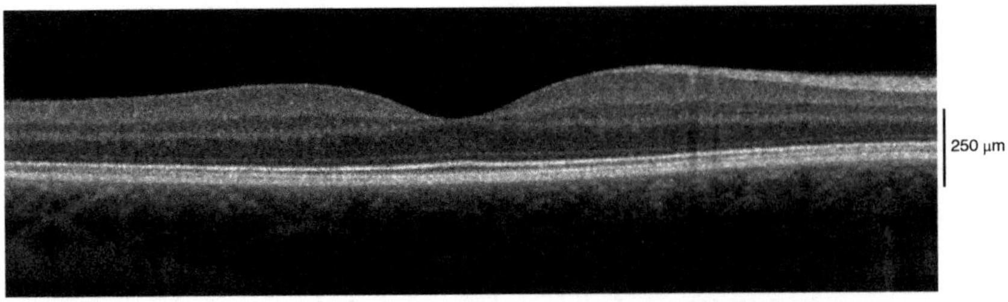

Figure 6.12 OCT image of retina. Source: Wies6014/Wikimedia Commons, CC BY-SA 4.0.

reflected from various layers of the retina is compared with the reference light. The interference is measured by a photodetector that produces a range of time delays for comparison. The interferometer integrates several data points to construct a few millimeters deep tomogram of the retinal structure. The OCT image of the retina can provide the integrity information of the retina and diagnostic signs for various eye diseases. In comparison with the confocal imaging, OCT has deeper tissue information with a lower resolution (Figure 6.12).

6.8.2 Fundus Photography for Blood Vessel Imaging in Retina

Another common retina imaging technique is fundus photography. Fundus camera is mainly composed of a microscope with flash light. Fundus photography can be performed with color filters or with fluorescent dyes such as fluorescein or ICG. Angiography is the main application of fundus photography, and the images can provide the blood vessel structure in normal or disease status. In probe-free fundus photography, usually a green filter of 540–570 nm is used for better contrast for blood vessels and associated hemorrhage, which is called as red-free fundus photography. Fluorescein is used to achieve even better signal for retinal vascular structure using proper excitation and emission filters. ICG, an NIR dye, is primarily used for deeper choroidal disease imaging (Figure 6.13).

6.8.3 Neuron Imaging on Retina

In addition to the general structure or blood vessel information, cell-specific probes can provide more detailed information on the retina. As the retina is a highly wired neural network, it could be considered as an extended central nervous (brain) system. NeuO has been developed as the first universal live-neuron-specific probe (Section 4.3.3), and NeuA is a covalent bond version of NeuO with a reactive acrylic group. NeuO and NeuA have been used for ex vivo imaging of brain slices and retina tissue, visualizing the neuronal structure in the partially viable tissue (Section 5.3.2).

In addition to the ex vivo tissue imaging capability, the in vivo neuron selectivity of NeuO was tested in mouse brain by intravenous injection. When brains and spinal cords were harvested and imaged, clear neuronal signals were obtained from the neurons of the hippocampus, cerebellum, and spinal cord. The optimal harvest

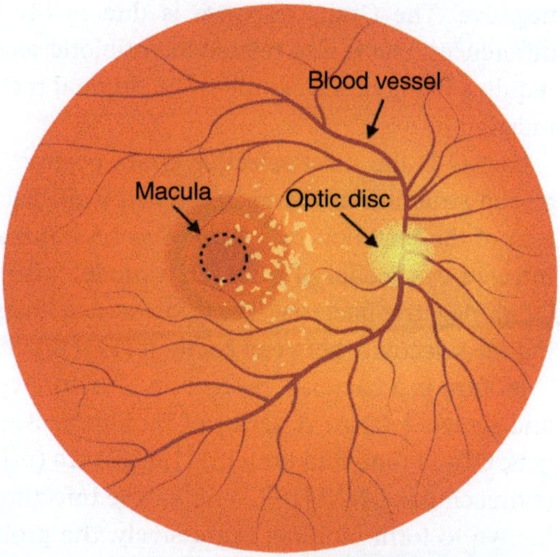

Figure 6.13 Fundus photography.

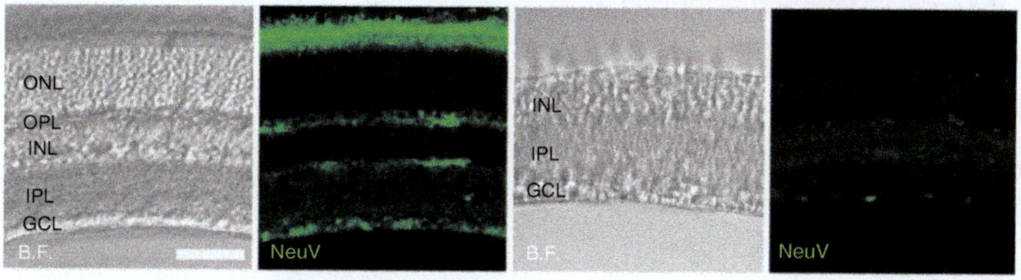

Figure 6.14 Discrimination of healthy and retinal disease by NeuA. Source: Ref. [11], from John Wiley & Sons.

time point was about one hour after injection, with the best neuron-to-glia signal ratio [4].

Through the intravenous injection of NeuO and NeuA in the tail vein of the mouse, a clear image of the neuronal structure was also obtained from the retina. As NeuO or NeuA stains only healthy neurons and the retina is an optically accessible tissue, NeuA is expected to be a noninvasive diagnostic probe. Retinitis pigmentosa (RP) is an inherited condition that is characterized by photoreceptor degeneration. NeuA strongly stained the outer segment layer of photoreceptor cells and some bipolar cells in the healthy retina, but there was only weak staining in the photoreceptor-degenerated retinas. Therefore, the NeuA probe can be used as an RP detection tool in preclinical conditions [11] (Figure 6.14).

6.8.4 Bacterial Infection on Cornea

Cornea is the outermost structure of the eye and is in direct contact with the environment. Therefore, it would be easy to make an eye infection model by scratching the cornea and inoculating it with bacteria. Bacteria are classified based on how they stain with the Gram reagent, namely Gram positive and Gram negative. The Gram response is due to the cell wall and membrane structure differences, and is also related to antibiotic sensitivity. As Gram staining procedure requires fixing of the samples, the original method is not applicable to whole-body imaging.

BacGO is the most selective Gram-positive probe reported so far, targeting the glycan structure of the bacterial cell wall. BacGO demonstrated a clear visualization of the infection in a *Staphylococcus aureus*, a type of Gram-positive bacteria, infected eye model of mice. In a model with *Pseudomonas aeruginosa* (a type of Gram-negative bacteria), the fluorescence imaging result was negative, demonstrating the selectivity between bacteria [12] (Figure 6.15).

Biofilm is a bacterial protection structure composed of proteins, carbohydrates, and secreted DNA. A series of biofilm sensors were developed with distinct targets: CDy11 (amyloid protein) [13], CDy14 (psl exopolysaccharide) [14], and CDr15 (extracellular DNA) [15]. With an eye infection model with *P. aeruginosa*, which is known to form biofilms aggressively, the probes demonstrated clear visualization of the bacterial colony hidden in the biofilm. The genetic knockout for each target abolished the staining capability, demonstrating the selectivity of each probe for the designated biomarker (Figure 6.16).

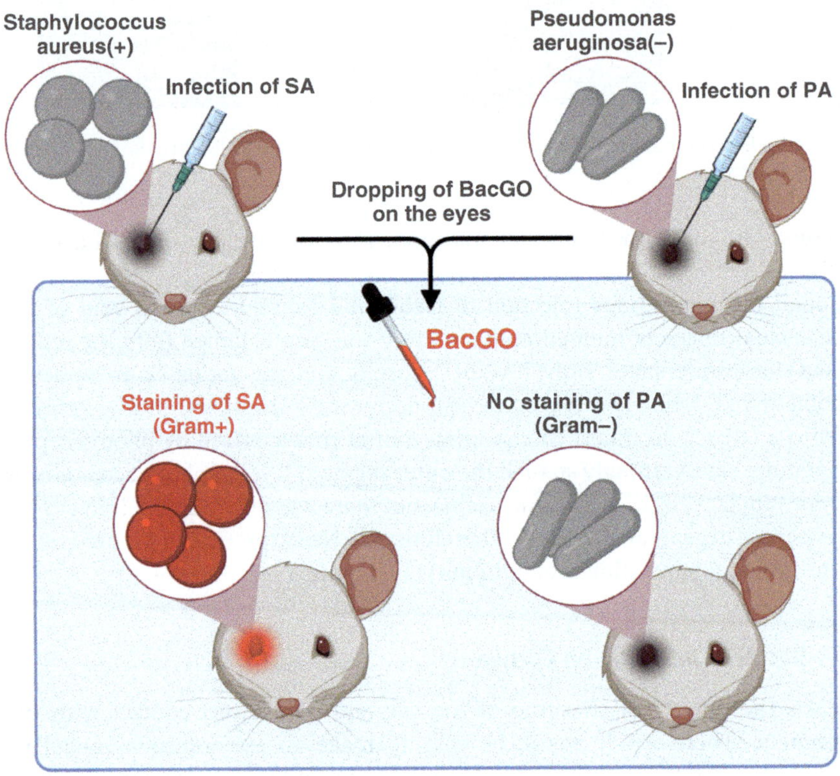

Figure 6.15 BacGo images on eyes.

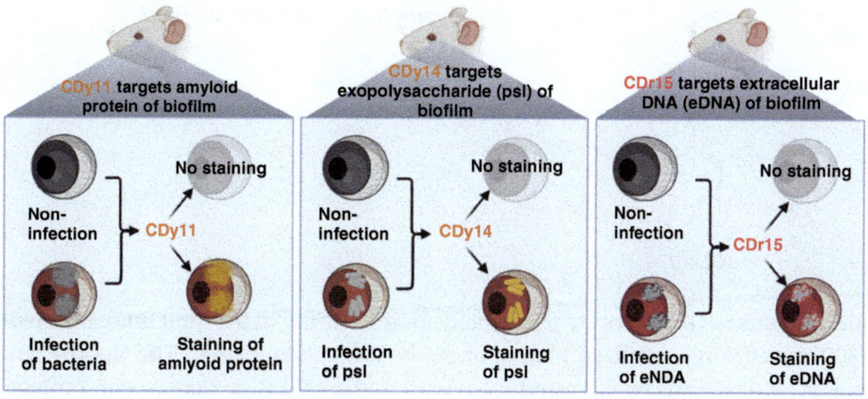

Figure 6.16 Infection images with CDy11, CDy14, and CDr15.

6.9 Macrophage Imaging in Ischemia and Inflammation

Macrophage is a representative innate immune cell responding to a broad range of stimuli, such as infection, cancer, and ischemia. The monocyte cells, circulating precursors of macrophages, are recruited to the inflammation site and mature into macrophages. Many fluorescent probes have been developed and used for the imaging of inflammation sites. For deep signal penetration, NIR probes would be preferable for whole body imaging. MF800 (Macrophage-specific NIR Fluorophore) is one of such probes, developed for in vivo imaging for macrophages [16] (Figure 6.17).

MF800 was discovered from unbiased cell-based screening using photostable NIR dye library [17]. MF800 visualized the macrophage accumulation in LPS (lipopolysaccharide isolated from Gram-negative bacteria)-induced mouse paw model of inflammation. The long wavelength of MF800 facilitated the penetration of the light through the tissue, providing a strong enough fluorescence signal for imaging. In addition, MF800 was applied for intravital microscopy in an ischemic

Figure 6.17 Structure of MF800.

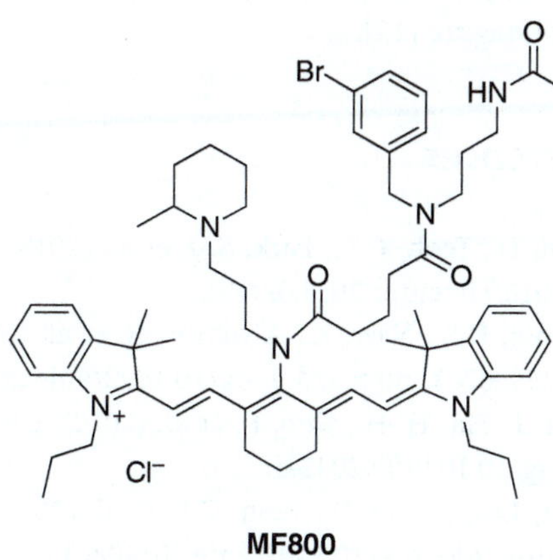

MF800

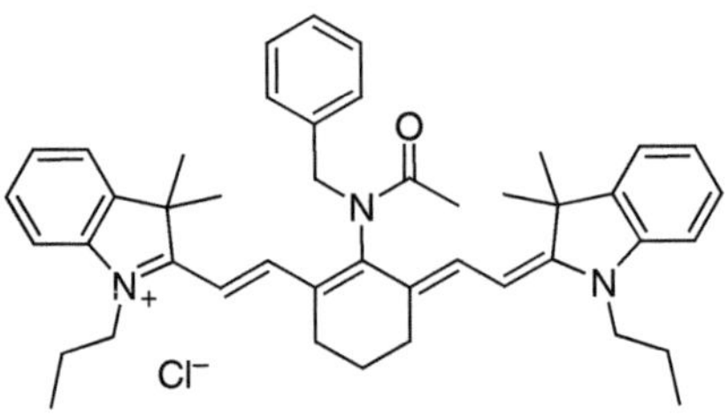

CDnir7

Figure 6.18 Structure of CDnir7.

model of mouse, which was created by a ligating an artery in the hind limb. Injected MF800 specifically visualized macrophages in the ischemic area. The fluorescent signal of MF800 was well overlapped with Csf1r-EGFP, a fluorescent protein reporter for macrophages.

A simpler structure of an NIR probe for the macrophage, CDnir7 (Compound of Designation near infrared 7), was also introduced in parallel. The long wavelength of CDnir7 ($\lambda_{ex}/\lambda_{em}$ = 806/821 nm) allowed the visualization of the inflammation site of LPS-treated paw model of mouse from the whole-body imaging. Furthermore, the fluorescence molecular tomographic (FMT) imaging generated the 3D distribution of CDnir7 in the paw model.

Orthotopic 4T1 breast tumor mouse model is known to recruit large numbers of tumor-associated macrophages. Through photoacoustic imaging, the pharmacokinetic behavior of CDnir7 could be obtained in 4D measurement (3D spatial resolution with time progress). Multispectral optoacoustic tomography (MSOT) allowed the differential imaging of oxygenated and deoxygenated hemoglobin distribution in contrast to CDnir7 localization in the cancer area [18] (Figure 6.18).

Alzheimer's disease (AD) is considered as a chronic inflammation-based neurodegenerative disease. When injected, CDnir7 showed the penetration through the blood–brain barrier and accumulation in the AD mouse brain. The deep tissue penetration of CDnir7 signal was further challenged to measure for imaging AD mouse brain using MSOT technique. During MSOT, CDnir7 showed higher fluorescent signal in the AD mouse cerebral cortex compared to that of the control mouse, demonstrating the potential of CDnir7 as a molecular imaging probe for AD brain imaging [19].

References

1 Su, D., Teoh, C.L., Park, S.-J. et al. (2018). *Chem* 4: 1128–1138. https://doi.org/10.1016/j.chempr.2018.02.016.

2 Choi, H.S., Nasr, K., Alyabyev, S. et al. (2011). *Angew. Chem. Int. Ed.* 50: 6258–6263. https://doi.org/10.1002/anie.201102459.

3 Li, J., Ha, H.-H., Guo, L. et al. (2010). *Chem. Commun.* 46: 2932–2934. https://doi.org/10.1039/b920432f.

4 Er, J.C., Leong, C., Teoh, C.L. et al. (2015). *Angew. Chem. Int. Ed.* 54: 2442–2446. https://doi.org/10.1002/anie.201408614.

5 Teoh, C.L., Er, J.C., Mukherjee, P., and Chang, Y.T. (2015). *Tomography* 1: 30–36. https://www.mdpi.com/2379-139X/1/1/30#.

6 Lee, J.H., So, J.-H., Jeon, J.H. et al. (2011). *Chem. Commun.* 47: 7500–7502. https://doi.org/10.1039/C1CC11253H.

7 Chun, H.-S., Han, S.-J., Shin, D.-S. et al. (2019). *Dyes Pigments* 171: 107720. https://doi.org/10.1016/j.dyepig.2019.107720.

8 Zaheer, A., Lenkinski, R.E., Mahmood, A. et al. (2001). *Nat. Biotech.* 19: 1148–1154. https://doi.org/10.1038/nbt1201-1148.

9 Kang, N.Y., Lee, S.C., Park, S.J. et al. (2013). *Angew. Chem., Int. Ed.* 52: 8557–8560. https://doi.org/10.1002/anie.201302149.

10 Kang, N.Y., Lee, J.Y., Lee, S.H. et al. (2020). *J. Am. Chem. Soc.* 142: 3430–3439. https://doi.org/10.1021/jacs.9b11173.

11 Park, S.J., Woon, Q.T.S., Er, J.C. et al. (2021). *ChemBioChem* 22: 1915–1919. https://doi.org/10.1002/cbic.202100011.

12 Kwon, H.Y., Liu, X., Choi, E.G. et al. (2019). *Angew. Chem. Int. Ed.* 58: 8426–8431. https://doi.org/10.1002/anie.201902537.

13 Kim, J.-Y., Sahu, S., Yau, Y.-H. et al. (2016). *J. Am. Chem. Soc.* 138: 402–407. https://doi.org/10.1021/jacs.5b11357.

14 Kwon, H.Y., Kim, J.Y., Lee, J.Y. et al. (2018). *Chem. Commun.* 54: 11865–11868. https://doi.org/10.1039/C8CC05544K.

15 Kwon, H.-Y., Kim, J.-Y., Liu, X. et al. (2019). *Biomater. Sci.* 7: 3594–3598. https://doi.org/10.1039/C9BM00152B.

16 Yoo, J.S., Das, R.K., Jow, Z.Y., and Chang, Y.-T. (2014). *PLoS One* 9: e103721. https://doi.org/10.1371/journal.pone.0103721.

17 Das, R.K., Samanta, A., Ha, H.-H., and Chang, Y.-T. (2011). *RSC Adv.* 1: 573–575. https://doi.org/10.1039/C1RA00498K.

18 Kang, N.-Y., Park, S.-J., Ang, X.W.E. et al. (2014). *Chem. Commun.* 50: 6589–6591. https://doi.org/10.1039/C4CC02038C.

19 Park, S.-J., Ho, C.J.H., Arai, S. et al. (2019). *Sci. Rep.* 9: 12052. https://doi.org/10.1038/s41598-019-48329-4.

7

Imaging for Biological Environment and Function

One cut image based on the structure of a biological system may belong to static and characteristic snapshot of life. However, life is dynamic and momentary, sustained only by the endless energy and material exchange, that is metabolism. To keep the harmonized and balanced function of life (so-called homeostasis), a constant adjustment of each metabolite and energy component is necessary, responding to environmental challenges. Cells are separated by a lipid bilayer from the outer space, and the functional status of cells could be measured by the quantitative changes of cellular components, signaling molecules, and metabolites. In addition, physical parameters such as temperature and viscosity are also important indicators of cellular function. This is similar to fever or swelling, which is an indication of the whole-body condition. Sensors and probes for these parameters would provide information on the functional status of cells and body.

The pioneer of the imaging for cellular functions is Roger Y. Tsien. In the late twentieth century, the discovery and development of cellular signal transduction was about to sprout. Through the cell membrane, a hydrophilic signaling molecule from outer space could not directly enter the cell, but indirectly transfer the signal by activating cell surface receptors. The activated receptors secrete the second messenger into the cytoplasm, and sometimes they further relay and amplify the signal by mobilizing a third messenger from intracellular storage. One of the best-studied signal transduction pathways was using IP3 (inositol 1,4,5-trisphosphate) as a second messenger and Ca^{2+} as the third messenger. Although the quantitative biochemical analysis of IP3 and Ca^{2+} was possible, the extraction and time-consuming analysis procedures made the real-time monitoring of the messengers impossible.

Tsien developed a series of fluorescent sensor for Ca^{2+} and opened up a bright window to directly look into the biological systems. Using the fast-responding Ca^{2+} sensors, a real-time monitoring of Ca^{2+} in live cells and bodies became available all of sudden. The amazing tools unveiled the dynamic nature of Ca^{2+} changes inside cells, not only in quantity, but also the frequency of the Ca^{2+} wave. The visualization of the third messenger made the connection of the first messenger to the biological function clear through time and space dimensions. The Ca^{2+} sensors could also visualize the neuronal firing in real time. It is no exaggeration that Tsien's Ca^{2+} sensors fully blossomed the signal transduction field. The impact of this work is reflected by

Sensors and Probes for Bioimaging, First Edition. Young-Tae Chang and Nam-Young Kang.

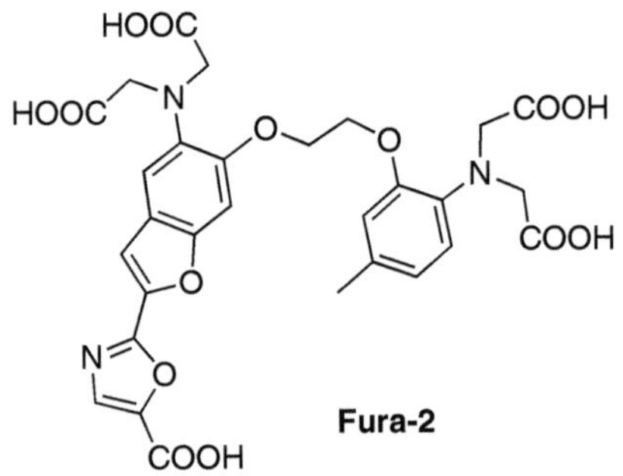

Figure 7.1 Structure of Fura-2.

the high number of citations (more than 20 000) of the paper on the representative Ca^{2+} sensor Fura-2 [1] (Figure 7.1).

7.1 pH

Our life forms are based on water environment. Life may have started from seawater, and the aquatic life is still surrounded by water. There would be merits for aquatic lives living in the water, such as easy access to the nutrients and less burden to sustain the body weight against gravity. Later, some life forms left the sea and evolved to live on the land. The land creatures live in the air, without directly contacting the water environment. The adaptation might have been difficult in the beginning as they needed to develop a new oxygen supply system, lungs instead of gills. Strong scaffold was also required to sustain the body against gravity, as buoyancy is no longer available on the land. Still, the rewards of the pioneering evolution were great, as the current dominant life forms of earth are land creatures, outnumbering the aquatic lives by enormous difference. Still, inside of all the life forms are based on water environment regardless where they live, and basically all the biological reactions occur in aqueous media.

We may feel water is a very familiar material to us. Water has no special taste and is not a harmful material to us. At room temperature, water is a liquid. We know the molecular formula of water is H_2O, a molecule composed of two hydrogen atoms and one oxygen atom. Although it looks like a simple molecule, water is quite unique and not so inert molecule. Liquid water is not just a collection of stable H_2O molecules. The hydrogen–oxygen bond of water is not static, but actually quite dynamic. Oxygen, the heavier atom in a water molecule, constantly exchanges hydrogen with neighboring water molecules. It means one isolated water molecule is difficult to define in liquid water. Even in the frozen state of water, that is ice, sometimes it is difficult to define which hydrogens belong to which water molecule. Therefore, the water is more likely a dynamic network of oxygens and hydrogens in an approximately 1 : 2 ratio. Still, if we take a snapshot of water, the majority of the molecules exist as H_2O, and small portions exist as dissociated form of H^+ and OH^-. By common convention, we use H^+ with the name proton, but it may be difficult for a proton to exist as a bare proton nucleus without any electron. So, it would be

$$H_2O + H_2O \rightleftharpoons H_3O^+ + HO^- \qquad [H_3O^+][OH^-] = 10^{-14}\ M$$

Figure 7.2 Dissociation of water and its equilibrium constant.

more reasonable to assume that two water molecules exchange a proton between them forming H_3O^+ (hydroxonium ion) and OH^- (hydroxide ion) pairs. The concentration of each ionic species is found to form an equilibrium and the value of $[H_3O^+][OH^-]$ is 10^{-14} M (Figure 7.2).

In pure water, the ratio of H_3O^+ and OH^- is 1, so each ion concentration would be 10^{-7} M. To the water, it is possible to add extra H_3O^+ and OH^- ions, then the equilibrium of water dissociation is shifted by Le Chatelier's principle. Still, the dissociating constant of water is not changed, displaying a fixed value of 10^{-14}. So, $[H_3O^+]$ is increased to 10^{-6} M, and $[OH^-]$ is decreased to 10^{-8} M. To make the exponential number easy to use, the concentration of $[H_3O^+]$ is described as $pH = -\log[H_3O^+]$. So, neutral water is pH7, and 10^{-6} M of $[H_3O^+]$ is pH6. When $[H_3O^+]$ is higher than $[OH^-]$, the environment is acidic, and the opposite situation is basic. Especially, the water-soluble basic material is called as alkali. Usually, acidic solutions taste sour, and alkaline solutions are bitter.

The proton concentration in the water environment affects the chemical reaction dramatically. Acids attract the electron and bases donate the electron, disturbing the molecule's stability. Therefore, either the acid or base can catalyze many chemical reactions in a biological system. In neutral water, the water molecule concentration is approximately 55 M (55 mol in 1 l water). The $[H_3O^+]$ is 0.1 μM, and the proton-to-water ratio is about 1 to 550 million, which implies the acid-catalyzed reaction could be quite rare. In a strong acid condition of pH1, $[H_3O^+]$ is 0.1 M, the proton-to-water ratio is about 1 to 550. Now, the environment would be quite dramatic with super high concentration of proton, and the acid-catalyzed reaction will dominate the system. The biological system may be in danger if the acid–base change is so dramatic, so an ionic buffer system prevents such a big swing of pH. The usual biological pH is maintained in the range of pH 4–8, with a slightly acidic environment preferred rather than a basic one, overall. In mammalian cells, the organelles have various pH: early endosomes (pH 6.3), late endosomes (<pH 6.0), lysosomes (<pH 5.5), endoplasmic reticulum (ER: pH 7.2–7.5), Golgi body (pH 6.2–6.7), peroxisomes (pH 6.9–8.2), nucleus (pH 7.3), cytoplasm (pH 7.3), and mitochondrial matrix (pH 7.9) [2]. Mitochondria have a lipid double layer, and the proton gradient accumulated in the intermembrane space is acidic (pH 6.8) than the matrix (pH 7.7) [3]. Due to the aberrant glycolysis and poor perfusion, tumors usually have a slightly acidic extracellular matrix condition (pH 6.2–6.9) than that of normal tissue [4]. Diseases or stress conditions may also change the pH of the cells and tissues, and pH monitoring would provide the readout for the progress of each event (Figure 7.3).

Based on the abovementioned situation, measuring pH would be one of the very fundamental processes to understand the environment and functional status of the biological system. Naturally, among many pH indicators, fluorescent sensors in the physiological range of pH4–8 would be of special interest for bioimaging. Among

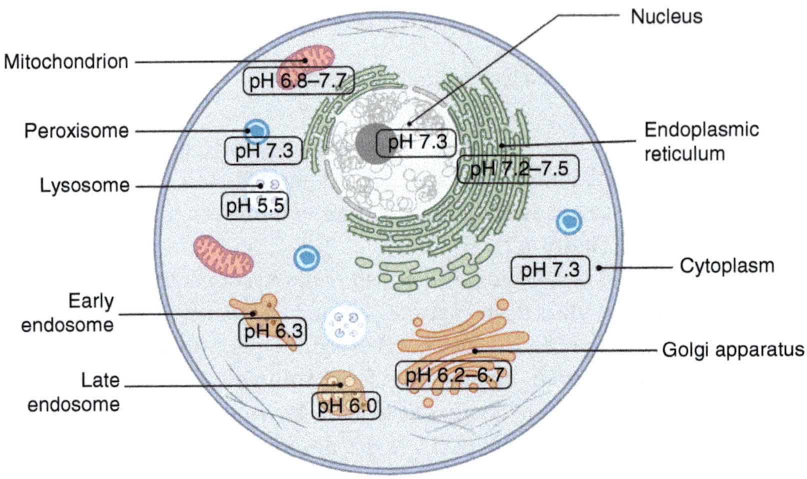

Figure 7.3 pH distribution in organelles of cell.

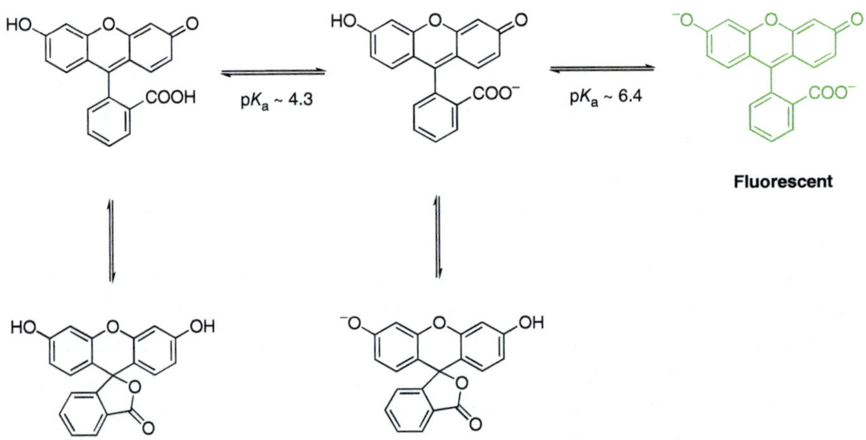

Figure 7.4 pH-dependent fluorescence of fluorescein ionization.

popular fluorophores, fluorescein has intrinsically pH-dependent fluorescence property. With the ionization equilibrium, fluorescein is fully fluorescent at a basic pH (above pH9) and quenched at an acidic pH (Figure 7.4).

The anionic character of fluorescein is not favorable for the cell entry. One common trick to make anionic dye for cell entry is the acetylation of the hydroxyl groups, which block the anion character of the dye. Once they are inside cell, esterase enzymes hydrolyze the ester, releasing the acetyl group and the original fluorescein in the cell (Figure 7.5).

SNARF-1 (seminaphthorhodafluors) has been a popular a pH sensor with the pK_a of 7.5 for a little basic environment monitoring [5]. SNARF-1 has dual-excitation and dual-emission properties, and a ratiometric imaging was commonly employed. SNARF-1 is also cell impermeable due to the anionic hydroxyl and carboxyl groups.

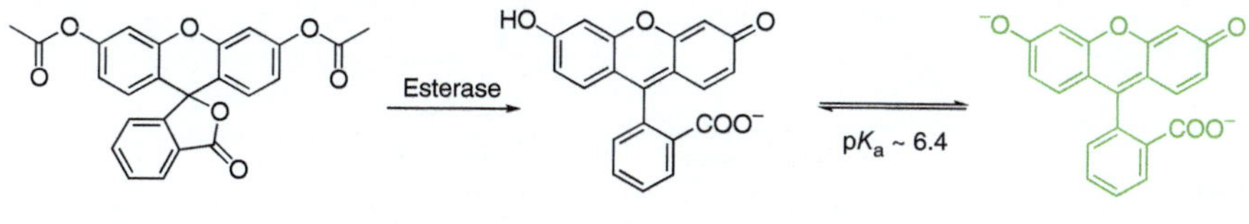

Figure 7.5 Acetylation of fluorescein and action of esterase inside the cell.

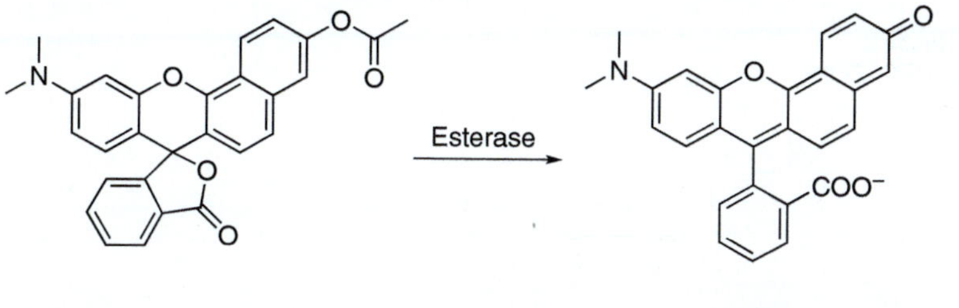

Figure 7.6 Structure of SNARF-1.

Similar to fluorescein case, the hydroxyl group can be blocked by the acetyl group, and the carboxyl group can be protected by an acetoxymethyl (AM) group. AM group is also cleavable by esterase, generating acetate and formaldehyde as the by-products. Acetyl and AM groups protected SNARF-1 can go inside the cell and upon cleavage of the protecting group, SNARF-1-stained mitochondria and was used for mitochondrial pH [6] (Figure 7.6).

The representative acidic pH sensors are summarized in Figure 7.7 with biological imaging application: Probe 2 (pK_a 4.0, lysosome probe) [7], PQ-Lyso (pK_a 4.2, lysosome probe) [8], Lyso-pH (pK_a 5.0, lysosome pH) [9], and Rlyso (pK_a 5.47, lysosome probe) [10].

For mitochondrial matrix pH measurement, pH sensors with higher pK_a were introduced and summarized in Figure 7.8: piperazine-naphthalimide (pK_a 6.18) [11], Mito-pH-1 (pK_a 7.33) [12], and CMP1 (two-photon probe, pK_a 7.95) [13].

7.2 Metal Ions

The extracellular space is a high NaCl environment mimicking seawater, with about 130 mM concentration. Through ion-selective channels, the intracellular space keeps comparable K+ ion concentration to keep osmotic pressure balances. Therefore Na^+ and K^+ are the highest concentration metal ions across the cell membrane. The Na^+/K^+ ratio in the outer space is about 10, and that in the inner space is opposite of 1/10. During the neuron cell activation, Na^+ pours into the cytoplasm, depolarizing the membrane potential. A follow-up K^+ flow to the outer space generates a temporary hyperpolarization before reaching the resting state. In addition to the most abundant Na^+ and K^+ ions, there are several biologically

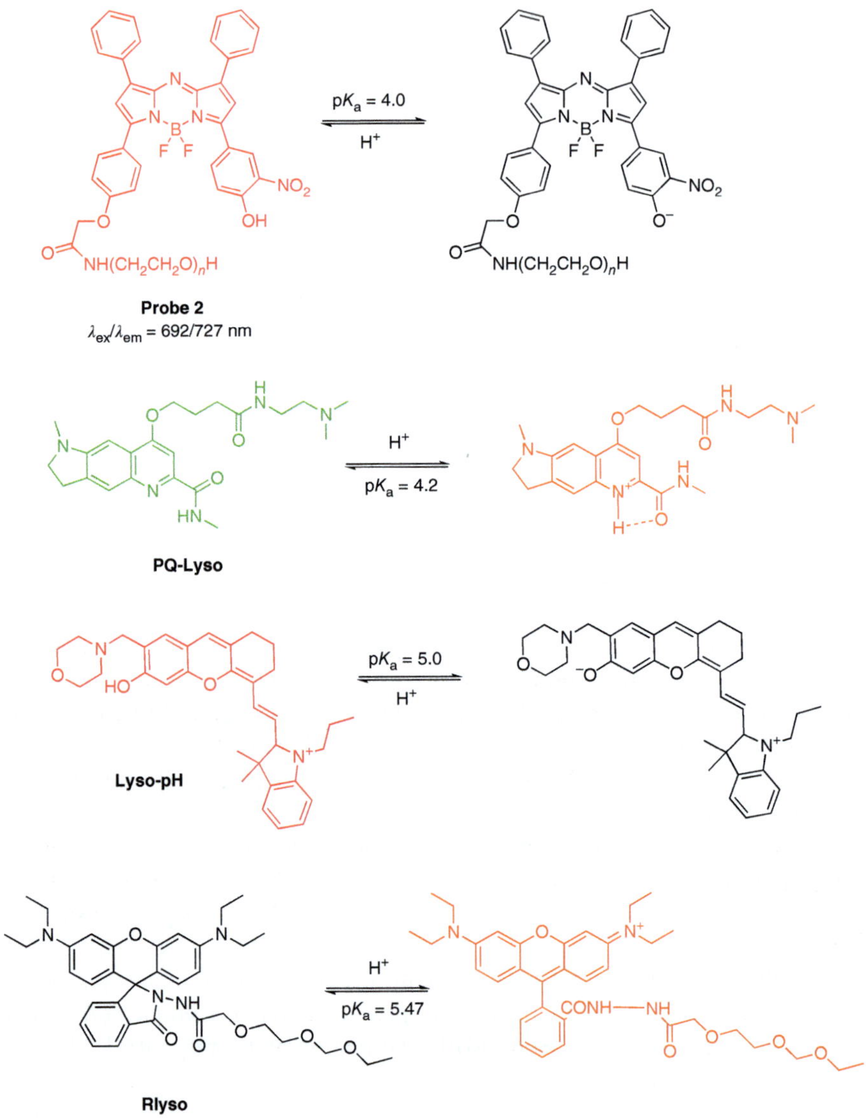

Figure 7.7 Acidic pH sensors.

important metal ions in the cell with different distribution among organelles, such as Ca^{2+}, Mg^{2+}, Zn^{2+}, Fe^{2+}, and Cu^{2+} (Figure 7.9) [14]. Therefore, monitoring the functional metal ions in and out of the cells would be important, and fluorescent metal sensors would play important roles.

7.2.1 Na^+ and K^+

The early generation of Na^+ and K^+ sensors are based on crown ethers conjugated to fluorophores. When the sensor is free from the metal ion, the lone pair of electrons

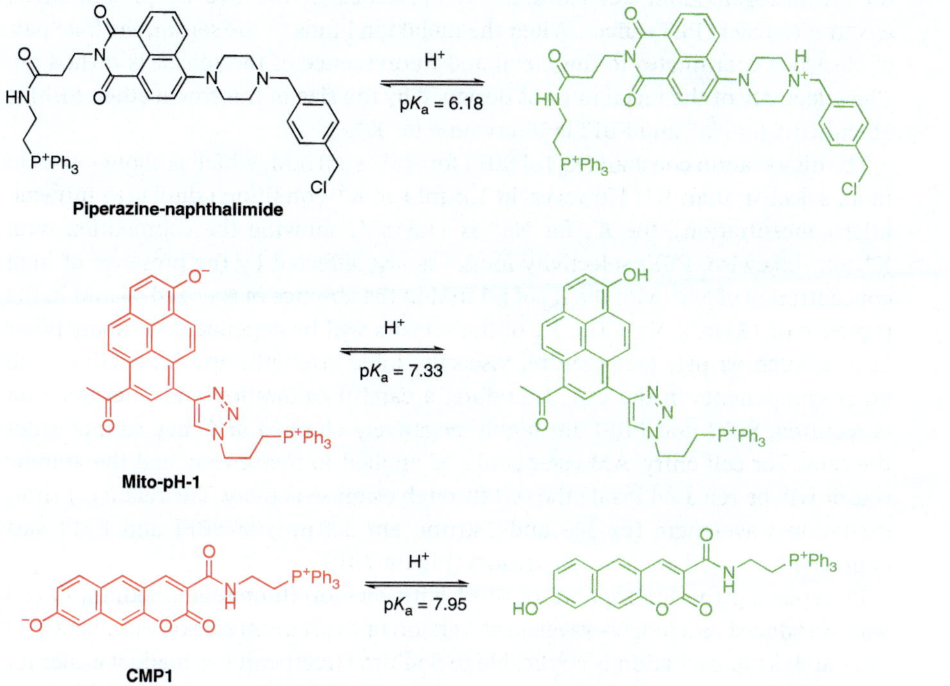

Figure 7.8 Neutral to slightly basic pH sensors.

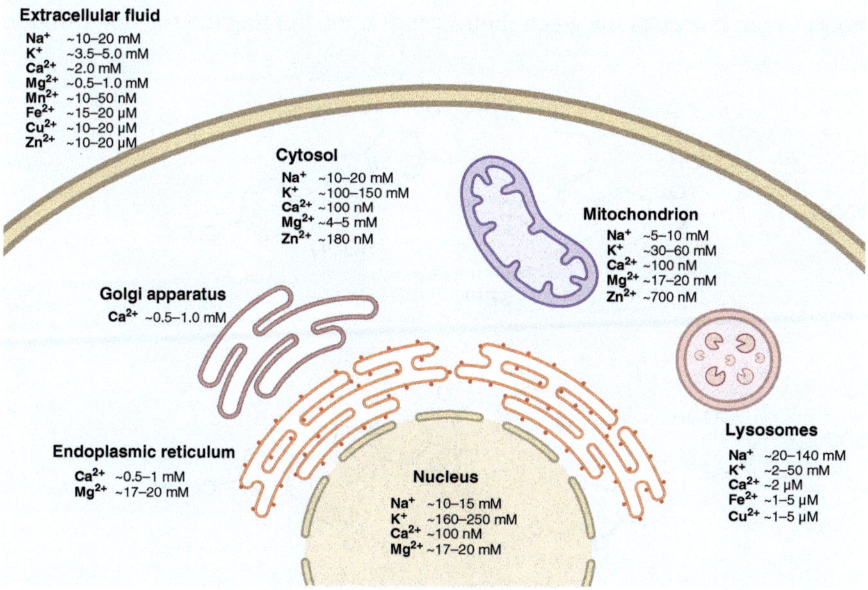

Figure 7.9 Major metal ion concentrations in and out of cells.

on the nitrogen atom quenches the fluorescence of the dye by photoinduced electron transfer (PET) effect. When the metal ion binds to the sensor, the lone pair of electrons coordinates to the metal and fluorescence of the sensor is turned on. The selectivity of the metal ion was designed by the size of the crown ether: SBFI is 15-crown-5 for Na^+ and PBFI is 18-crwon-6 for K^+.

The dissociation constant (K_d) of SBFI for Na^+ is 3.8 mM, which is about ~18-fold more selective than K^+. However, in 135 mM of K^+ condition (similar to intracellular concentration), the K_d for Na^+ is 11.3 mM, showing the competition with K^+ ion. Likewise, PBFI selectivity for K^+ is also affected by the presence of high concentration of Na^+ with the K_d of 5.1 mM in the absence of Na^+ and 44 mM in the presence of 135 mM Na^+. The K_d of the sensors will be dependent on many other factors, such as pH, temperature, viscosity, ionic strength, and interaction with other components in the cell. Therefore, a careful calibration and interpretation is required. SBFI and PBFI are highly negatively charged and they cannot enter the cells. For cell entry, AM ester could be applied to the sensors and the anionic sensor will be released inside the cell through esterase activity. The relatively short excitation wavelength (ex 340 and 380 nm, em 500 nm) of SBFI and PBFI cast demands for longer-wavelength sensors (Figure 7.10).

By replacing the fluorophore of SBFI with dichlorofluorescein, Sodium Green was introduced as a longer-wavelength version of the sodium sensor. Popular argon laser at 488 nm excitation is applicable to Sodium Green and has made it easier for confocal microscopy and flow cytometric analysis. The cell permeability of Sodium Green could be achieved by acetylation of the four hydroxyl groups as in fluorescein. A similar size of Na^+ capturing 15-crown-5 ether could be attached to different positions of difluorofluorescein, generating CoroNa Green (ex/em = 492/516 nm). CoroNa Green increases the green fluorescence upon binding the Na^+ ion through

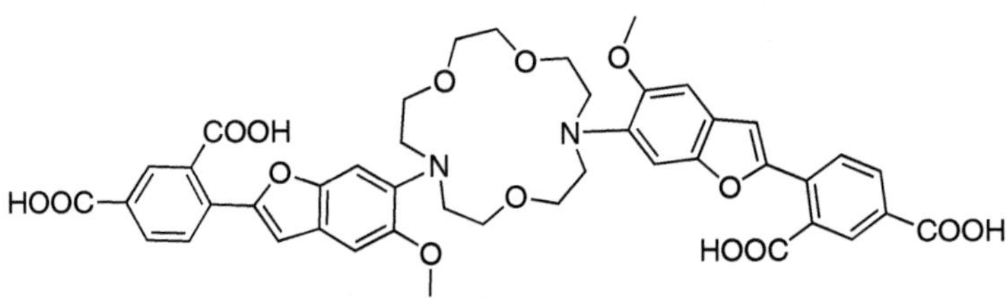

SBFI

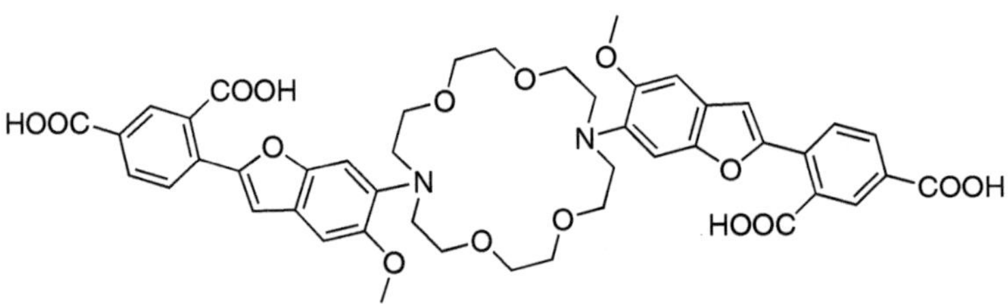

PBFI

Figure 7.10 Structure of first-generation Na^+ and K^+ sensors.

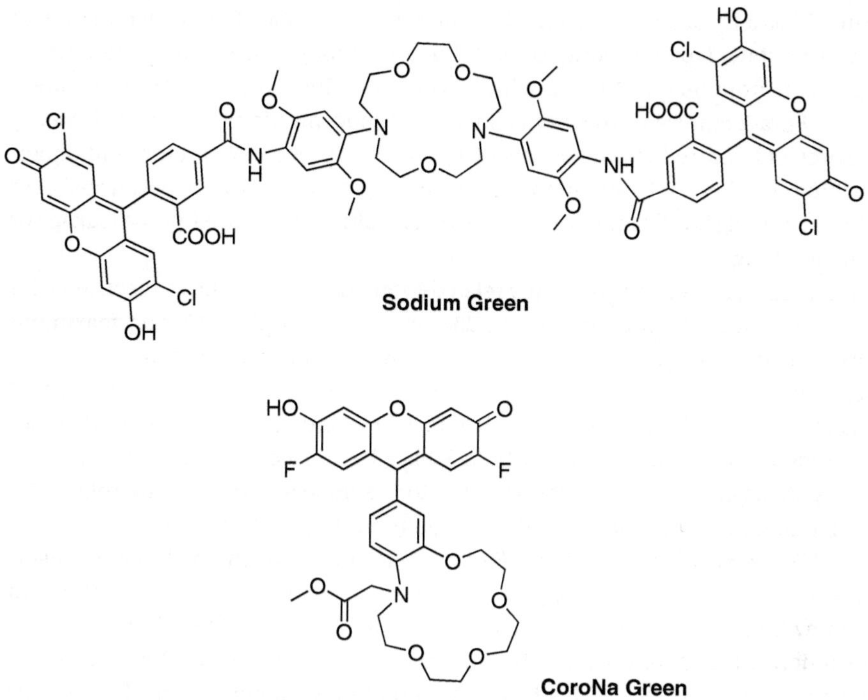

Figure 7.11 Structure of Sodium Green and CoroNa Green.

the blocked PET effect. The AM ester of CoroNa Green has better cell permeability than Sodium Green tetraacetate, and CoroNa Green responds to a broader range of Na^+ concentration. With the strong electron withdrawing fluorine group, CoroNa Green has a lower pK_a than fluorescein, and works better in neutral conditions [6] (Figure 7.11).

7.2.2 Ca^{2+}

The cytosolic Ca^{2+} concentration is about 100 nM at the resting state, while extracellular Ca^{2+} concentration is in the ~ mM range. The dramatic difference in Ca^{2+} concentration in and out of the cells seemed designed by life to utilize Ca^{2+} as a signaling messenger. The main Ca^{2+} storage organelles are ER and Golgi body with a lumen Ca^{2+} concentration of 0.5–1 mM. When cells get an activation signal from outside, through the signal transduction pathways and Ca^{2+} channel opening on the storage organelle membrane, the cytosolic Ca^{2+} concentration spikes up to hundred to thousand times in seconds. The secreted Ca^{2+} ions regulate the functions of many enzymes and receptors, and eventually affect the gene expressions of specific proteins. The Ca^{2+} secretion is often in wave forms, and the temporal and spatial information on Ca^{2+} concentration is important to understand the signal transduction events. The success of Fura-2 for real-time monitoring of Ca^{2+} in live cells and tissues literally revolutionized the signal transduction field.

Fura-2 has a K_d of 140 nM, which is similar to the cytosolic Ca^{2+} concentration at the resting state. Upon binding to Ca^{2+}, Fura-2 exhibits a strong absorption wavelength shift from 380 to 340 nm with 505 nm emission, so 340/380 nm excitation ratio for Fura-2 represents the intracellular Ca^{2+} concentration [15]. The ratiometric measurement can reduce the effects of the uneven probe distribution and photobleaching. The Fura-2-based fluorescence imaging usually could be performed over an hour without significant fluorescence loss, resulting from either probe leakage or photobleaching.

In contrast to Fura-2, a structurally related compound, Indo-1, has a spectrum shift in emission rather than excitation. At 338 nm excitation, the emission maximum shifts from 485 to 405 nm. Therefore, the 405/485 nm emission ratio for Indo-1 represents the intracellular Ca^{2+} concentration. In a later study, the K_d of Indo-1 was measured as 250 nM and was not much affected by binding to proteins or other biological molecules [16]. Fura-2 and Indo-1 have been used by many scientists and cited by thousands of papers, contributing to the understanding of the role of the Ca^{2+} ion in signal transduction and cell regulation (Figure 7.12).

The UV range of excitation of Fura-2 and Indo-1 urged to develop longer-wavelength Ca^{2+} probes. Among many probes developed, Calcium Green-1 and Calcium Orange are the representative successful Ca^{2+} probes [17] (Figure 7.13).

As a derivative of iminodiacetate, calcein (ex/em = 495/515 nm) is a relatively low-affinity Ca^{2+} chelator with a K_d of 0.1 ~ 1 mM range at physiological conditions [18]. With the low affinity, calcein may not be proper to measure the cytosolic Ca^{2+} concentration, but is a good indicator for the high Ca^{2+} storage of the sarcoplasmic reticulum, which is found in muscle cells and is similar to ER of other cells. With high concentration Ca^{2+} ability, calcein is useful for detecting mineralized Ca^{2+} such as in bone [19]. Calcein self-quenches at above 70 mM and is commonly used as an indicator of lipid vesicle leakage [20]. Calcein is negatively charged and cell impermeable. Acetyl- and AM-protected calcein can be loaded in the cell, and with the active esterase enzyme in healthy cells, the free calcein is released and retained in cells for a long time, compared to the analogues with lower numbers of carboxylate [21]. Highly negatively charged molecule is difficult to penetrate the cell membrane,

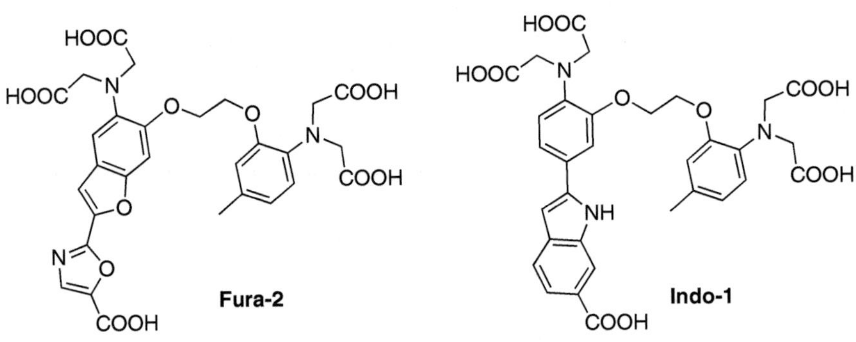

Figure 7.12 Structure of Fura-2 and Indo-1.

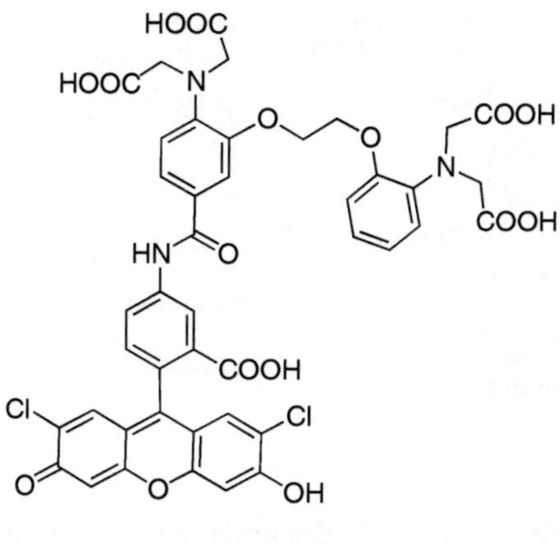

Calcium Green-1

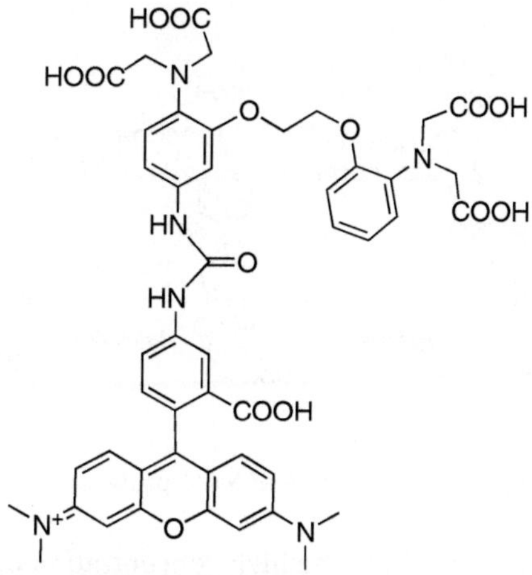

Calcium Orange

Figure 7.13 Structure of Calcium Green-1 and Calcium Orange.

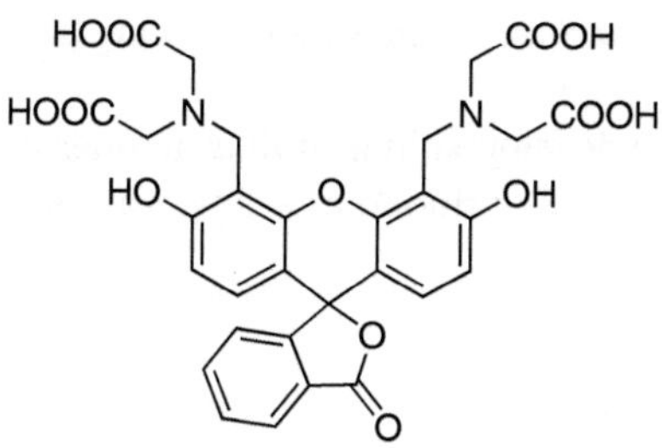

Calcein

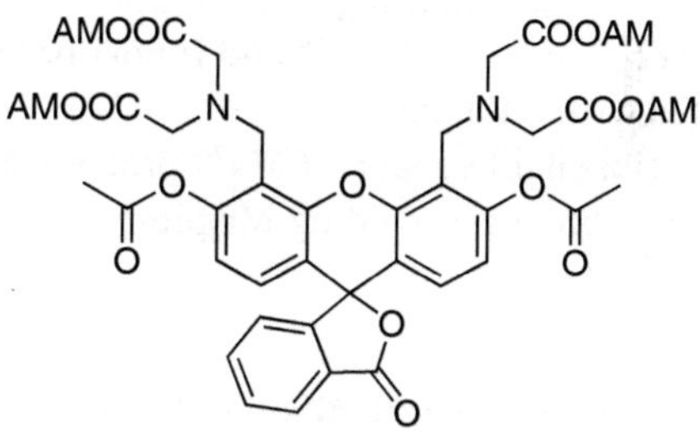

Calcein-AM
AM: acetoxymethyl
$-CH_2OAc$

Figure 7.14 Structure of calcein- and AM-protected cell-permeable derivative.

and once formed in the cytosol, the probe is difficult to exit the cell, acting as a long-term probe in the cells (Figure 7.14).

7.2.3 Mg^{2+}

Unlike Ca^{2+}, Mg^{2+} concentration is higher inside the cell compared to the environment (0.5∼1 mM), implying the constitutional role of Mg^{2+} in the biological system. The cytosolic concentration of Mg^{2+} is about 5 mM, and Mg^{2+} is involved in regulation of enzyme reactions, hormonal secretion, and muscular contraction. Mg^{2+} is the main counter anion of DNA, forming the bridge between abundant phosphate groups. Due to the high content of DNA in the nucleus, the Mg^{2+} concentration in the nucleus is ∼20 mM, even higher than that of the cytosol.

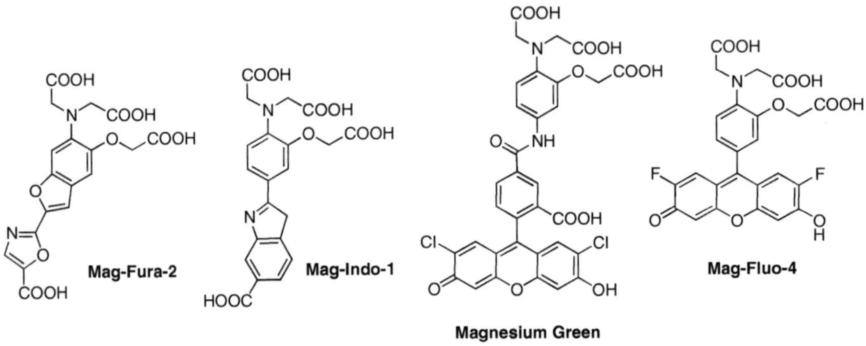

Figure 7.15 Structure of Mg^{2+} probes.

For matching the high concentration of Mg^{2+} in the cell, the probe needs to have a weak binding affinity. By modifying the Ca^{2+} probe, Fura-2 and Indo-1 were modified with smaller chelating groups to provide Mag-fura-2 and Mag-indo-1, which have a K_d of 1.9 and 2.7 mM each for Mg^{2+} [6]. As an application, Mag-indo-1 was used for intracellular Mg^{2+} concentration change in mouse skeletal muscle contraction [22]. Similar to Fura-2 and Indo-1 probes, Mag-fura-2 and Mag-indo-1 are also excited with UV light.

A visible light version of Ma^{2+} probes was also developed in a similar manner to the Ca^{2+} probes, providing Magnesium Green and Mag-fluo-4, as a few examples (Figure 7.15).

7.2.4 Metal Ion Selectivity of Fluorescent Sensors

As described above, the current metal ion sensors are designed based on metal coordination by multidentate chelators. Usually, the lone pair of electrons on the nitrogen atom of the chelator quenches the fluorescence of the sensor by PET. When coordination occurs to the analyte metal ion, the lone pair of electrons is donated to the metal ion and the PET effect is abolished, turning on the fluorescence. The representative chelators are described in Figure 7.16 and extensive motifs are summarized in the literature [23] (Figure 7.16).

The selectivity of the sensors for each metal ion should be achieved by differential affinity of the chelator to different metal ions. Naked metal ion is considered as a zero-dimensional positively charged dot or at most a small symmetric sphere. Chelators can be modified by several strategies. One is the selection of electron-donating atoms, such as O, N, and S, which defines the hard and soft electronic characters. Increasing the dentate number may increase the affinity to the metal ion. If the chelator is a cyclic structure, the ring size may have preference over the size of metal ions. Altogether, still the combination of the options may not be broad enough to cover similar metal ions.

When Ca^{2+} and Mg^{2+} sensors were tested against many other metal ions, including heavy transition metal ions, not surprisingly, there was significant

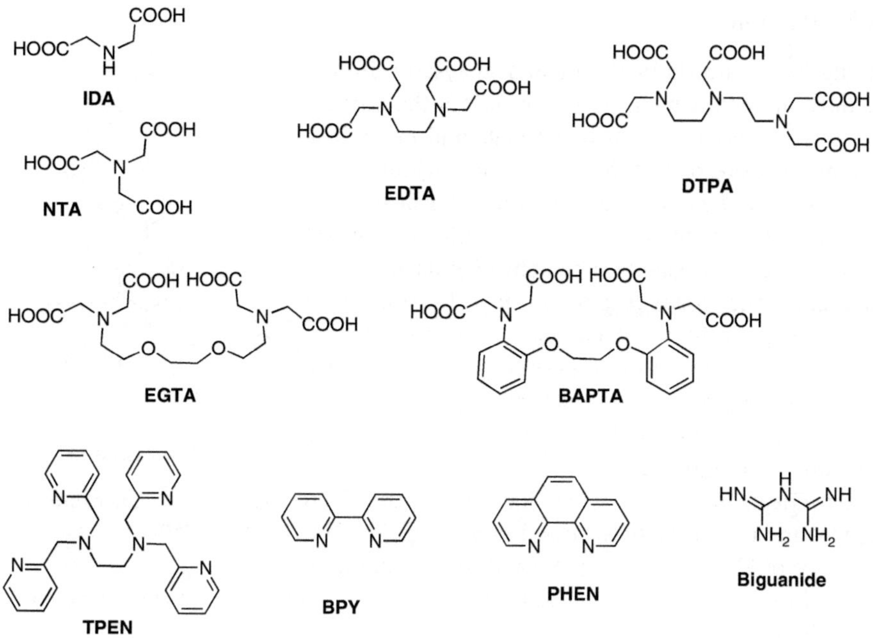

Figure 7.16 Structure of representative metal ion chelators.

cross-reactivity and the development of a single metal-ion-selective sensor seems intrinsically challenging [6]. Therefore, in the biological system, monitoring a target metal ion is always difficult due to the competition with other existing metal ions. One way is designing a sensor that has distinctively higher affinity to the target metal ion, by matching the character of the metal ion and the recognition motif. Another way is to use an existing probe for a new application, taking advantage of the given situation. For example, Ca^{2+} concentration in the cytosol is much lower than Mg^{2+} concentration, and a sensor that responds for both ions to a similar degree could still be used for Mg^{2+} monitoring as the Ca^{2+} response would be negligible. Also, the affinity of the sensor to the target metal ion is dependent on many factors in the biological system, such as pH, temperature, viscosity, binding proteins, and other competing ions. The background of the sensor reflects all the factors in the environment, and if the target ion change is the single dominant event, it would be possible to extract out the information on the change.

Although the selective ion sensor is difficult to achieve especially for the heavy metal ions, the cross-reactivity to multiple sensors allows pattern analysis of the fluorescence change for metal distinction. In an artificial tongue format, an array of sensors is used, and the optical changes (either absorption or fluorescence) are analyzed by principal component analysis (PCA) or linear discriminant analysis (LDA) to qualitatively and also quantitatively determine the metal ion identity and concentrations [24].

7.2.5 Iron Ion

Among heavy metals in the biological system, iron is the most abundant metal ion. Iron ion is the component of hemoglobin in red blood cells and exists in a dynamic mixture of Fe^{2+} and Fe^{3+}. Hemoglobin is an oxygen-transport protein, and the red color of blood is due to the iron ion complex in the hemoglobin. Iron ion is also the key component of cytochromes as an important regulator for redox reaction in the cells. While iron ions are involved in many biological functions, most of them are sequestered by binding either to functional proteins or storage proteins such as ferritin. Therefore, free iron ion detection is usually considered as a pathogenic situation of cell injury [25].

Phen Green is an iron metal ion sensor, based on phenacetin-binding motif for iron ion binding. The fluorescence of Phen Green is quenched upon binding to Fe^{2+} and Fe^{3+}, and has been used for the iron ion change in various biological systems (Figure 7.17).

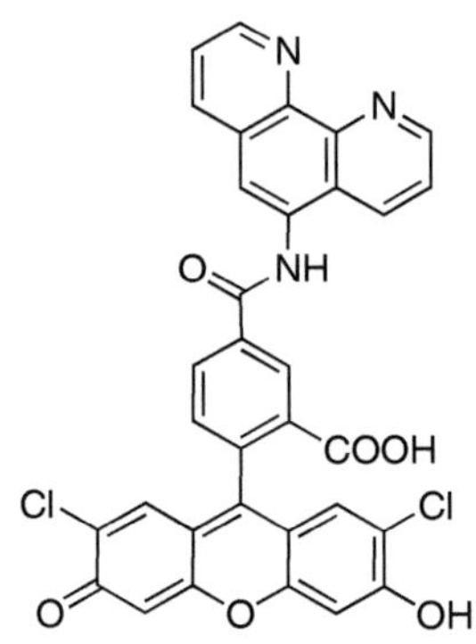

Phen Green

Figure 7.17 Structure of Phen Green.

7.2.6 Zn^{2+}

Zn^{2+} is the second most abundant transition metal ion in the biological system after iron. Many Zn^{2+}-binding proteins are involved in transcriptional regulation of genes. Zn^{2+} is also important for pancreatic insulin secretion, forming a crystalline high Zn^{2+}–insulin complex in beta cells [26]. There is evidence that stress-induced Zn^{2+} release of presynaptic neurons is a contributing factor for neurological disorders such as Alzheimer's disease[27]. Concentration of Zn^{2+} in the extracellular space is about 10–20 μM, and the cytosolic concentration is about 180 nM. Still, the free Zn^{2+} concentration is very low in most cells (<1 nM), with the majority of it being bound to proteins and nucleic acids [28].

TSQ is one of the earliest fluorescent probes developed for Zn^{2+} detection. TSQ is a cell-membrane-permeable Zn^{2+} probe [29] and forms a 2 : 1 complex with free Zn^{2+} [30]. With a metalloprotein the complex may be 1 : 1. TSQ was used to detect Zn^{2+} translocation from the presynaptic nerve terminal to the postsynaptic terminal during the ischemic events [31].

The successful story of Ca^{2+} probes also inspired to develop a Zn^{2+}-selective probe. By removing one acetate group from BAPTA of Ca^{2+}-selective probe, the Ca^{2+} affinity was reduced, maintaining the original high affinity to Zn^{2+} [32]. The resulting FluoZin-3 showed high affinity to Zn^{2+} (K_d = 15 nM) and was used for Zn^{2+} release from pancreatic beta cells. The rhodamine derivative RhodZin-3 was also designed based on the same strategy and has been used for mitochondrial Zn^{2+} homeostasis [33].

Newport Green has a moderate binding affinity to Zn^{2+} ($K_d \sim 1$ μM) and is useful for monitoring Zn^{2+} influx through neurons through Zn^{2+} channels [34]. Newport

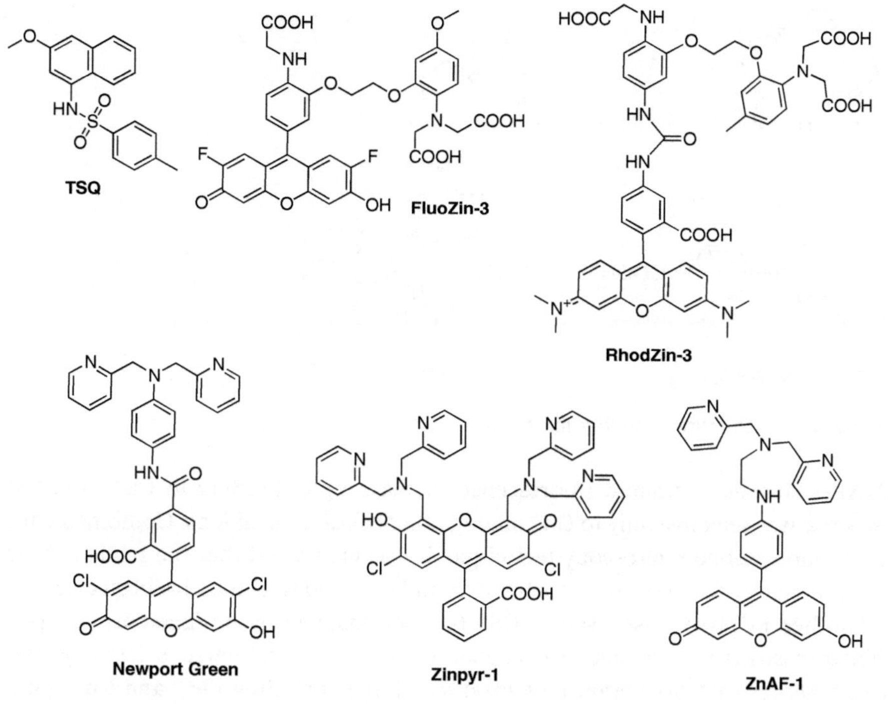

Figure 7.18 Structure of Zn^{2+} probes.

Green is also used to detect human pancreatic beta cells based on their high Zn^{2+} concentration complexed to insulin [35].

Utilizing di-2-picolylamine as the Zn^{2+}-capturing motif, a series of high-affinity Zn^{2+} probes were introduced by Lippard group. The first probe Zinpyr-1 is based on dichlorofluorescein and has extremely high affinity to Zn^{2+} ($K_d = 0.7$ nM) [36]. By careful modulation of the electron transfer property between the bottom aryl donor and the xanthene acceptor, ZnAF-1 was developed with a low background and large turn-on response to Zn^{2+} [37] (Figure 7.18).

7.2.7 Copper Ion

Copper is the third most abundant heavy metal ion in the biological system after iron and zinc. The reversible reduction of Cu^{2+} to Cu^{+} makes copper ion an important biological regulator, and also makes the detection more complicated. Due to the active redox property, unregulated copper ion can cause oxidative stress and damages in aging-related diseases such as Alzheimer's disease [38].

The iron sensor, Phen Green, has been used for copper ion detection with its broad response to heavy metal ions including Hg^{2+}, Pd^{2+}, Cd^{2+}, and Ni^{2+}. Similar to iron ion effect, the binding of the copper ion quenches the fluorescence of Phen Green [6].

The first turn-on sensor for Cu^{+} was developed by coupling thiacrown ether as the binding motif to pyrazoline dye, yielding CTAP-1 [39]. With 365 nm excitation,

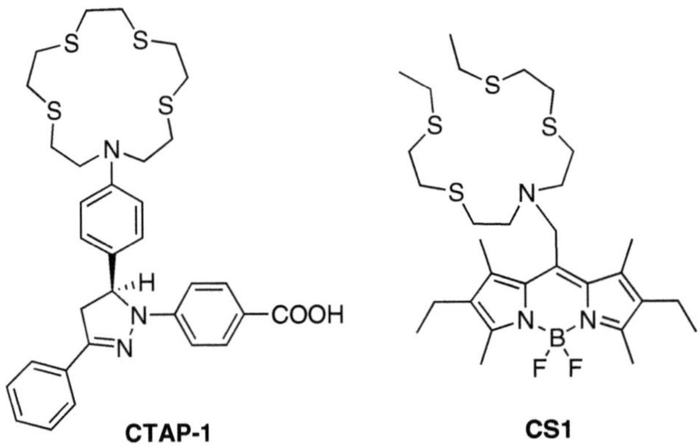

Figure 7.19 Structure of copper ion sensors.

CTAP-1 showed a 4.6-fold fluorescence increase upon binding to Cu^+, and the response was selective only to Cu^+ among other cellular metal ions. Combined with X-ray fluorescence microscopy technique, it was elucidated that the labile copper ions are mainly localized in mitochondria and Golgi body in the fibroblast cells.

Another BODIPY-based senor, CS1 (coppersensor-1), is designed using open thioether structure for selective copper binding. CS1 exhibited selectivity and sensitivity for Cu^+ over other biological metal ions including Zn^{2+} and Cu^{2+} [40]. CS1 has very high affinity to Cu^+ (K_d = 3.6 pM) and showed a 10-fold increase of fluorescence upon biding to Cu^+. In HEK293 cells, CS1 could respond to the changes of labile intracellular copper ions in the living condition (Figure 7.19).

7.3 Metabolites

7.3.1 ATP

ATP is the most widely used biological energy molecule and the amount of ATP may represent the energy status of cell and body. As mitochondria are the power plants to generate ATP through oxidative phosphorylation, the ATP quantity measurement in mitochondria reflects the energy production in the cell. ATP-Red 1 was developed as a turn-on ATP sensor through a multisite-binding design [41]. ATP-Red 1 showed a selective response to ATP compared to other nucleotides and localized in mitochondria in the cell. Thus, ATP-Red 1 could be used for mitochondrial ATP level monitoring. When the cells were starved, the ATP-Red 1 signal decreased marginally. With the inhibitor of oxidative phosphorylation, KCN, the sensor signal dropped dramatically. Apoptosis inducer camptothecin increased the sensor signal, reflecting the increased ATP production, which is a known requisite for apoptotic cell death process [42] (Figure 7.20).

In an independent study, ATP-Red 1 has been utilized to isolate a subpopulation of ATP-high cancer cells, which are aggressive with increased proliferation, stemness, anchorage independence, cell migration, invasion, multidrug resistance, and

Figure 7.20 Structure of ATP-Red 1.

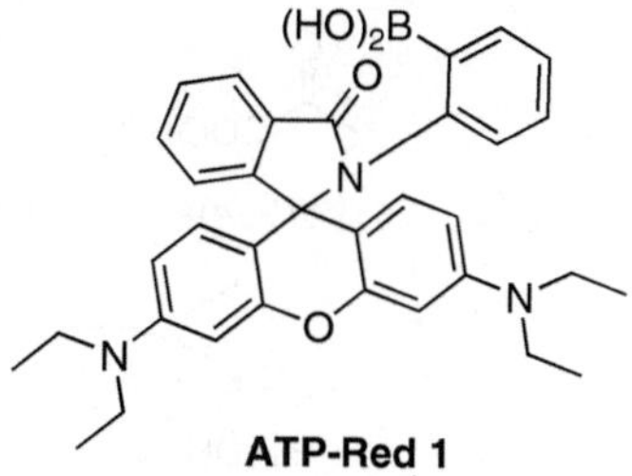

ATP-Red 1

antioxidant capacity [43]. These results show that live-cell-compatible ATP sensors can distinguish and also isolate ATP-high cell populations, which can be utilized for further biological studies, such as population-selective anticancer drug testing.

7.3.2 NADH

NAD^+ and its reduced form NADH are indispensable cofactors in all living cells. The NADH/NAD^+ ratio reflects the oxidoreductive status of cells. NADH is also the fuel for mitochondrial ATP production, where one NADH generates three ATP molecules. Therefore, a selective live-cell NADH probe would be of great utility. Through a bio-inspired strategy, enzyme-mimicking NADH probe, BA-Resa, and its advanced form, RA-Resa, have been developed [44]. The boronic acid moieties of the probes act as binding motifs for NADH, and resazurin acts as a reduction reporter by NADH (Figure 7.21).

In live cells, the RA-Resa signal was dose-dependently increased by glucose addition, reflecting the production of NADH. Lactate is the reduced form of pyruvate, and during anaerobic glycolysis, the accumulated pyruvate cannot enter the tricarboxylic acid (TCA) cycle but is converted into lactate by using NADH. An excess amount of lactate reverses the reaction by converting NAD^+ into NADH. As a result, lactate addition into cells induces a higher signal of RA-Resa, and the signal is quenched by additional pyruvate addition. With this example, RA-Resa would be a useful imaging tool for oxidoreductive status of cells (Figure 7.22).

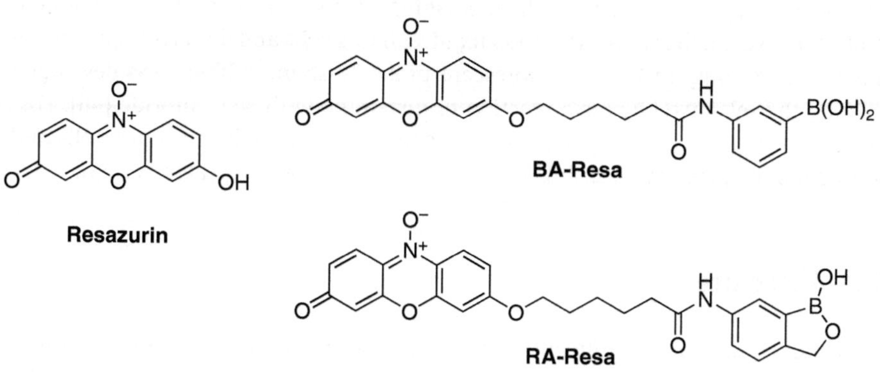

Figure 7.21 Structure of BA-Resa and RA-Resa.

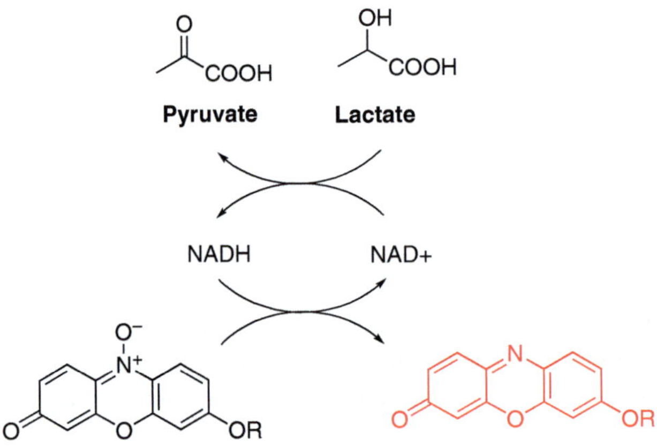

Figure 7.22 Biochemical scheme of RA-Resa action and lactate cycle.

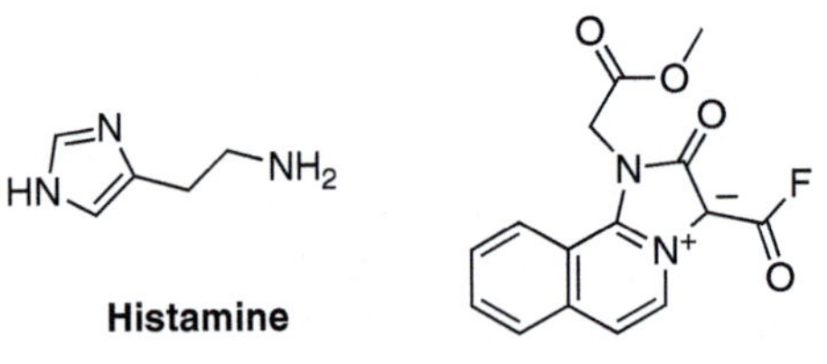

Figure 7.23 Structure of histamine and histamine blue.

7.3.3 Histamine

Histamine is an alkaloid molecule involved in local immune responses and also acts as a neurotransmitter [45]. Histamine is mainly synthesized in granules in mast cells and in basophils. Mast cells are abundant in exposed tissues such as the nose and mouth, and basophils are one type of granulocyte in the bloodstream. When these cells are activated by antigen, release the histamine to the extracellular space, as the first line of innate immune defense from injury and infection. Overproduction and release of histamine causes allergic symptoms. The conventional Giemsa reagents stain the granules in strong blue, distinguishing basophils from eosinophils or neutrophils. However, the Giemsa method requires fixing cells and limits the application to live-cell staining. In 2012, a fluorescent probe "histamine blue" was developed utilizing the selective chemical reaction of histamine with acid fluoride [46]. Histamine blue was successfully applied to image histamine in live immune cells and to discriminate primary basophils from other granulocytes [47] (Figure 7.23).

7.4 Viscosity

Intracellular viscosity significantly influences the material transport and interactions of biomolecules. It is known that the abnormal changes of intercellular viscosity are related to diseases such as Alzheimer's disease, atherosclerosis, and diabetes. Imaging tool to monitor the viscosity change of organelles would be a great tool to understand the cellular status. Intracellular organelles have different viscosity due to the building blocks and environmental differences. Dynamic viscosity (absolute viscosity) has the dimension of force × time/area, and the unit is poise, defined as 0.1 Pa s or 0.1 kg/m s. For convenience, cP (centipoise, one-hundredth of a poise) is commonly used for biological applications. By this scale, each organelle's reported viscosity is as follows: cytoplasm (14.5 cP) [48], lipid droplet (34–57 cP) [49], lysosome (50–90 cP) [50], nucleus (34–59 cP) [51], mitochondrion (63–68 cP) [52], ER (130 cP) [53], and plasma membrane (193–307 cP) [54] (Figure 7.24).

One of the important and difficult problems in viscosity measurement is the influence by the temperature change; in general, an increased temperature decreases the viscosity. If the stimulus is chemical or biological, rather than physical temperature change, it is difficult to isolate or distinguish if the change is from viscosity or temperature changes. An ideal probe would be a biological temperature-independent (maybe in 25–50 °C range) viscosity sensor or a biological viscosity-independent (15–300 cP range) temperature sensor. However, it is extremely difficult to isolate the two factors in reality. Therefore, always a careful caution should be applied to the analysis of these sensors for proper interpretation.

Molecular rotor has been a popular motif for viscosity sensors. The rotor motif rotates in a low-viscosity condition, and when connected to a fluorophore, the

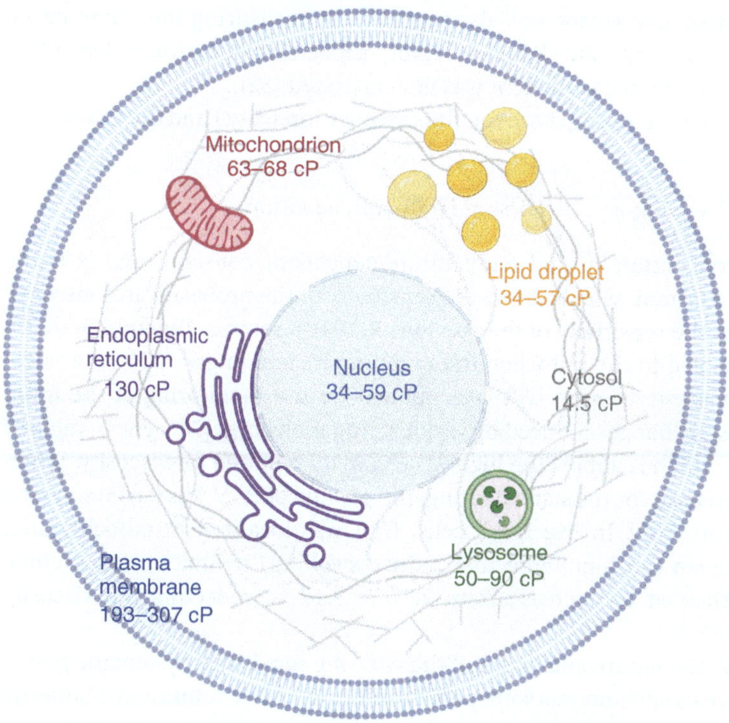

Figure 7.24 Summary of the viscosity of intracellular organelles.

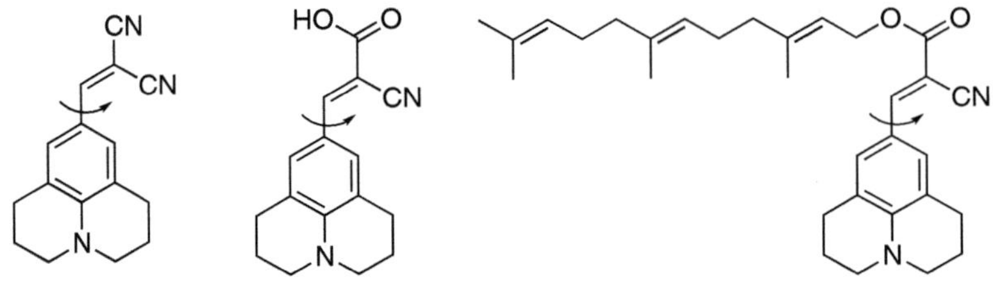

DCVJ CCVJ FCVJ

Figure 7.25 Structure of DCVJ and its derivatives for membrane viscosity measurement.

fluorescence is quenched due to nonradiative decay induced by the free rotation. The quenched fluorescence is recovered when the sensor is in a highly viscous condition by the restricted rotation. DCVJ with dicynovinyl is a first-generation rotor-based viscosity sensor and its derivatives have been used for membrane viscosity measurement [55] (Figure 7.25).

Phenyl group attached to the meso position of BODIPY is another popular rotor motif. With medium size of hydrophobic chain, the BODIPY rotor has been used for the fluorescence lifetime imaging in live cell and the intracellular viscosity was estimated as 140 cP [56]. Combined together with a viscosity-insensitive coumarin molecule and a mitochondria-targeting motif, another BODIPY-based self-calibrating viscosity sensor was also reported. By measuring the fluorescence ratio between coumarin and BODIPY rotor, probe-concentration-independent mitochondrial viscosity measurement was demonstrated [52].

The quantitative relationship between fluorescence intensity I and the solvent viscosity η is

$$\log I = C + \chi \log \eta \qquad \text{(Förster–Hoffmann equation)}$$

where C is a concentration- and temperature-dependent constant, and χ is the dye-dependent constant, which shows the sensitivity of the probe towards viscosity [57]. To improve the sensitivity of the previous BODIPY rotor, a dimethyl BODIPY motif was connected to the mitochondria-targeting moiety to give BTV as a novel mitochondrial viscosity sensor. BTV was applied for the monitoring of the mitochondrial viscosity changes induced by various drug molecules [58]. The χ value of BTV was 0.718, which is almost the highest sensitivity reported for viscosity. When lifetime has been used for the same plotting, the χ value of BTV was 0.605, showing a little lower sensitivity. In live HeLa cells, BTV was localized in mitochondria, and with monensin (an ionophore that induce swelling) treatment, the lifetime measurement showed an increase from 2.77 to 3.17 ns, reflecting the viscosity increase (Figure 7.26).

To understand the nature and origin of the viscosity sensitivity, systematic preparation of derivatives and analysis were performed for BODIPY rotors and a butterfly conformation model was proposed [59]. A further study for viscosity sensitivity and temperature dependency elucidated the relationship providing a model for the new

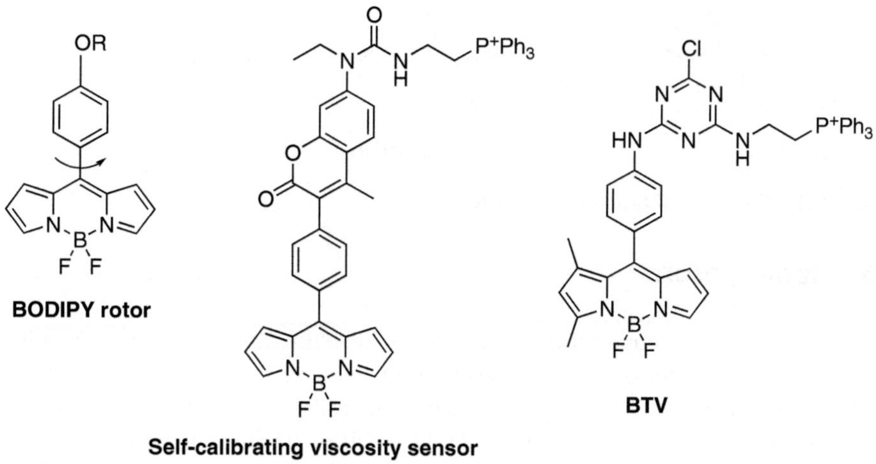

Figure 7.26 Structure of BODIPY rotor, self-calibrating viscosity sensor, and BTV.

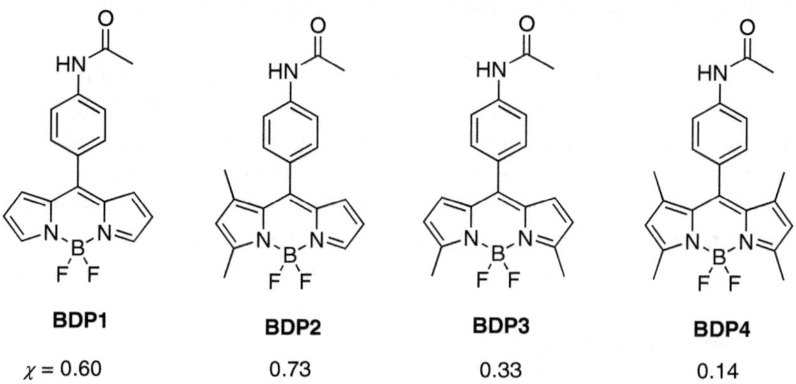

Figure 7.27 Structure of BDP1-4.

sensor design. Based on the results, the record-high viscosity-sensitive probe, BDP2, with a χ value of 0.73 was elucidated [60] (Figure 7.27).

Naphthalene dyes with an electron donor and acceptor are usually polarity sensors with solvatochromic properties. Prodan is the representative dye and shows red shift emission in polar solvents by stabilization of the excited state in a polar environment. Laurdan is a hydrophobic derivative of prodan with a longer hydrocarbon chain that can anchor in the plasma membrane of cells. In addition to the polarity sensitivity, laurdan also exhibits a viscosity sensor property that acts as a rotor and is used for lipid membrane phase recognition [61] and fluidity monitoring [62] (Figure 7.28).

The viscosity change is one of the important characteristics in various diseases and the related probe development and application is further discussed in Section 8.1.4.

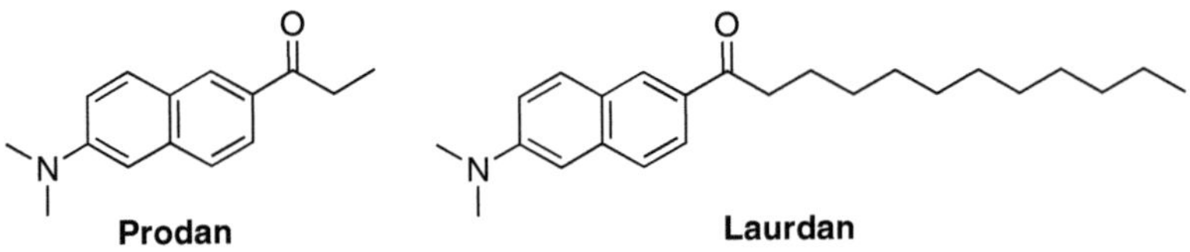

Figure 7.28 Structure of prodan and laurdan.

7.5 Temperature

Keeping an optimum temperature is essential to keep the functions of life. Usually, the chemical reactions get accelerated with an increase in temperature, approximately two times per 10 °C. Many of the biological reactions are catalyzed by enzymes, and the enzyme activity is highly sensitive to the temperature. Mammals try to keep the body temperature constant (37 °C in humans) by spending a lot of energy, while reptiles do not have such a heating system. As a result, at low temperatures, the metabolism of reptile becomes slower, and eventually they enter a sleeping mode. This is the trade-off of self-control and high energy expenditure!

Fluorescent molecules usually show high quantum yield when they keep structural rigidity with proper orientation of energy transition at the excited state. At higher temperature, the molecular motion becomes more dynamic, and usually the quantum yield of fluorescence decreases. The lifetime of fluorescent molecule is determined the prolonged time period at the excited state. At higher temperature, lifetime also usually gets short. If the fluorescence intensity or lifetime change is big enough against temperature change, such molecules have potential to be used as a temperature sensor. If the molecules are live-cell compatible, they can be temperature-sensitive imaging probe.

7.5.1 ER Thermometer

Through a systematic temperature sensitivity screening of DOFL and cell localization test, the first ER-specific small-molecule thermometer, ER thermo yellow was developed [63]. ER thermo yellow is selectively localized in ER in live HeLa cells, and showed 3.9%/°C fluorescent intensity changes, which is the highest among reported small-molecule thermometers. In the same condition, the control compound ER tracker showed only a 0.3% change. The intracellular temperature gradient was generated by a laser-heating metal nanoparticle located in outside of cells, and the distance from the nanoparticle and the temperature gradient was linearly correlated.

ER thermo yellow was successfully applied to the measurement of intracellular temperature change in SH-SY5Y neuron cells by piezoelectric nanoparticle as a reporter [64] and the intracellular temperature monitor during the photothermal release of cargo from liposomes in live HeLa cells [65]. Fluorescence intensity depends not only on the quantum yield, but also on the localized concentration of the probe in the region of interest. To avoid the artifact generated from the different amount of probe, fluorescence lifetime imaging microscopy (FLIM) is a good option. ER thermo yellow was successfully adapted in C2C12 myotube using the FLIM mode for temperature change monitoring, showing that the heat production in myotube is highly localized and is coupled to a Ca^{2+} burst [66]. A remote myotube activation by gold nanoshells was monitored by ER thermo yellow, and about 5 °C of temperature change was accompanied by myotube contraction [67].

Mammals have a unique heat generation method through brown adipose tissue (BAT), which is abundant in newborn animals and also inducible during sustained cold exposure. ER thermo yellow has been tried to monitor the temperature change in BAT and a significant photostability problem was observed. ER thermo yellow is equipped with a chloroacetyl group and was originally designed to form a covalent bond with the thiol group in the cell. To improve the photostability, a modification of ER thermo yellow by changing the chloroacetyl group into an acetyl group generated ERthermAC (Figure 7.29).

When ERthermAC was tested in a fixed WT-1 (a mouse brown preadipocyte cell line) the temperature sensitivity was about 1.07%/°C between 18 and 35 °C, and 4.76%/°C between 36 and 43 °C. When WT-1 cell differentiated to BAT, UCP1 (uncoupling protein 1) is increased. UCP1 is localized in the inner mitochondrial membrane and uncouples oxidative phosphorylation from respiration, resulting in increased heat production. Isoproterenol (also known as isoprenaline) is a β-adrenergic agonist which is known to upregulate UCP-1 expression and thermogenesis. When isoproterenol was treated in WT-1 cells, the ER temperature was increased from 25 to 42 °C based on the ERthermAC signal changes, which is surprisingly high. During the isoproterenol treatment, the mitochondrial membrane potential was decreased due to the uncoupling, confirmed by JC-1 measurement [68]. By Tseng group, a CRISPR-engineered human brown-like (HUMBLE) adipocyte was generated and a similar temperature increase was observed by forskolin treatment using ERthermAC [69].

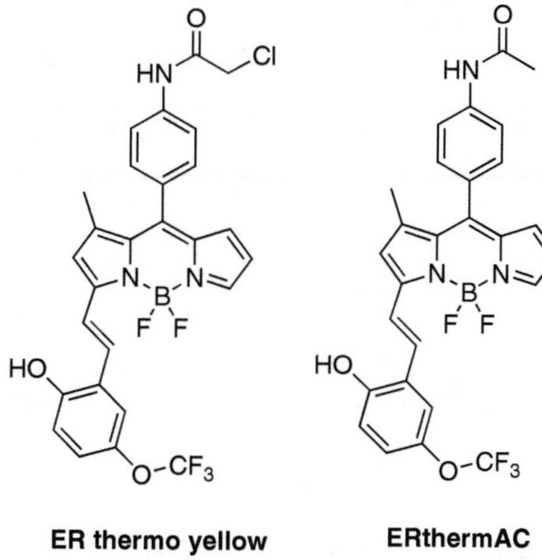

Figure 7.29 Structure of ER thermo yellow and ERthermAC.

7.5.2 Mitochondrial Thermometer

In parallel to ER thermometer, a mitochondrial thermometer MTY (Mito Thermo Yellow) was developed by temperature response and localization screen of DOFL. In live NIH3T3 mouse fibroblast cells, MTY showed 2.7%/°C of sensitivity. The negative control compound A15 has a very low sensitivity of 0.3%/°C, and a slightly modified I30 has 1.6%/°C, implying that the piperidine motif plays an important role. MTY has been successfully applied to different types of cells, such as C2C12 myotube, mESC, Chang liver cells, BAT, and HeLa cells with comparable temperature sensitivity [70].

An astounding observation was reported using MTY for mitochondrial temperature. Both in HEK293 cells (human embryonic kidney) and primary skin fibroblasts, the temperature of mitochondria was near 50 °C when the mitochondrial respiratory chain was fully functional. This is much higher than the scientists believed and even claimed an impossible range. Interestingly, the activity of the enzymes in the respiratory chain was maximal at or slightly above 50 °C, implying that the local mitochondrial temperature may be indeed so high. This provocative results obtained by MTY may need further validation by independent methods or prompt critical re-examination of the literature on mitochondria [71].

MTY has also been used for the temperature monitoring in thermoresponsive nanocarrier for mitochondria-targeted drug delivery [72]. A covalent binding version of MTY, MTX (MTY for crosslinking), was prepared for potential long-term labeling of mitochondria, and thus expecting the longer monitoring of temperature change. Through a careful analysis of both MTY and MTX, a significant cell dependency and impact by the mitochondrial membrane potential were cast as potential caveat of these charged probes [73] (Figure 7.30).

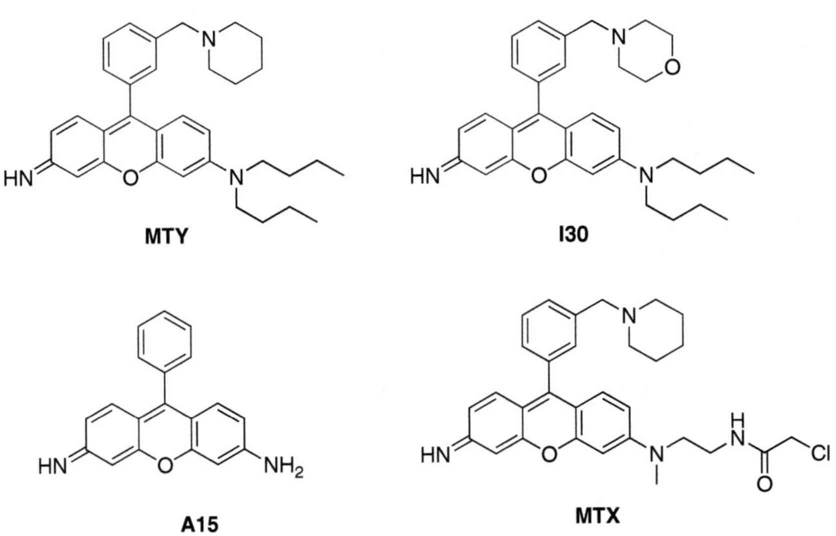

Figure 7.30 Structure of MTY, I30, A15, and MTX.

7.5.3 Organelle-Specific Fluorescent Thermometers

In addition to the ER and mitochondria, cells are composed of many more compartmentalized organelles. Intracellular local temperature is closely related to cellular process, and monitoring the temperature at each organelle would be desirable. In 2022, a full set of organelle-specific thermometers are introduced as a unified palette, covering the plasma membrane, ER, Golgi body, lysosomes, mitochondria, nucleus, and lipid droplets. Utilizing the fact that viscosity and temperature effect to fluorescence of the thermometer is reversely correlated, the most sensitive viscosity motif, BDP2 (Section 7.4), is chosen for the temperature reporter of site-specific organelle thermometers, Thermo Greens (TGs).

For specific delivery of the thermometer to each organelle, BDP2 is conjugated with various organelle-targeting motifs, such as n-undecanoyl group (Lipid Droplet Thermo Green, DTG), triphenylphosphonium group (Mitochondria Thermo Green, MTG), 4-(2-aminoethyl)morpholine group (Lysosome Thermo Green, LTG), 3-(dodecyl(methyl)amino)propane-1-sulfonate (Plasma Membrane Thermo Green, PTG), D-sphingosine group (Golgi Thermo Green, GTG), and Hoechst 33258 (Nucleus Thermo Green, NTG). Through fluorescence lifetime-based cell imaging, TGs were demonstrated to monitor the organelle-specific temperature gradients created by external heating. The fluorescence lifetime-based thermometry shows that each organelle experiences a distinct temperature increment that depends on the distance away from the heat source (Figure 7.31).

TGs are further demonstrated in the quantitative imaging of heat production at different organelles in brown adipocytes. Brown adipocytes are thermogenic cells by utilizing the proton gradient of mitochondria through UCP1 (uncoupler protein) rather than ATP production. Therefore, mitochondria are the hot spots in the cell through the heat generation. In a study with TG, it was found that the temperature change was the highest, followed by ER, and then other remote organelles, such as lipid droplets. To date, TGs are the first palette batch of small-molecular fluorescent thermometers that can cover almost all typical organelles.

7.6 Reactive Oxygen Species and Reactive Nitrogen Species

Reactive oxygen species (ROS) are a series of radical or reactive oxygen-containing species that are generated during the oxidation reaction involving oxygen. Oxidation is basically the same process with burning, and chemical oxidation is just slowed down burning process. In biological systems, mitochondria are the specialized organelle for the slow burning, handling the dangerous oxygen. If the oxidation of organic molecules (composed of C and H) is completed, the products are relatively stable as of H_2O and CO_2. If the oxidation is not fully finished in a short amount of time, reactive intermediates, represented by ROS, are accumulated in the local area of the body. Not surprisingly, mitochondria release most of the ROS in the cells. Due to the high activity, ROS inevitably affect the biomolecules and often play important roles both for normal and pathological processes [74].

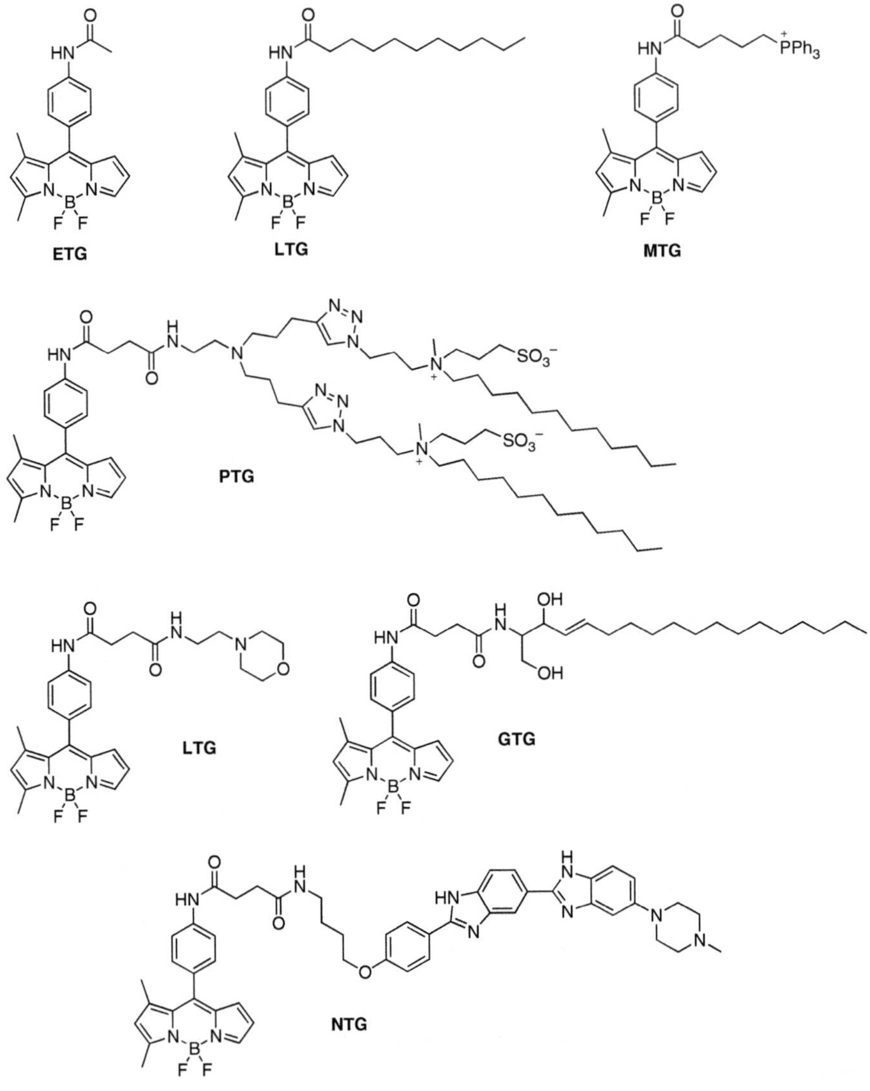

Figure 7.31 Structure of TGs.

As the first step of oxidation, molecular oxygen forms superoxide ($^{\bullet}O_2^-$) by taking up an electron. Dismutation of two superoxide radicals makes hydrogen peroxide (H_2O_2) by abstracting protons from the environment. It is noteworthy that one molecule of oxygen is regenerated during the reaction. As superoxide is super reactive and toxic, superoxide dismutase (SOD) enzyme catalyzes the reaction to form the less toxic H_2O_2 in the biological system. H_2O_2 takes up another electron to form hydroxide and hydroxyl radical, and the latter is also a highly reactive species. One more electron take-up finishes the reaction forming water as the final stable end product. Alternatively, H_2O_2 can be decomposed to water and molecular oxygen by catalase enzyme. The ROS can react with other biological partners to form various

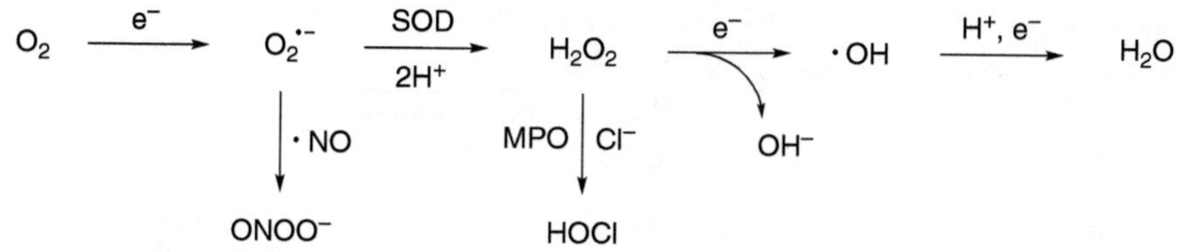

Figure 7.32 Reaction schemes for ROS production and transformation.

reactive materials. For example, superoxide can react with nitric oxide (•NO) to form peroxynitrite ($ONOO^-$), a representative reactive nitrogen species (RNS). Hydrogen peroxide reacts with chloride to generate hypochlorous acid (HOCl), a chemical weapon used by immune cells. Industrially, chlorine is commonly used for water disinfection. When chlorine is dissolved in water, hypochlorous acid (HOCl) is formed as the primary active disinfection ingredient. The salt form NaOCl is the common bleach material in daily life and HOCl is more reactive than the ionic form. The pK_a of HOCl is 7.53 and it exists in about half ionized form in physiological condition [75]. The representative ROS/RNS and their transformation are summarized in Figure 7.32 [76].

The ROS generated in the biological system are essential to normal function of life, and they are involved in signal transduction, muscle relaxation, plate aggregation, blood pressure control, immune regulation, and so on. [77]. However, the situation of too high or too low ROS is harmful to the biological system, leading to irreversible damages to cells and tissues. It is not surprising to observe the reports that oxidative stress is related to various diseases [78] and aging [79]. It is genuinely challenging to study ROS, due to their short lifetime and various reaction partners in the biological system such as antioxidants. For the efficient study of ROS and their impact on the biological system, a versatile sensor or probe with high sensitivity and biocompatibility would be of great importance.

Dichlorodihydrofluorescein (DCFH) is the widely used general ROS or oxidative stress sensor. The reduced form of the dye is oxidized by various oxidants to convert DCFH into the oxidized form, DCF, which is a bright fluorescent dye. For intracellular application, DCFH-DA (diacetate form of DCFH) is used for better cell membrane penetration. Inside the cell, the acetate group may be removed by esterase, and the released DCFH reacts with ROS. DCFH was originally developed as an H_2O_2 sensor [80], but it was turned out not to discriminate different ROS [81]. This is a common observation for most of the dihydro form of reduced dyes' behavior, such as rhodamine 123 [82]. So, DCFH is useful to gauge the sum of oxidative stress in the cell, but for a specific ROS study, a matching individually selective sensor development is necessary (Figure 7.33).

7.6.1 Superoxide

Superoxide is the first ROS converted from molecular oxygen and the estimated physiological concentration is in pM range [83]. The pK_a of the conjugated acid is 4.8 and exists mainly in the anionic form under physiological condition. Superoxide is

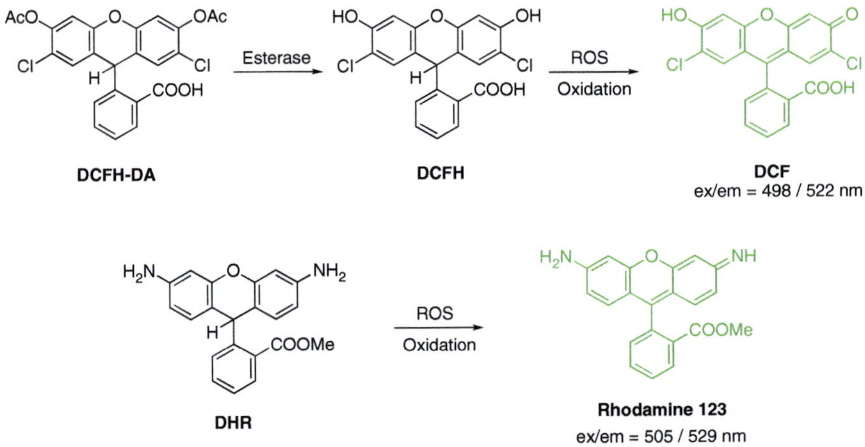

Figure 7.33 Reaction of dihydro dyes with ROS to form fluorescent product.

a weak oxidant and a potent reductant especially for heavy metal ions such as Fe^{3+}. The primary ROS, superoxide generates various secondary ROS such as H_2O_2 and $ONOO^-$ by dismutation or reacting with other biomaterials.

Dihydroethidium has been developed for superoxide sensor and a fluorescent product ethidium is formed by reaction with superoxide with moderate specificity [84]. With accumulated data, it seems dihydroethidium also reacts with various ROS (such as $ONOO^-$ and hydroxyl radical) not only with superoxide and may provide the information for more general oxidation status of cells, rather than superoxide specific one [85]. Interestingly, the reaction products of dihydroethidium are not identical with different color material generation, which casts the clues for selective sensor development [86]. Another potential problem of using ethidium in cell imaging is its intercalating activity to DNA double helix, and the resulting fluorescence increase makes the interpretation of the data more complicated [87].

To suppress the DNA intercalation, a bulky neopentyl group was introduced to the same fluorophore, generating a more selective probe MitoNeoD. The introduced deuterium also helps to increase the chemical selectivity of the probe to singlet oxygen with the kinetic isotope effects at the hydrogen abstraction step [88].

Using the same hydrogen abstraction mechanism, pyridyl benzothiazoline was developed as more selective superoxide sensor turning on the fluorescence. The probe did not react with H_2O_2 or increased fluorescence with the hydroxyl radical, showing higher selectivity than the previous sensors [89]. With the similar idea, dimeric thiazoline derivative, DBZTC, was developed for superoxide sensor [90].

The nucleophilicity of superoxide provided another possibility of selective sensor design. A fluorophore with sulfonyl or phosphinyl group could react with superoxide, releasing hydroxyl product which is fluorescent. PF-1, a fluorescein molecule with two phosphinyl groups, showed moderate selectivity over other ROS and fluorescence increase upon reaction with superoxide [91]. Further improved probe, HKSOX-1 contains more electron-withdrawing Tf (triflate or

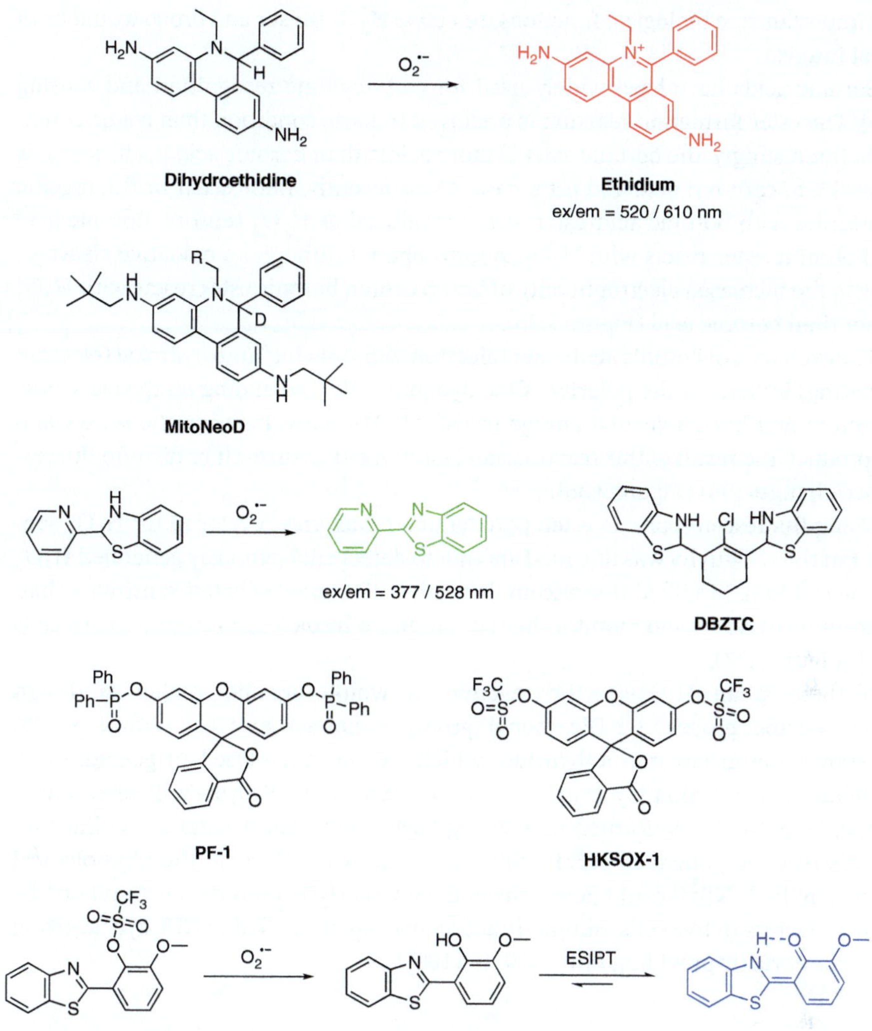

Figure 7.34 Superoxide sensors.

trifluoromethanesulfonyl) leaving group, showed much better superoxide selectivity compared to PF-1 [92]. The same chemistry was applied to the benzothiazole motif for a superoxide sensor and the mechanism was suggested that the resulting phenol compound undergoes excited state intramolecular proton transfer (ESIPT) to turn on the fluorescence [93] (Figure 7.34).

7.6.2 H_2O_2

Hydrogen peroxide is mainly in neutral state under physiological condition with a pK_a of 11.6. As a relatively stable molecule, it can diffuse across the cell membrane to regulate many biological functions [74]. Thiol group of cysteine in proteins can be modified by H_2O_2 through forming sulfenic, sulfinic, or sulfonic acid [94]. With the importance in biological functions, selective H_2O_2 sensor and probe would be of great interest.

Boronic acids have been widely used for carbohydrate recognition and sensing [95]. The ester formation reaction is preferred in basic condition than acidic condition. Interestingly, the boronic ester is more acidic than boronic acid itself, which is opposite to carboxyl acid and ester case. More recently, fluorescent or fluorogenic molecules with boronic acid/ester were introduced as H_2O_2 sensors. Boronic acid and boronic ester reacts with H_2O_2 to form phenol through an oxidative cleavage. Due to the increased electrophilicity of boron center, boronic ester reacts with H_2O_2 faster than boronic acid (Figure 7.35).

The exchange of boronic acid/ester (electron withdrawing) into hydroxyl (electron donating) influences the polarity of the dye molecule. Depending on the electronic structure and intramolecular charge transfer (ICT) status between the sensor and its product, the result of this reaction can result in either turn-off or turn-on fluorescence changes [96] (Figure 7.36).

Using fluorescein boronic ester, peroxyfluor-1 was first developed for H_2O_2 sensor, but the sensitivity was not good enough to detect endogenously generated H_2O_2 in the cell [97]. Modified fluorescein derivative PG1 showed better sensitivity than peroxyfluor-1 and could monitor the H_2O_2 induced by biological stimuli in live cells [98] (Figure 7.37).

If there is an H_2O_2-specific reaction, it would provide a clue to design H_2O_2-specific probes. 1,2-Dicarbonyl group is known to react with H_2O_2 to rearrange the group into anhydride, which can be hydrolyzed to generate carboxylate. This reaction is a named reaction as known as Baeyer–Villiger reaction, which is originally performed in a strong basic condition. A careful modification of this reaction generated NBzF which can react with H_2O_2 in the physiological condition [99]. NBzF could detect the endogenous H_2O_2 production stimulated by growth factors in live cells. Similar functional group is applied to NIR dye targeting mitochondria by providing Mito-NIRHP [100].

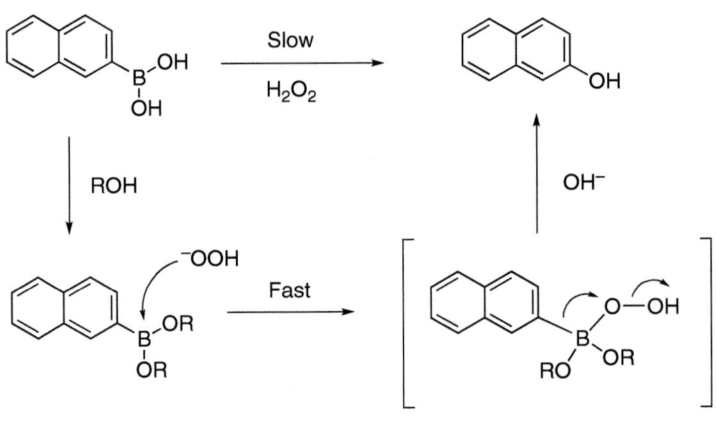

Figure 7.35 Reaction of boronic acid/ester with H_2O_2.

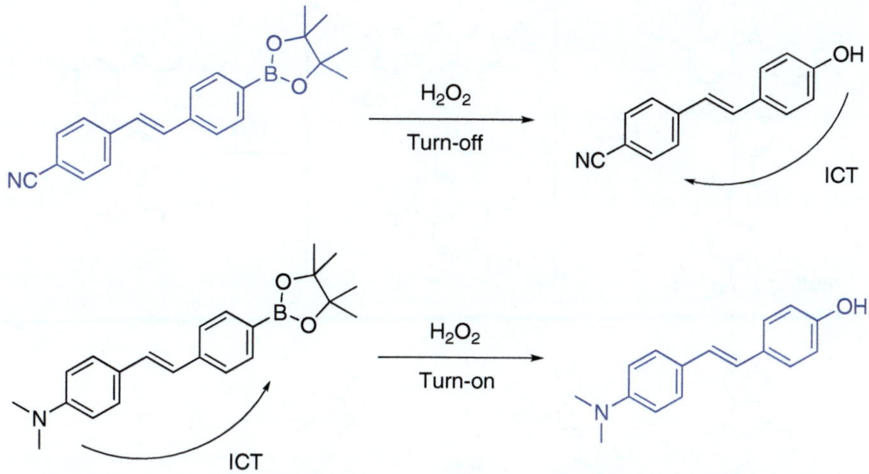

Figure 7.36 Hydrogen peroxide reaction with boronic acid with turn-off and turn-on changes.

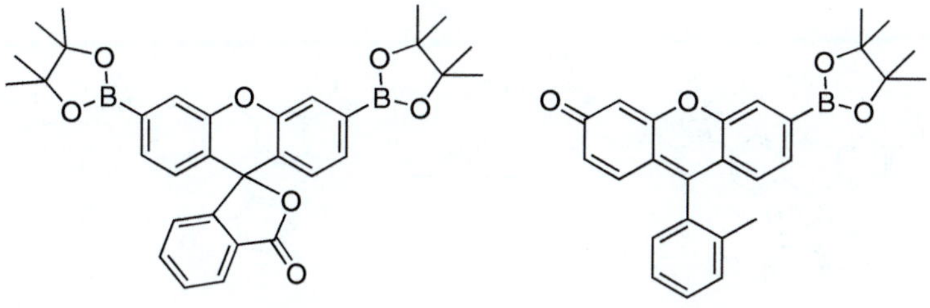

Figure 7.37 Boronic ester-based H_2O_2 sensor for biological application.

Another example of H_2O_2-specific reaction is the tandem Payne/Dakin reaction. The salicylaldehyde can be converted to catechol in the presence of H_2O_2, and the subsequent cleavage of carbamate is applied to a rhodamine structure to generate HKPerox-1 [101]. HKPerox-1 could monitor the H_2O_2 burst in macrophages induced by a signaling molecule, phorbol myristate acetate (PMA) (Figure 7.38).

7.6.3 $ONOO^-$

Peroxynitrite ($ONOO^-$) is formed by a superoxide reaction with nitric oxide. The conjugated acid peroxynitrite (ONOOH) has a pKa of 6.8, so under physiological conditions, both neutral and anionic forms exist, exhibiting complex reaction properties. Depending on the reactants, peroxynitrite can act as one or two electron oxidants and also as a potent nucleophile. Various reaction-based sensors have been developed for detection of $ONOO^-$.

Electrophilic trifluoromethyl ketone can be oxidized by $ONOO^-$ to form dioxirane intermediate, which in turn releases the fluorescent dye by rearrangement. Utilizing

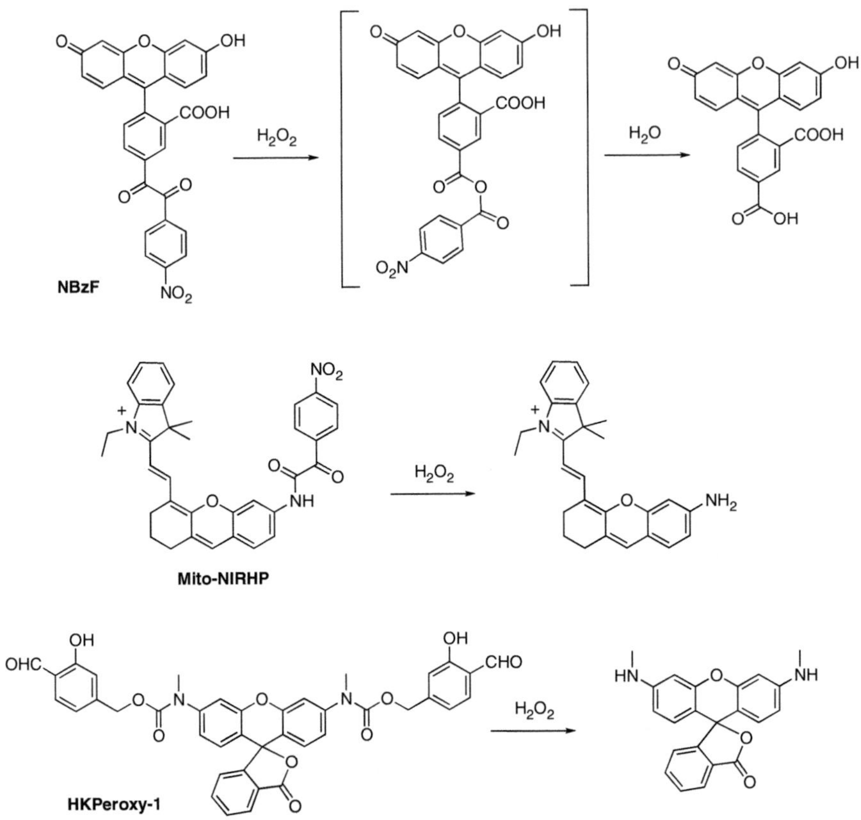

Figure 7.38 Reaction-based H_2O_2 sensor for biological application.

trifluoromethyl ketone, a series of HKGreen probes were developed [102]. Among the series, HKGreen-3 showed the best selectivity to $ONOO^-$ and enabled the visualization of endogenous $ONOO^-$ production in macrophage cells by various biological stimuli [103].

N-dearylation was another strategy used for $ONOO^-$ selective sensor development. Fluorescent dyes with arylamine is quenched by the electron-rich arylamino group. If $ONOO^-$ oxidizes the aryl group and releases the free amine group, the quenched fluorescence could be released to act as a turn-on $ONOO^-$ sensor. HKGreen-4 is the first kind of the probe and could detect endogenous $ONOO^-$ production in macrophages attacked by bacteria [104]. Using the same strategy, a longer wavelength sensor HKYellow was developed and applied to detect $ONOO^-$ generation in hepatic inflammation of mouse model [105].

$ONOO^-$ reaction with organoselenium or organotellurium was another successful strategy of selective $ONOO^-$ sensor design. The selenium and tellurium quench the fluorescence of the dye, and they once react with $ONOO^-$ form oxide and induce the fluorescence increase of the sensor. Interestingly, the oxidation reaction is reversible and can go back to the reduced form by glutathione or ascorbate. The

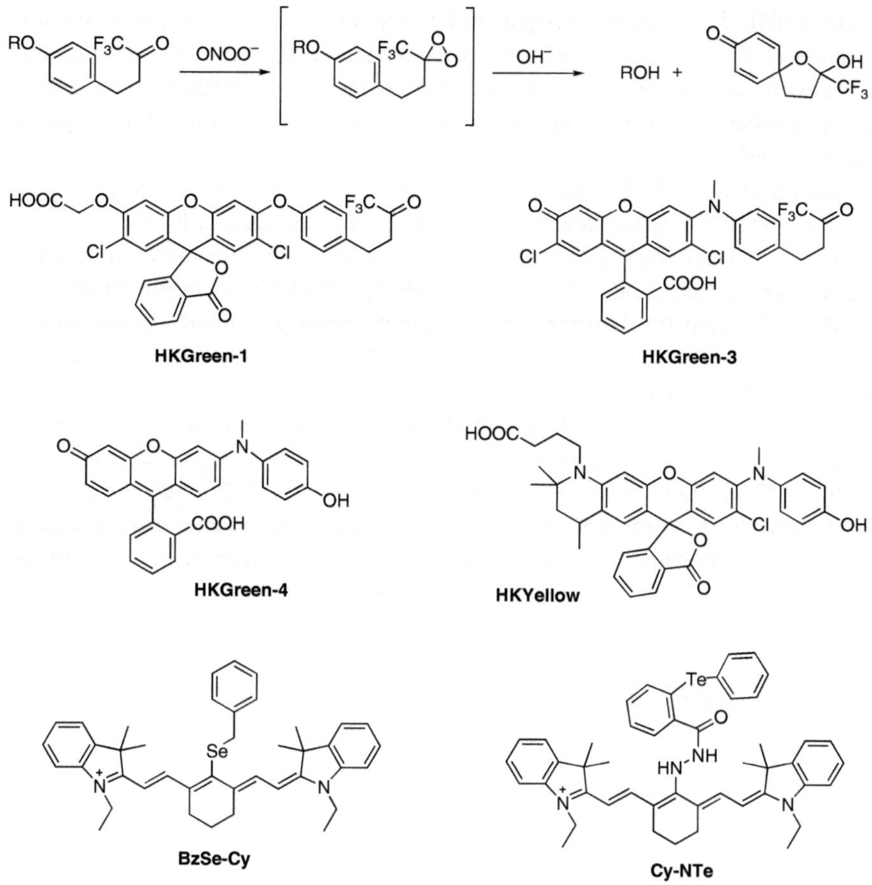

Figure 7.39 Structures of $ONOO^-$ sensors.

endogenous $ONOO^-$ could be monitored by the NIR selenium sensor, BzSe-Cy, in macrophage cells [106]. Also, the redox cycles in living cells could be monitored by the tellurium-based sensor Cy-NTe [107] (Figure 7.39).

7.6.4 HOCl and Hypochlorite

HOCl is a powerful oxidizing agent with a pK_a of 7.4, which implies the coexistence of neutral (hypochlorous acid) and anion (hypochlorite) forms under the physiological condition. HOCl is formed by the reaction of H_2O_2 and chloride catalyzed by myeloperoxidase (MPO) in immune cells as a common chemical weapon against microbial infection. Due to its high chemical toxicity, excess production of HOCl can be potentially harmful to the host cells and body. A series of HOCl sensors were developed based on selective reaction with HOCl.

Electron-rich phenols can be readily oxidized by HOCl to from benzoquinones. By fine tuning the reaction, the first HOCl-selective sensor, HKOCl-1, was developed in 2008 [108]. The *para*-methoxyphenol group in HKOCl-1 quenches the fluorescence of the molecule, and the HOCl oxidation abolished the ability, and in turn induced fluorescence turn-on. This strategy was further optimized to generate a versatile HOCl sensor for live-cell imaging of HOCl with an example of HKOCl-3 [109].

Oxidation of C=C double bond by HOCl was a well-known reaction and has been used for HOCl sensor development. One of the examples is PMN-TPP and it was used for ratiometric fluorescence measurement for HOCl generation in acute inflammation mouse model [110]. HOCl selectively oxidizes diacylhydrazine to diacyldiamide, then the intermediate further decomposes to carboxylate. Utilizing the chemistry, a rhodamine sensor, Rh-TPP, was developed to measure the mitochondrial HOCl concentration [111].

As another strategy, a thiocarbamate group was selected to develop a HOCl-specific reactant to develop a HOCl sensor. This reaction was selective only to HOCl, without response to $ONOO^-$, H_2O_2, $^{\bullet}OH$, NO, and so on. [112]. The fluorescence turn-on mechanism is ESIPT, and the same chemistry has been applied to generate various emission wavelength HOCl sensors with the rainbow color range [113] (Figure 7.40).

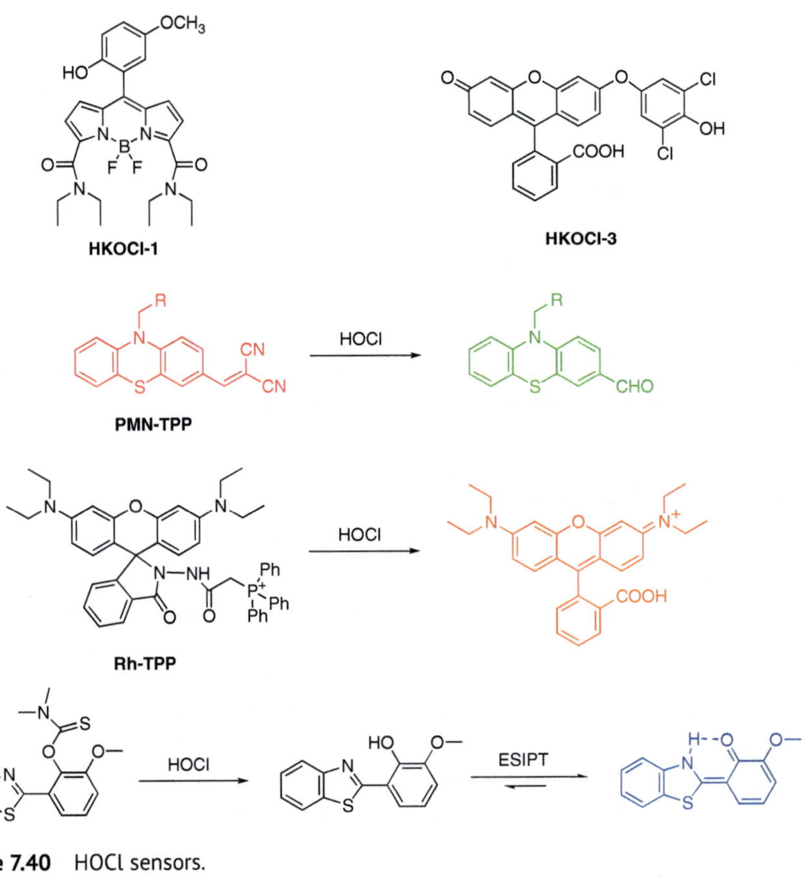

Figure 7.40 HOCl sensors.

7.6.5 Hydroxyl Radical

With a super high reduction potential of +2.3 V, hydroxyl radical is the most dangerous one among all ROS. While the real-time monitoring of the hydroxyl radical in biological system would be important, the short lifetime and low steady-state concentration in biological system, mainly due to the high reactivity, makes it challenging to develop a useful sensor.

Tetramethylpiperidine-1-yl-oxyl (TEMPO) group is a relatively stable radical and has been widely used as a fluorescence quencher through its paramagnetic property. Hydroxyl radical directly or indirectly reacts with TEMPO and can remove the radical character and the fluorescence quenching effect. By combining TEMPO with a BODIPY scaffold, TEMPO-BDP was developed as a hydroxyl radical sensor [114]. In the application of TEMPO-BDP, DMSO was used as the first hydroxyl radical trap, generating a methyl radical. Subsequently, the methyl radical reacted with TEMPO forming a N-methoxy group and removed the quenching effect of TEMPO. Although it was used for intracellular hydroxyl radical measurement, its limitations included the poor selectivity over superoxide and the requirement for high concentration of DMSO.

O-dearylation is a useful strategy for highly reactive oxygen species (hROS), and the selectivity among hROS is the main issue. By a careful optimization with the introduction of flanking iodine, HKOH-1 was developed for selective hydroxyl radical sensor, which suppresses the reaction with a common competitor, HOCL or $ONOO^-$. HKOH-1 was successfully used for monitoring endogenous hydroxyl radical production in stimulated macrophages and UV-irradiated HeLa cells [115] (Figure 7.41).

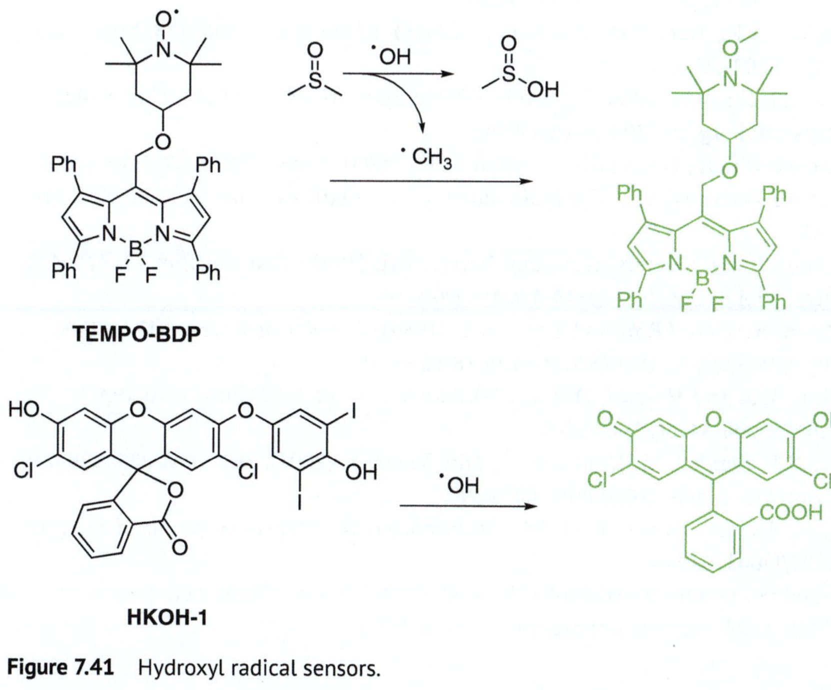

Figure 7.41 Hydroxyl radical sensors.

References

1 Grynkiewicz, G., Poenie, M., and Tsien, R.Y. (1985). *J. Biol. Chem.* 260: 3440–3450. https://doi.org/10.1016/S0021-9258(19)83641-4.

2 Shen, J., Zeng, Y., Zhuang, X. et al. (2013). *Mol. Plant* 6: 1419–1437. https://doi.org/10.1093/mp/sst079.

3 Santo-Domingo, J. and Demaurex, N. (2011). *J. Gen. Physiol.* 139: 415–423. https://doi.org/10.1085/jgp.201110767.

4 Boedtkjer, E. and Pedersen, S.F. (2020). *Annu. Rev. Physiol.* 82: 103–126. https://doi.org/10.1146/annurev-physiol-021119-034627.

5 Whitaker, J.E., Haugland, R.P., and Prendergast, F.G. (1991). *Anal. Biochem.* 194: 330–344. https://doi.org/10.1016/0003-2697(91)90237-n.

6 Wiederschain, G.Ya. (2010). *The Molecular Probes Handbook*, 11e. Life Technologies.

7 Grossi, M., Morgunova, M., Cheung, S. et al. (2016). *Nat. Commun.* 7: 10855. https://doi.org/10.1038/ncomms10855.

8 Li, G., Zhu, D., Xue, L., and Jiang, H. (2013). *Org. Lett.* 15: 5020–5023. https://doi.org/10.1021/ol4023547.

9 Wan, Q., Chen, S., Shi, W. et al. (2014). *Angew. Chem. Int. Ed.* 53: 10916–10920. https://doi.org/10.1002/anie.201405742.

10 Zhu, H., Fan, J., Xu, Q. et al. (2012). *Chem. Commun.* 48: 11766–11768. https://doi.org/10.1039/C2CC36785H.

11 Lee, M.H., Park, N., Yi, C. et al. (2014). *J. Am. Chem. Soc.* 136: 14136–14142. https://doi.org/10.1021/ja506301n.

12 Cao, L., Zhao, Z., Zhang, T. et al. (2015). *Chem. Commun.* 51: 17324–17327. https://doi.org/10.1039/C5CC07118F.

13 Sarkar, A.R., Heo, C.H., Xu, L. et al. (2016). *Chem. Sci.* 7: 766–773. https://doi.org/10.1039/C5SC03708E.

14 Bischof, H., Burgstaller, S., Waldeck-Weiermair, M. et al. (2019). *Cells* 8: 492. https://doi.org/10.3390/cells8050492.

15 Barreto-Chang, O.L. and Dolmetsch, R.E. (2009). *J. Vis. Exp.* 1067. https://doi.org/10.3791/1067 and *The Molecular Probes Handbook*, Life Technologies, 11e, 2010.

16 Ikenouchi, H., Peeters, G.A., and Barry, W.H. (1991). *Cell Calcium* 12: 415–422. https://doi.org/10.1016/0143-4160(91)90067-O.

17 Minta, A., Kao, J.P.Y., and Tsien, R.Y. (1989). *J. Biol. Chem.* 264: 8171–8178. https://doi.org/10.1016/S0021-9258(18)83165-9.

18 Chiu, V.C. and Haynes, D.H. (1977). *Biophys. J.* 18: 3–22. https://doi.org/10.1016/S0006-3495(77)85592-6.

19 Du, S.J., Frenkel, V., Kindschi, G., and Zohar, Y. (2001). *Dev. Biol.* 238: 239–246. https://doi.org/10.1006/dbio.2001.0390.

20 Tsao, Y.S. and Huang, L. (1985). *Biochemistry* 24: 1092–1098. https://doi.org/10.1021/bi00326a004.

21 Izumi, S., Urano, Y., Hanaoka, K. et al. (2009). *J. Am. Chem. Soc.* 131: 10189–10200. https://doi.org/10.1021/ja902511p.

22 Csernoch, L., Bernengo, J.C., Szentesi, P., and Jacquemond, V. (1998). *Biophys. J.* 75: 957–967. https://doi.org/10.1016/s0006-3495(98)77584-8.

23 Tosato, M. and Di Marco, V. (2019). *Biomolecules* 9: 269. https://doi.org/10.3390/biom9070269.

24 Xu, W., Ren, C., Teoh, C.L. et al. (2014). *Anal. Chem.* 86: 8763–8769. https://doi.org/10.1021/ac501953z.

25 Reardon, T.F. and Allen, D.G. (2009). *J. Physiol.* 587: 4705–4716. https://doi.org/10.1113/jphysiol.2009.173005.

26 Tayaramma, T., Ma, B., Rohde, M., and Mayer, H. (2006). *Stem Cells* 24: 2858–2867. https://doi.org/10.1634/stemcells.2006-0109.

27 Datki, Z.L., Hunya, Á., and Penke, B. (2007). *Brain Res. Bull.* 74: 183–187. https://doi.org/10.1016/j.brainresbull.2007.06.001.

28 Krężel, A. and Maret, W. (2006). *J. Biol. Inorg. Chem.* 11: 1049–1062. https://doi.org/10.1007/s00775-006-0150-5.

29 Frederickson, C.J., Kasarskis, E.J., Ringo, D., and Frederickson, R.E. (1987). *J. Neurosci. Methods* 20: 91–103. https://doi.org/10.1016/0165-0270(87)90042-2.

30 Prat, M.D., Guiteras, J., Compañó, R., and Beltrán, J.L. (1991). *J. Fluoresc.* 1: 267–272. https://doi.org/10.1007/BF00865250.

31 Varea, E., Ponsoda, X., Molowny, A. et al. (2001). *J. Neurosci. Methods* 110: 57–63. https://doi.org/10.1016/S0165-0270(01)00417-4.

32 Gee, K.R., Zhou, Z.-L., Qian, W.-J., and Kennedy, R. (2002). *J. Am. Chem. Soc.* 124: 776–778. https://doi.org/10.1021/ja011774y.

33 Sensi, S.L., Ton-That, D., Weiss, J.H. et al. (2003). *Cell Calcium* 34: 281–284. https://doi.org/10.1016/S0143-4160(03)00122-2.

34 Suh, S.W. (2009). *J. Neurosci. Methods* 177: 1–13. https://doi.org/10.1016/j.jneumeth.2008.09.016.

35 Lukowiak, B., Vandewalle, B., Riachy, R. et al. (2001). *J. Histochem. Cytochem.* 49: 519–528. https://doi.org/10.1177/002215540104900412.

36 Walkup, G.K., Burdette, S.C., Lippard, S.J., and Tsien, R.Y. (2000). *J. Am. Chem. Soc.* 122: 5644–5645. https://doi.org/10.1021/ja000868p.

37 Hirano, T., Kikuchi, K., Urano, Y. et al. (2000). *J. Am. Chem. Soc.* 122: 12399–12400. https://doi.org/10.1021/ja002467f.

38 Tougu, V., Karafin, A., and Palumaa, P. (2008). *J. Neurochem.* 104: 1249–1259. https://doi.org/10.1111/j.1471-4159.2007.05061.x.

39 Yang, L., McRae, R., Henary, M.M. et al. (2005). *Proc. Natl. Acad. Sci. U.S.A.* 102: 11179–11184. https://doi.org/10.1073/pnas.0406547102.

40 Zeng, L., Miller, E.W., Pralle, A. et al. (2006). *J. Am. Chem. Soc.* 128: 10–11. https://doi.org/10.1021/ja055064u.

41 Wang, L., Yuan, L., Zeng, X. et al. (2016). *Angew. Chem., Int. Ed.* 55: 1773–1776. https://doi.org/10.1002/anie.201510003.

42 Zamaraeva, M.V., Sabirov, R.Z., Maeno, E. et al. (2005). *Cell Death Differ.* 12: 1390–1397. https://doi.org/10.1038/sj.cdd.4401661.

43 Fiorillo, M., Scatena, C., Naccarato, A.G. et al. (2021). *Cell Death Differ.* 28: 2797–2817. https://doi.org/10.1038/s41418-021-00788-x.

44 Wang, L., Zhang, J., Kim, B. et al. (2016). *J. Am. Chem. Soc.* 138: 10394–10397. https://doi.org/10.1021/jacs.6b05810.

45 Nieto-Alamilla, G., Márquez-Gómez, R., García-Gálvez, A.-M. et al. (2016). *Mol. Pharmacol.* 90: 649–673. https://doi.org/10.1124/mol.116.104752.

46 Kielland, N., Vendrell, M., Lavilla, R., and Chang, Y.-T. (2012). *Chem. Commun.* 48: 7401–7403. https://doi.org/10.1039/C2CC32292G.

47 Gao, M., Lee, S.H., Park, S.H. et al. (2021). *Angew. Chem. Int. Ed.* 60: 23743–23749. https://doi.org/10.1002/anie.202108536.

48 Ghosh, S., Chattoraj, S., Mondal, T., and Bhattacharyya, K. (2013). *Langmuir* 29: 7975–7982. https://doi.org/10.1021/la400840n.

49 Jüngst, C., Klein, M., and Zumbusch, A. (2013). *J. Lipid Res.* 54: 3419–3429. https://doi.org/10.1194/jlr.M042515.

50 Wang, L., Xiao, Y., Tian, W., and Deng, L. (2013). *J. Am. Chem. Soc.* 135: 2903–2906. https://doi.org/10.1021/ja311688g.

51 Sun, W., Cui, J.-X., Ma, L.-L. et al. (2018). *Analyst* 143: 5799–5804. https://doi.org/10.1039/C8AN01643G.

52 Yang, Z., He, Y., Lee, J.-H. et al. (2013). *J. Am. Chem. Soc.* 135: 9181–9185. https://doi.org/10.1021/ja403851p.

53 Yang, Z., He, Y., Lee, J.H. et al. (2014). *Chem. Commun.* 50: 11672–11675. https://doi.org/10.1039/C4CC04915B.

54 Kubánková, M., Summers, P.A., López-Duarte, I. et al. (2019). *ACS Appl. Mater. Interfaces* 11: 36307–36315. https://doi.org/10.1021/acsami.9b10426.

55 Haidekker, M.A., Ling, T., Anglo, M. et al. (2001). *Chem. Biol.* 8: 123–131. https://doi.org/10.1016/s1074-5521(00)90061-9.

56 Kuimova, M.K., Yahioglu, G., Levitt, J.A., and Suhling, K. (2008). *J. Am. Chem. Soc.* 130: 6672–6673. https://doi.org/10.1021/ja800570d.

57 Förster, Th. and Hoffmann, G. (1971). *Phys. Chem.* 75: 63–76. https://doi.org/10.1524/zpch.1971.75.1_2.063.

58 Su, D., Teoh, C.L., Gao, N. et al. (2016). *Sensors* 16: 1397. https://doi.org/10.3390/s16091397.

59 Liu, X., Chi, W., Qiao, Q. et al. (2020). *ACS Sensors* 5: 731–739. https://doi.org/10.1021/acssensors.9b01951.

60 Liu, X., Chi, W., Gómez-Infante, A.deJ. et al. (2021). *Bull. Korean Chem. Soc.* 42: 91–94.

61 Osella, S. and Knippenberg, S. (2019). *ACS Appl. Bio Mater.* 2: 5769–5778. https://doi.org/10.1021/acsabm.9b00789.

62 Brignac-Huber, L.M., Reed, J.R., Eyer, M.K., and Backes, W.L. (2013). *Drug Metab. Dispos.* 41: 1896–1905. https://doi.org/10.1124/dmd.113.053611.

63 Arai, S., Lee, S.-C., Zhai, D. et al. (2014). *Sci. Rep.* 4: 6701. https://doi.org/10.1038/srep06701.

64 Marino, A., Arai, S., Hou, Y. et al. (2015). *ACS Nano* 9: 7678–7689. https://doi.org/10.1021/acsnano.5b03162.

65 Arai, S., Lee, C.-L.K., Chang, Y.-T. et al. (2015). *RSC Adv.* 5: 93530–93538. https://doi.org/10.1039/C5RA19729E.

66 Itoh, H., Arai, S., Sudhaharan, T. et al. (2016). *Chem. Commun.* 52: 4458–4461. https://doi.org/10.1039/C5CC09943A.

67 Marino, A., Arai, S., Hou, Y. et al. (2017). *ACS Nano* 11: 2494–2508. https://doi.org/10.1021/acsnano.6b08202.

68 Kriszt, R., Arai, S., Itoh, H. et al. (2017). *Sci. Rep.* 7: 1383. https://doi.org/10.1038/s41598-017-00291-9.

69 Wang, C.H., Lundh, M., Fu, A. et al. (2020). *Sci. Transl. Med.* 12: eaaz8664. https://doi.org/10.1126/scitranslmed.aaz8664.

70 Arai, S., Suzuki, M., Park, S.-J. et al. (2015). *Chem. Commun.* 51: 8044–8047. https://doi.org/10.1039/C5CC01088H.

71 Chrétien, D., Bénit, P., Ha, H.-H. et al. (2018). *PLoS Biol.* 16: e2003992. https://doi.org/10.1371/journal.pbio.2003992.

72 Wang, D., Huang, H., Zhou, M. et al. (2019). *Chem. Commun.* 55: 4051–4054. https://doi.org/10.1039/C9CC00603F.

73 Chrétien, D., Bénit, P., Leroy, C. et al. (2020). *Chemosensors* 8: 124. https://doi.org/10.3390/chemosensors8040124.

74 Winterbourn, C. (2008). *Nat. Chem. Biol.* 4: 278–286. https://doi.org/10.1038/nchembio.85.

75 Harris, D.C. (2009). *Exploring Chemical Analysis*, 4e, 538.

76 Bai, X., Ng, K.K.-H., Hu, J.J. et al. (2019). *Annu. Rev. Biochem.* 88: 605–633. https://doi.org/10.1146/annurev-biochem-013118-111754.

77 Azzi, A., Davies, K.J.A., and Kelly, F. (2004). *FEBS Lett.* 8: 3–6. https://doi.org/10.1016/S0014-5793(03)01526-6.

78 Vendemiale, G., Grattagliano, I., and Altomare, E. (1999). *Int. J. Clin. Lab. Res.* 29: 49–55. https://doi.org/10.1007/s005990050063.

79 Beckman, K.B. and Ames, B.N. (1998). *Physiol. Rev.* 78: 547–581. https://doi.org/10.1152/physrev.1998.78.2.547.

80 Keston, A.S. and Brandt, R. (1965). *Anal. Biochem.* 11: 1–5. https://doi.org/10.1016/0003-2697(65)90034-5.

81 Crow, J.P. (1997). *Nitric Oxide* 1: 145–157. https://doi.org/10.1006/niox.1996.0113.

82 Gomes, A., Fernandes, E., and Lima, J.L.F.C. (2005). *J. Biochem. Biophys. Methods* 65: 45–80. https://doi.org/10.1016/j.jbbm.2005.10.003.

83 Murphy, M.P. (2009). *Biochem. J.* 417: 1–13. https://doi.org/10.1042/bj20081386.

84 Münzel, T., Afanas'ev, I.B., Kleschyov, A.L., and Harrison, D.G. (2002). *Arterioscler. Thromb. Vasc. Biol.* 22: 1761–1768. https://doi.org/10.1161/01.atv.0000034022.11764.ec.

85 Murrant, C.L. and Reid, M.B. (2001). *Microsc. Res. Tech.* 55: 236–248. https://doi.org/10.1002/jemt.1173.

86 Zhao, H., Kalivendi, S., Zhang, H. et al. (2003). *Free Radic. Biol. Med.* 34: 1359–1368. https://doi.org/10.1016/s0891-5849(03)00142-4.

87 Zhao, H., Joseph, J., Fales, H.M. et al. (2005). *Proc. Natl. Acad. Sci. U.S.A.* 102: 5727–5732. https://doi.org/10.1073/pnas.0501719102.

88 Shchepinova, M.M., Cairns, A.G., Prime, T.A. et al. (2017). *Cell Chem. Biol.* 24: 1285–1298. https://doi.org/10.1016/j.chembiol.2017.08.003.

89 Tang, B., Zhang, L., and Zhang, L-l. (2004). *Anal. Biochem.* 326: 176–182. https://doi.org/10.1016/j.ab.2003.11.023.

90 Gao, J.J., Xu, K.H., Tang, B. et al. (2007). *FEBS J.* 274: 1725–1733. https://doi.org/10.1111/j.1742-4658.2007.05720.x.

91 Xu, K., Liu, X., Tang, B. et al. (2007). *Chem. Eur. J.* 13: 1411–1416. https://doi.org/10.1002/chem.200600497.

92 Hu, J.J., Wong, N.-K., Ye, S. et al. (2015). *J. Am. Chem. Soc.* 137: 6837–6843. https://doi.org/10.1021/jacs.5b01881.

93 Wu, L., Liu, L., Han, H.-H. et al. (2019). *New J. Chem.* 43: 2875–2877. https://doi.org/10.1039/C8NJ05656K.

94 Burgoyne, J.R., Madhani, M., Cuello, F. et al. (2007). *Science* 317: 1393–1397. https://doi.org/10.1126/science.1144318.

95 Bull, S.D., Davidson, M.G., van den Elsen, J.M.H. et al. (2013). *Acc. Chem. Res.* 46: 312–326. https://doi.org/10.1021/ar300130w.

96 Lampard, E.V., Sedgwick, A.C., Sun, X. et al. (2018). *ChemistryOpen* 7: 262–265. https://doi.org/10.1002/open.201700189.

97 Chang, M.C.Y., Pralle, A., Isacoff, E.Y., and Chang, C.J. (2004). *J. Am. Chem. Soc.* 126: 15392–15393. https://doi.org/10.1021/ja0441716.

98 Miller, E.W., Tulyathan, O., Isacoff, E.Y., and Chang, C.J. (2007). *Nat. Chem. Biol.* 3: 263–267. https://doi.org/10.1038/nchembio871.

99 Abo, M., Urano, Y., Hanaoka, K. et al. (2011). *J. Am. Chem. Soc.* 133: 10629–10637. https://doi.org/10.1021/ja203521e.

100 Xie, X., Yang, X., Wu, T. et al. (2016). *Anal. Chem.* 88: 8019–8025. https://doi.org/10.1021/acs.analchem.6b01256.

101 Ye, S., Hu, J.J., and Yang, D. (2018). *Angew. Chem. Int. Ed.* 57: 10173–10177. https://doi.org/10.1002/anie.201805162.

102 Yang, D., Wang, H.-L., Sun, Z.-N. et al. (2006). *J. Am. Chem. Soc.* 128: 6004–6005. https://doi.org/10.1021/ja0603756.

103 Peng, T. and Yang, D. (2010). *Org. Lett.* 12: 4932–4935. https://doi.org/10.1021/ol102182j.

104 Peng, T., Wong, N.-K., Chen, X. et al. (2014). *J. Am. Chem. Soc.* 136: 11728–11734. https://doi.org/10.1021/ja504624q.

105 Peng, T., Chen, X., Gao, L. et al. (2016). *Chem. Sci.* 7: 5407–5413. https://doi.org/10.1039/C6SC00012F.

106 Yu, F., Li, P., Li, G. et al. (2011). *J. Am. Chem. Soc.* 133: 11030–11033. https://doi.org/10.1021/ja202582x.

107 Yu, F., Li, P., Wang, B., and Han, K. (2013). *J. Am. Chem. Soc.* 135: 7674–7680. https://doi.org/10.1021/ja401360a.

108 Sun, Z.-N., Liu, F.-Q., Chen, Y. et al. (2008). *Org. Lett.* 10: 2171–2174. https://doi.org/10.1021/ol800507m.

109 Hu, J.J., Wong, N.-K., Lu, M.-Y. et al. (2016). *Chem. Sci.* 7: 2094–2099. https://doi.org/10.1039/C5SC03855C.

110 Xiao, H., Li, J., Zhao, J. et al. (2015). *J. Mater. Chem. B* 3: 1633–1638. https://doi.org/10.1039/C4TB02003K.

111 Hou, J.-T., Wu, M.-Y., Li, K. et al. (2014). *Chem. Commun.* 50: 8640–8643. https://doi.org/10.1039/C4CC02673J.

112 Wu, L., Yang, Q., Liu, L. et al. (2018). *Chem. Commun.* 54: 8522–8525. https://doi.org/10.1039/C8CC03717E.

113 Shi, D., Chen, S., Dong, B. et al. (2019). *Chem. Sci.* 10: 3715–3722. https://doi.org/10.1039/C9SC00180H.

114 Li, P., Xie, T., Duan, X. et al. (2010). *Chem. Eur. J.* 16: 1834–1840. https://doi.org/10.1002/chem.200901514.

115 Bai, X., Huang, Y., Lu, M., and Yang, D. (2017). *Angew. Chem. Int. Ed.* 56: 12873–12877. https://doi.org/10.1002/anie.201705873.

8

Imaging for Disease

8.1 Introduction

Disease is an abnormal state of life due to an infection or an imbalance of the normal functions and regulations. A biomarker is a measurable indicator of a biological state or condition, and for proper disease detection or diagnosis, elucidation of relevant biomarkers is essential. In general, the ideal biomarkers are prominently different between disease and normal status, and the measurement of biomarkers is noninvasive and convenient. Depending on the sampling methods, the procedure is classified as ex vivo and in vivo approaches. In the ex vivo approach, the test sample is taken from the body and the measurement is performed outside of the body (ex vivo). Unless the sample is reintroduced into the body after labeling or treatment, usually ex vivo samples are used only for in vitro measurement. The used samples may be discarded or stored for a long-term study or as a reference. The ex vivo samples include urine, blood, tear, nasal mucus, sweat, hair, epithelial cells, and biopsy tissue.

The fluidic samples are mainly used for biochemical analysis, targeting metabolites, proteins, and nucleic acids such as DNA or RNA. Metabolites could be analyzed by gas chromatography (GC) or high-performance liquid chromatography (HPLC) coupled with mass spectrometry detector, or colorimetric chemical reactions. The quantitative measurement of proteins is usually performed by immunoassays using specific antibody. For high sensitivity, signal amplification by enzyme-linked immunosorbent assay (ELISA) is the representative method. For quick response, lateral flow test chips with nanoparticle-based sensing are the most common. If the protein biomarkers are functional enzymes, a color-generating biochemical reaction could be utilized. Nucleic acid biomarkers are the best among any biochemical tests for the high sensitivity due to their intrinsic capability of amplification through polymerase chain reaction (PCR), which is mimicking the biological proliferation in terms of replication of the genetic material. Recently, circulating exosomes have emerged as a new source of enriched biomarkers. The representative biomarkers for disease are listed in Table 8.1.

The biochemical analysis of biomarkers is useful, but the sample is consolidated into one batch, losing the localization information where it was sampled from. Therefore, bioimaging is an ideal approach to maintain the spatial information

Sensors and Probes for Bioimaging, First Edition. Young-Tae Chang and Nam-Young Kang.

Table 8.1 Representative biomarkers for disease.

Biomarker	Target disease
Glucose	Diabetes
Human chorionic gonadotropin (HCG)	Pregnancy
Prostate-specific antigen (PSA)	Pancreatic cancer
Carcino embryonic antigen (CEA)	Colorectal cancer
Alpha FetoProtein (AFP)	Prostate, liver cancer
Circulating tumor cell CTC)	Tumor metastasis

of the biomarkers in disease cell or tissue. The imaging data would be applicable both for diagnosis (detection of disease) and prognosis (response monitoring of treatment). Ultrasound, X-ray, and magnetic resonance imaging (MRI) provide structural information of the body and thus visualize the abnormal shape or size of the disease area. With the development of computed tomography (CT) techniques, three-dimensional (3D) imaging is also commonly available to increase the abnormal area detection. Still, for the molecular information of the disease area, sensors or probes are widely used for bioimaging. In most cases, matching biomarkers for each imaging probe provide a mechanistic understanding of the probe's mode of action (MOA) and a molecular-level diagnosis of the disease.

8.2 Cancer Imaging

Tumor tissue is composed of the fast-growing cancer cells and unusually big size of the tissue results in low oxygen and nutrients supply in the inner area of the tumor chunk. To overcome the supply problem, tumor tissue recruits new blood vessels (neoangiogenesis) and the new vessels characteristically have enhanced permeability. As a result, various materials exhibit a favored penetration and residing tendency in the tumor area. This effect is called enhanced permeability and retention (EPR) property of tumors. Depending on the looseness of the vessel, even high-molecular-weight polymer probes or nanoparticles including quantum dots can also leak out and image the tumor site in a passive manner [1].

The EPR property could also be applied to the drug delivery system for selective cancer treatment. Nanoparticles loaded with drug molecules could be delivered to the tumor site by EPR effect and by sensing the tumor environment, the drug molecules could be released only in the tumor site. The smart drug delivery system using nanoparticles combined with EPR effect have been widely studied [2]. While useful and convenient, the EPR effect of tumor tissue is often inhomogeneous in the same tumor and among different tumors, generating serious signal variations. Also, the physical property changes of tumor area are difficult to distinguish from swelling induced by inflammation. For specific imaging of cancer, a molecular biomarker and targeting probe may be necessary (Figure 8.1).

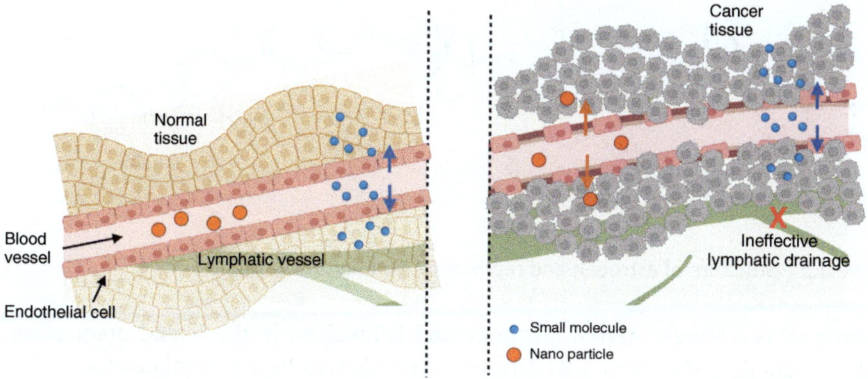

Figure 8.1 Enhanced permeability and retention (EPR).

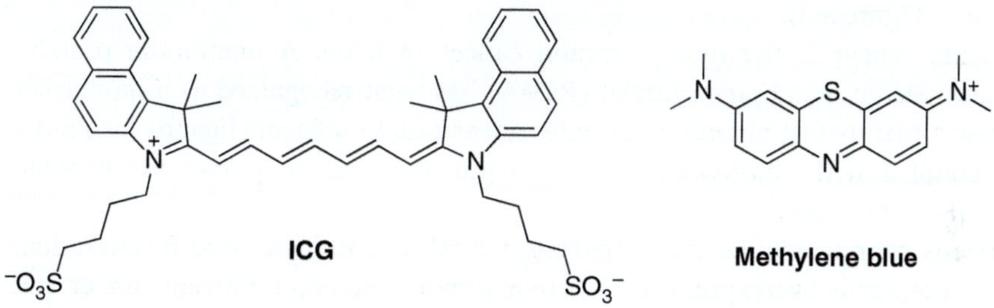

Figure 8.2 Structure of ICG and methylene blue.

IndoCyanine Green (ICG) is a near-infrared (NIR) range fluorescence dye with 800–850 nm emission on excitation at 700–800 nm and a representative tumor imaging agent working by EPR. ICG is the first US-FDA-approved fluorescent imaging probe for clinical usage and has been used for various applications including ocular blood flow, hepatic function, and sentinel lymph node imaging during tumor resection [3]. Intravenously injected ICG binds to serum albumin and accumulates in the tumor through the EPR effect, enabling tumor imaging [4]. Methylene blue is another FDA-approved long-wavelength fluorescent probe (ex/em = 665/685 nm) and has been used for pancreatic and breast cancer visualization [5] (Figure 8.2).

8.2.1 Imaging by Cancer-Specific Biomarker Binding

Tumor cells may overexpress specific proteins such as receptors or signaling molecules as the tumor biomarkers. If the proteins are expressed on the cell membrane, fluorescently labeled antibodies would be an option for tumor imaging through the biomarker. Instead of a fluorescent reporter, a drug molecule could be coupled to the antibody and forms a therapeutic agent. The approach is called antibody–drug conjugate (ADC) and is getting more and more popular [6]. As the same antibody is used for diagnosis and also for therapy with small-molecule modification, this could be an example of theranostics (therapy + diagnosis). In this book, the contents will be focused on small-molecule probes.

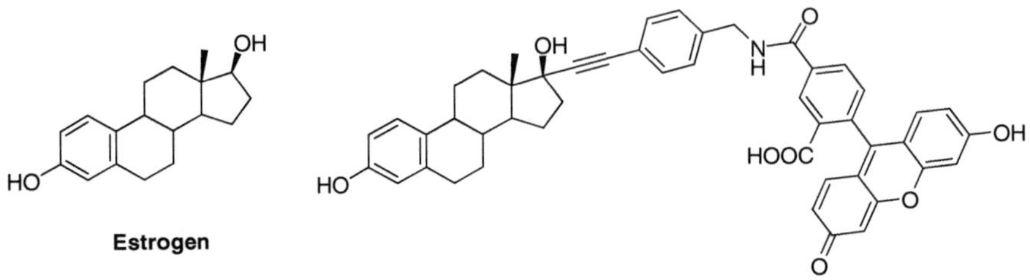

Figure 8.3 Structure of estrogen and representative fluorescent probe.

Estrogen is a female steroid hormone and is involved in the sexual maturation and reproduction of women. ER (Estrogen Receptor) is a ligand-regulated transcription factor and plays an important role in breast cancer growth. Estrogen derivatives coupled with fluorescence motifs [7] have been developed and used for breast cancer detection (Figure 8.3).

Prostate cancer is the most common cancer in male. A membrane protein, prostate-specific membrane antigen (PSMA) has been recognized as a biomarker and also a therapeutic target for prostate cancer. Small-molecule ligands for PSMA were coupled with fluorescent motifs to provide imaging probes for prostate cancer [8] (Figure 8.4).

Folate is vitamin B9 and it is important for cell maintenance and regeneration. Folate Receptor is overexpressed in various tumors, including ovarian cancer and myeloid leukemia. Naturally, folate-coupled fluorescent probes are developed and used for the tumor imaging [9] (Figure 8.5).

Cyclooxygenase-2 (COX-2) is the enzyme involved in inflammation progress by generating prostaglandins. With COX-1, COX-2 is one of the targets of aspirin.

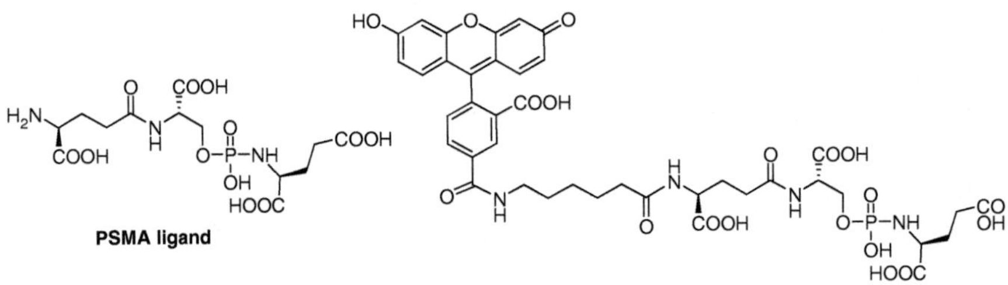

Figure 8.4 Structure of PSMA ligand and its fluorescent probe.

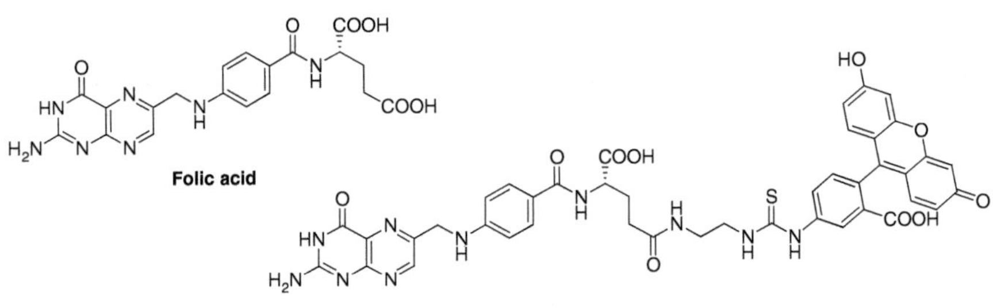

Figure 8.5 Structure of folate and representative fluorescent probe.

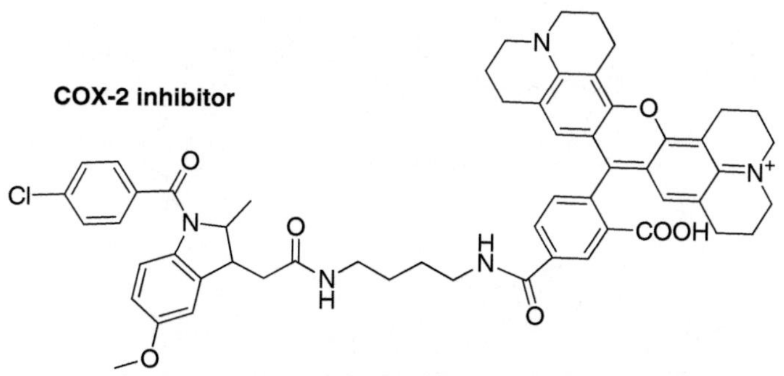

Figure 8.6 Structure of COX-2 inhibitor and representative fluorescent probe.

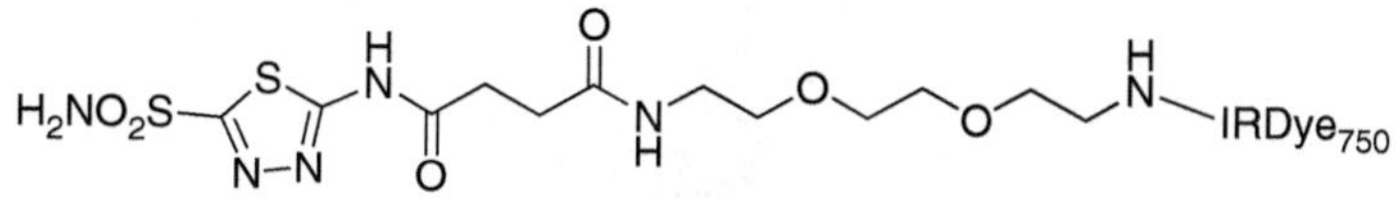

Figure 8.7 Structure of CAIX ligand and representative fluorescent probe.

By reducing the stomach ulcer probe (induced by COX-1), a COX-2-selective inhibitor attracted hot attention for heart attack prevention, with the nickname "super aspirin" in the corner of twenty-first century. With other unexpected side effects, the super aspirin heat has subsided. Instead, COX-2 showed up the potential as a tumor biomarker, for colorectal adenomas and adenocarcinomas. COX-2 inhibitors were coupled to fluorescent motifs and used for tumor imaging [10] (Figure 8.6).

Carbonic anhydrase is an enzyme that catalyzes the conversion of carbon dioxide into carbonic acid by reaction with water. One subtype, carbonic anhydrase IX (CAIX) was known to be associated with tumor progression and invasion, and overexpressed in solid tumors, such as colorectal tumors. Ligands for CAIX coupled with fluorescent motifs were developed into tumor imaging probes [11] (Figure 8.7).

Chlorotoxin is a peptide composed of 36 amino acids and is found in the venom of the deathstalker scorpion, which blocks small-conductance chloride channels [12]. Chlorotoxin is known to preferentially bind to various cancers via MMP2 and annexin A2. BLZ-100 is a chlorotoxin conjugated with ICG and has been tested in skin cancer patients [13].

Among histone deacetylase (HDAC) subtypes, HDAC6 is upregulated in hepatocellular carcinoma. FITC-SAHA was designed by linking an FDA-approved HDAC6 inhibitor, SAHA (suberoylanilide hydroxamic acid) to FITC. A NIR version of IRDye800CW-SAHA could be used for fluorescence-guided surgery in a hepatocellular carcinoma mouse model [14] (Figure 8.8).

8.2.2 Imaging by Cancer-Specific Metabolism

Tumors have fast growing and invading characteristics. When tumors invade neighboring tissues, they reorganize the extracellular matrix through various

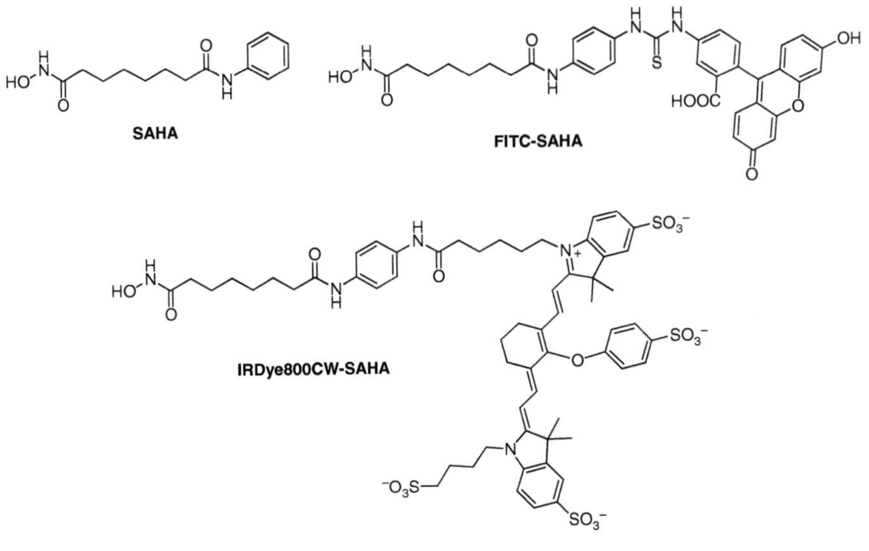

Figure 8.8 Structure of SAHA and representative fluorescent probe.

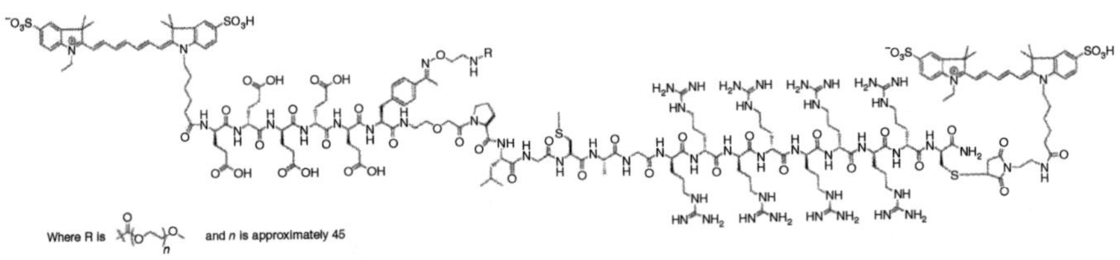

Figure 8.9 Structure of AVB-620. Source: [16].

enzymes [15]. Matrix metalloproteinases (MMP) is such a popular target for cancer imaging as a calcium-dependent zinc-containing peptidase. AVB-620 is designed with Cy5 and Cy7 as the fluorescence resonance energy transfer (FRET) pair on the substrate peptide. This probe has been used for breast cancer metastasis to lymph node through the ratiometric change of fluorescence [16] (Figure 8.9).

Cathepsin is another tumor-associated protease and is usually activated at a low pH such as in lysosomes. Cathepsin substrate peptide was used to design a specific probe. The probe LUM015 was built using FRET of Cy5 and QSY21 (quencher) pair on each side of the peptide, and when the peptide is cleaved by cathepsin, the fluorescence of Cy5 is restored from QSY21 quenching effect [17]. Similar concept was applied to cathepsin with NIR dyes to provide 6QC-NIR. The probes exhibited higher retention in tumor due to the accumulation of the fluorescent product in lysosome [18]. The same research group optimized the probe by screening the peptide sequence by library including unnatural amino acids. As a result, they developed probe 5 by the best sequence to replace the peptide of 6QC-NIR. Probe 5 was more reactive to cathepsin than 6QC-NIR and showed better tumor tissue contrast [19]. To further increase the selectivity, two independent peptides were introduced into one probe as an AND logic gate. Only when both of the peptides

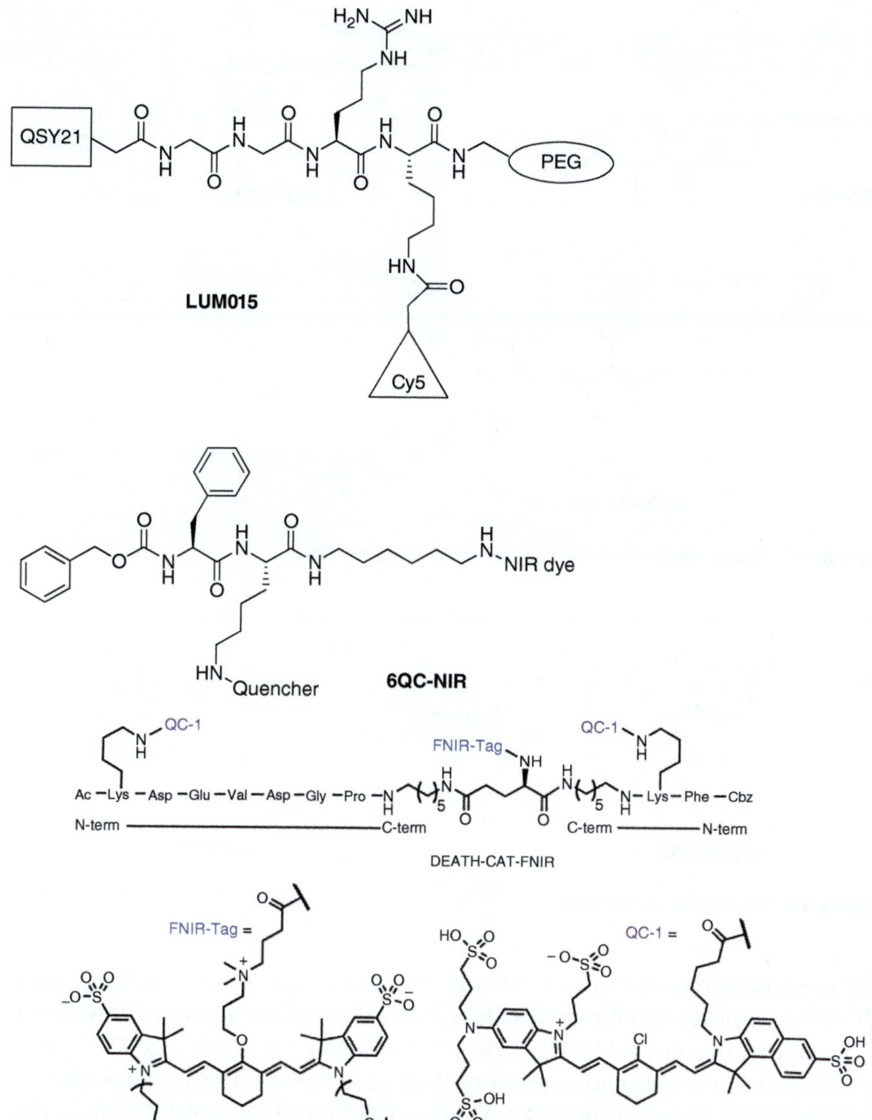

Figure 8.10 Structure of LUM015, 6QC-NIR, and DEATH-CAT-FNIR.

are cleaved, the probe (DEATH-CAT-FNIR) would show off the fluorescence [20] (Figure 8.10).

For further development of this approach, the key factor is the elucidation of the unique enzymes as the novel biomarker for tumor. Aminopeptidase N (APN) is elucidated as such a target and YH-APN is designed with dicyanoisophorone as the fluorophore and alanine as the APN substrate [21]. YH-APN visualized liver cancer xenograft and microcarcinoma in a mouse model. Similar to APN,

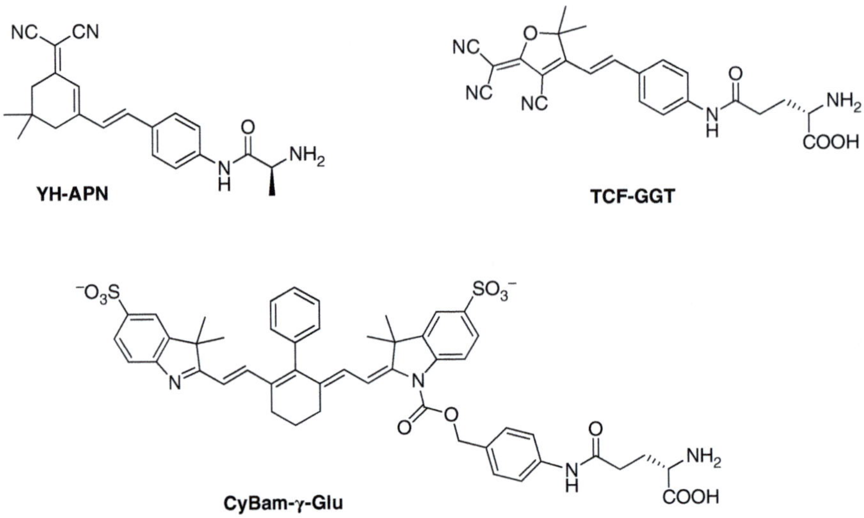

Figure 8.11 Structure of YH-APN, TCF-GGT, and CyBam-γ-Glu.

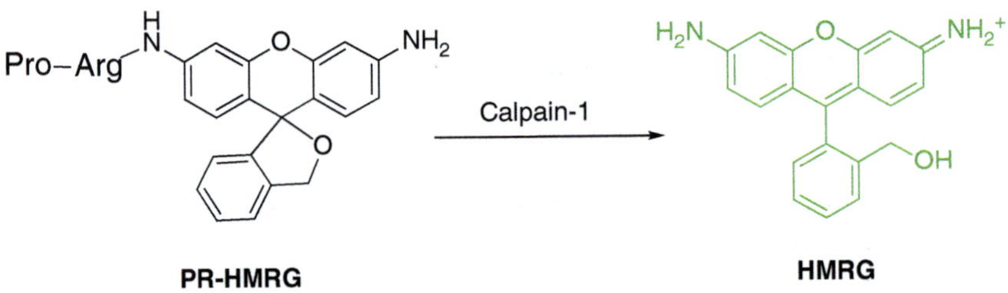

Figure 8.12 Structure of PR-HMRG.

GGT (γ-glutamyltraspeptidase) is another emerging tumor-associated peptidase. TCF-GGT was developed for GGT with TCF as a fluorophore and γ-glutamyl group as the substrate. Catalyzed by GGT, the probe releases the fluorescent TCF (em = 575–700 nm) and the excitation could be done either by one photon (ex = 559 nm) or two photons (ex = 800 nm) [22]. Another GGT-targeting probe CyBam-γ-Glu was independently developed using norcyanine and γ-glutamate. After metabolized by GGT, the probe fluoresced in lysosomes and visualized ovarian cancer in the metastatic tumor mouse model [23] (Figure 8.11).

Calpain-1 was selected as a novel tumor biomarker, and a specific probe was developed. Through a fluorescence-labeled dipeptide library screening, PR-HMRG was selected as the best-performing probe and visualized a brain tumor in an animal model [24] (Figure 8.12).

β-Galactosidase (β-Gal) is usually used as senescence biomarker of cells. But the enzyme was also found to be highly expressed in gastric cancer, and SPiDER-β-Gal was developed for the probe. Upon catalysis of β-galactosidase, the probe generates the fluorescent product with intracellular retention [25] (Figure 8.13).

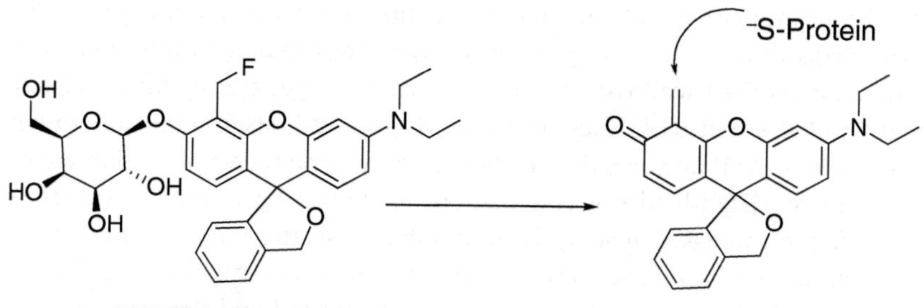

SPiDER-β-Gal

Figure 8.13 Structure of SPiDER-β-Gal.

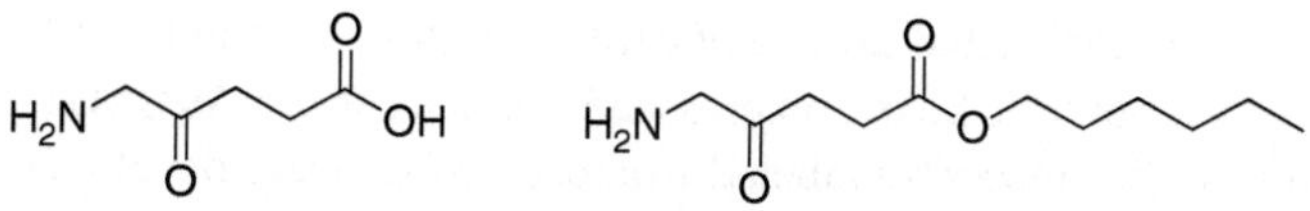

5-ALA **H-ALA**

Figure 8.14 Structure of 5-ALA and H-ALA.

5-Aminolevulinic acid (5-ALA) is an endogenous non-proteinogenic amino acid and is the key intermediate of porphyrin biosynthesis. Porphyrin is the main component of heme in animal [26] and chlorophyll in plant [27]. Chlorophyll absorbs light and converts the light energy into biomaterials through photosynthesis, and porphyrin is the light antenna. While part of the excited electrons in porphyrin contribute to fluorescence emission, others undergo to triplet state via the intersystem crossing pathway. The long-lived triplet state electrons can generate singlet oxygen or other reactive oxygen species (ROS), so porphyrin can be used as a photodynamic therapy (PDT) agent. So, porphyrin is a dual-function material of imaging and therapy, that is theranostic material. Therefore, 5-ALA has been used as a metabolic precursor for porphyrin and imaging probe for various cancers including urological [28] and gastrointestinal cancers [29]. While porphyrin is difficult to penetrate blood–brain barrier (BBB), 5-ALA can, so the unique application of 5-ALA is in brain tumor imaging and PDT [30]. H-ALA (hexyl ester of 5-ALA) is a derivative of 5-ALA and is an FDA-approved agent along with 5-ALA itself. H-ALA has been used for bladder cancer diagnosis [31] and for increasing the detection accuracy of PSMA-based prostate cancer detection [32] (Figure 8.14).

8.2.3 Imaging by Cancer-Specific Transporter

Tumor has altered metabolic pathways from normal tissue to sustain their selfish fast growth, and the phenomena is explained by Warburg effect. As a result, cancer cells prefer different nutrients compared to normal cells. For the inclined taste, cancer cells overexpress special transporters in higher level than the normal cells. If a probe can hijack the tumor-specific transporters, it would be a good choice as a tumor-specific probe.

One of the prominent metabolic changes in tumor cells is aerobic glycolysis. Glycolysis breaks down glucose (C6 sugar) to pyruvate/lactate (C3 intermediate), and pyruvate enters the tricarboxylic acid (TCA) cycle for generating high ATPs by full oxidation of the fuel molecules. In contrast, glycolysis itself generates much smaller amount of ATP and usually considered as an emergency situation event under low oxygen. High-load exercise without enough oxygen supply is such a situation, and the increased glycolysis accumulates lactate, which is known to induce muscle pain. Considering the low-oxygen environment, this is called as anaerobic glycolysis. In tumors, the glycolysis is accelerated and the accumulated pyruvate and lactate are used for anabolic metabolism, rather than energy generation through TCA cycle. Therefore, tumors in general overexpress carbohydrate transporters to meet their needs for high fuel materials. Glucose transporters (Gluts) are responsible for the facilitated uptake of carbohydrates. Among their 13 subtypes, at least, GLUT5 is known to be overexpressed in breast cancer. As GLUT5 has high affinity to fructose, fluorescently labeled fructose probes were developed and used for cancer imaging.

Thyroid is a gland organ which secretes thyroid hormones. The representative thyroid hormones (T3 and T4) contain iodine in the molecule, and thyroid tumor tends to take up abnormally high iodide. In a similar way, the aberrant metabolism of tumors may take up specific amino acids and other metabolites. These metabolites are relatively small and modification with larger fluorescent motif may alter their cellular uptake through designated transporters. As the result, radioactive iodide or amino-acid-based contrast agents for single-photon emission computed tomography (SPECT) or positron emission tomography (PET) are commonly used, but the fluorescently labeled probes are relatively rare.

8.2.4 Imaging by the Changed Environment of Cancer

Due to the aberrant glycolysis, most tumor cells accumulate high lactic acid in the cells. To maintain the intracellular pH slightly alkaline (~pH 7.4), several proton extrusion mechanisms are activated and induces extracellular acidosis in tumor microenvironment [33]. For example, human MCF-7 mammary carcinoma exhibited the extracellular pH as 6.44 [34]. Accordingly, pH sensors which are turned on in acidic condition have been used for tumor area imaging and there is an excellent review published in 2021 [35]. As a recent example, germanium-rhodamine pH sensor was developed through fine-tuning of the electrophilicity of the fluorescent molecule [36]. The spirocyclized form of the probe is nonfluorescent induced by nucleophilic addition of hydroxyl group. Under acidic condition, the spiro structure is opened, turning on the fluorescence with long wavelength (λ_{ex} 679 nm, λ_{em} 693 nm) with a pK_a of 6.24. With modification of HER2 antibody (Herceptin), the pH probe was internalized by endocytosis in HER2 overexpressed cancer cells and visualized the tumor site (Figure 8.15).

Not only extracellular matrix, the lysosomes of cancer cells have lower pH and higher viscosity than those of normal cells. Using the environmental changes, a dual probe of pH and viscosity could be used for cancer cell imaging [37]. While dual

Figure 8.15 Structure of germanium-rhodamine pH sensor.

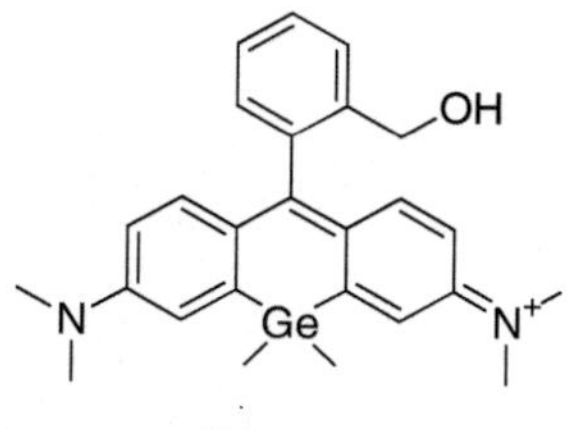

probe provides higher selectivity as the AND gate only works with two conditions exist together, sometimes selective single parameter-dependent sensor is required. An NIR probe was developed that is only sensitive to viscosity not to other factors such as pH, polarity, or ROS [38]. The probe demonstrated viscosity-dependent fluorescence increase in cells and worms, and also in an animal tumor model (Figure 8.16).

8.2.5 Circulating Tumor Cell (CTC)

Cancer cells can migrate from the original primary tumor site to remote locations of the body and form a secondary tumor. This process is called metastasis. During the metastasis, the cancer cells need to leave the original site, and penetrate into the blood vessel (intravasation) to travel to a remote site. When they reach the target area, they need to leak out through blood vessel wall to the surrounding tissue (extravasation) and form a new tumor colony. Therefore, if tumor cells are detected in the blood stream, it would be a biomarker for metastasis. Such a circulating tumor cell in the blood stream is called a circulating tumor cell (CTC) and has been used as a biomarker of tumors. Currently, CTC is isolated by its larger size compared to blood cells (that is, filtration through small holes) or antibody-based capturing method. However, the small number of CTC compared to overwhelmingly high number of blood cells make it difficult to isolate and characterize the property of the cells. Specific probes for CTC and more advanced isolation/analysis techniques are highly required.

8.2.6 Cancer Cell Line for Imaging

Cancer cells keep dividing and growing in principle forever. Compared to primary cells, which are harvested from tissues and have limited division capacity, the immortal property of cancer cells bestows researchers a lot of flexibility. In 1961, Leonard Hayflick reported that human skin cells can divide in laboratory conditions approximately 50 times before they stop dividing and enter senescence or apoptosis [39]. This division number of 50 has been considered the natural biological limit for cell growth and known as "Hayflick limit" following the discoverer's name.

Cell line is a population of cells that live for a long period, stably maintaining their characteristics and functions. Due to their immortal characteristics, majority of the established cell lines are cancer cell lines or carrying cancer properties to some extents. Cancer cell lines are widely used for cancer biology and treatment

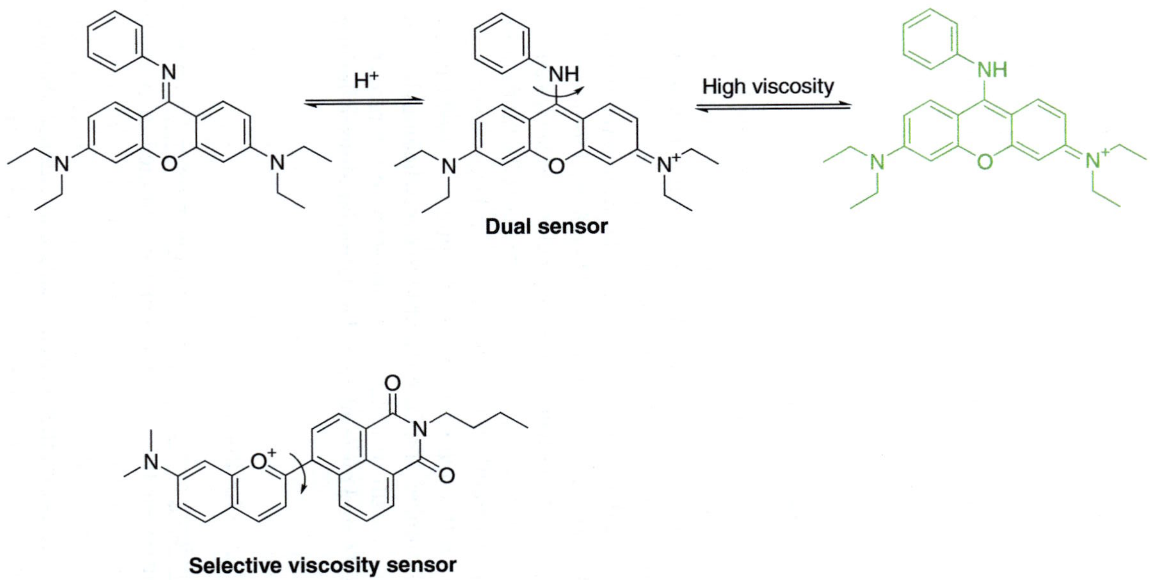

Figure 8.16 Dual sensor and viscosity sensor.

as surrogates for primary cancer cells. As different cells may have different features, for sharing the experimental results with other researchers, usage of common cell lines would be convenient. As a result, there are several dominatingly popular cancer cell lines widely used in cancer biology and imaging.

HeLa was the first successfully cultured cancer cell line, which was derived from the cervical cancer of Henrietta Lacks in 1951 [40]. At the moment, there was vigorous competition for cancer researchers to develop the first human cancer cell line. There was mouse cancer cell line established, but by then there was no available human cancer in a test tube, which can be cultured for long [41]. After the failure of the culture with chicken and calf embryonic plasma, the culture was finally successful with human placental blood-containing media [42]. Although the patient could not survive for long after her first diagnosis, the cultured cells have survived with the name of HeLa following the patient's name.

Since then numerous copies of HeLa cells were generated from laboratories all over the world until now. It is estimated that the research papers including HeLa cells are more than 60 000 by 2009 and the weight of HeLa cells generated could be more than 50 million tons and still growing [43]. HeLa cells have been used in various research fields including polio vaccine development, telomere discovery, genome project, radiation resistance, and so on. Through the study using HeLa cells, human papilloma virus (HPV) was uncovered as one of the causes of cervical cancer, the origin of HeLa Cells. HeLa cells were even sent to space to study human cell biology under zero gravity and used in experiments with nuclear bomb explosions, contributing to almost every aspect of human cell biology and cancer [44].

One of the mechanisms to make HeLa an into immortalized cell is activated telomerase, which keeps the telomer length from shortening during the cell division. Telomere is a repetitive DNA sequences localized at the end of chromosome and known to be shorten in the DNA replication during the cell division. The telomere shortening is considered as the molecular mechanism of Hayflick limit. As a stable human cell line, HeLa cells also contributed to develop nuclear staining probe in live cells [45]. While the genome of HeLa cells is remarkably stable even after many years of culturing [46], most of HeLa cells have 70–90 chromosomes and more than 20 translocations [47]. Compared to normal human chromosome of 46, the heterogenous and translocated HeLa cell chromosomes are highly complex. This kind of aberrant DNA structure is commonly observed in immortalized cell lines.

Since HeLa cells, more than hundreds of human cell lines are developed both for cancer and nonmalignant cells, but the standardization and identification of each cell line has been challenging due to the change of cell character and also contamination by other cell types [48]. Currently, American Type Culture Collection (ATCC) provides the global standard and authentic samples for the cell lines. The representative human and animal cell lines are summarized in Table 8.2.

8.2.7 Animal Models of Tumor Imaging

Tumor cells are usually cultured on 2D plastic surface with an optimized culture media. While the majority of high-throughput screening (HTS) for drug efficacy

Table 8.2 Representative human and animal cell lines.

Cell line	Organism	Tissue origin	Cell type
3T3-L1	Mouse	Embryo	Fibroblast
A549	Human	Lung	Lung carcinoma
C2C12	Mouse	Muscle	Myoblast
Caco-2	Human	Colon	Colorectal carcinoma
CHO	Chinese hamster	Ovary	Epithelium
COS-7	Monkey	Kidney	Fibroblast
HEK 293	Human	Embryonic kidney	Epithelium
HeLa	Human	Cervix epithelium	Cervical carcinoma
Hep G2	Human	Liver	Hepatoblastoma
HL-60	Human	Blood	Myeloblast
Jurkat	Human	White blood cells	T-cell leukemia
MCF-7	Human	Breast	Breast carcinoma
NIH-3T3	Mouse	Embryo	Fibroblast
NK-92	Human	Lymphoma	Natural killer cell
PC12	Rat	Adrenal medulla	Pheochromocytoma
PC-3	Human	Bone metastasis	Prostate carcinoma
SH-SY5Y	Human	Bone marrow	Neuroblastoma
THP-1	Human	Monocyte	Monocytic leukemia
U2OS	Human	Osteosarcoma	Epithelial
U87	Human	Glioblastoma	Epithelial-like
U937	Human	Leukemia	Monocytic leukemia

Source: Adapted from https://en.wikipedia.org/wiki/Cell_culture.

or imaging has been performed on the 2D cell culture, the cell cultures in the nonphysiological condition may not represent the cells residing in the complex microenvironment of a tissue. In vivo animal models would represent closer to real human tumors with 3D structure, and various animal models have been developed for cancer biology and treatment.

The most common tumor animal model is a xenograft of human cancer cells into mouse. The Greek prefix "xeno-" means foreign and "graft" means implant, so xenograft transplantation is transplanting cells or tissues of one species into a different species. Human cancer cells are adapted to survive in human body, escaping the human immune system. If they are transplanted into a mouse, they can be easily wiped out by the mouse immune system. So, usually the xenograft is exercised in an immunodeficient mouse. Mice with severe combined immunodeficiency (SCID) are commonly used for this purpose. SCID mice have mutations for B and T lymphocyte developments, so even though they have lymphocyte progenitors, usually they do not survive to maturity. Without B cells, they cannot produce antibodies, and without T cells, they cannot attack foreign invaders through cellular adaptive immune response. Under the conditions, xenograft cells have better chance to survive, avoiding the transplant rejection. The popular animal model, nude mice, have mutations for thymus development and as the result they have impaired T-cell system. Still, the B-cell system is functional, and the immunodeficiency of nude mouse is partial. For more complete immunodeficiency both to T- and B-cell systems, NOD (Non-obese Diabetic) or NSG (NOD scid Gamma) mice model is more commonly used (https://www.jax.org).

Tumor microenvironment is composed of mutually interacting immune cells and tumor cells. In the xenograft model in immunodeficient mice, it is not possible to study the immune cells in the tumor site. As a solution, humanized mouse is a more advanced model, which is built on the SCID mice by introducing human immune cells. As all the SCID mice requires clean environment and food to avoid any infection, the cost of maintenance is usually high. Considering the high cost of the animal model and maintenance, alternative option is the transplantation of mouse cancer cells into normal mouse. If the donor and receiver mice are inbred mice, their genetic backgrounds can be assumed to be almost the same. In this context, this type of procedure is called syngeneic graft transplantation in contrast to xenograft transplantation. From the aspect that they are in the same species, but different individual, this can be also called as allogeneic transplantation. If the donating tissue is from the same receiving individual, it is called as autologous transplantation. While much cheaper, in this case, the relevance of mouse tumor biology to that of human tumor remains as a question.

The source of cancer cells could be an already-established cancer cell line or primary cancer cells from a patient. Patient-derived xenograft (PDX) may present a more realistic model, assuming the character changes accumulated in the cancer cell line. Also, PDX can be used for individualized study for patient-specific treatment. Therefore, in a sense, the PDX model is an avatar of the patient at least for the tumor aspect.

Depending on the site of cancer cell injection, there are subcutaneous (most common ectopic choice) and orthotopic models. In the subcutaneous model, the cancer cells are transplanted under the skin regardless of the origin of the cancer. As the transplantation method is simple and the tumor growth is easy to monitor, the subcutaneous model is a popular first choice. However, tumors with different tissue origin may not show off the original characteristics under the skin. For example, liver cancer under the skin may have a totally different environment from being surrounded by liver cells. Therefore, a more sophisticated model would be transplantation of liver cancer into the liver, the original site of the tumor. In the same token, lung cancer in the lungs, kidney cancer in the kidney, and so on. This orthotopic model often requires trained surgical skills and more complicated procedures. Also, if the site of transplantation is deep inside the body, the visual monitoring of the growth and optical imaging with fluorescent probes would be more challenging.

Another important transplantation model is the metastasis model. For spontaneous metastasis, after orthotopic transplantation, monitoring the distal susceptible organ for the secondary tumor generation is the common practice. For the successful

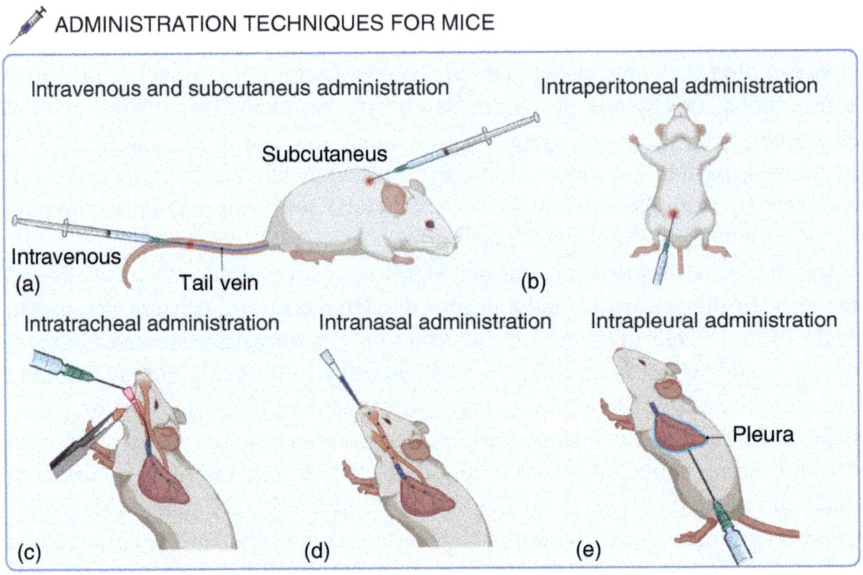

Figure 8.17 Animal model of tumor by transplantation.

metastasis, primary tumor formation, migration to secondary site, and secondary tumor formation should be completed. While mimicking the natural process, it would take a long time, and the occurrence of the events could be low. An alternative approach is experimental metastasis. The ectopic cancer cells are introduced into the bloodstream either through the vein or artery or into the abdominal cavity through intraperitoneal injection. The settlement and growth of tumor in the target organ would be monitored. In this case, at least for the primary tumor formation could be omitted and the migration of CTC is enhanced by the large number of cancer cell injection. As a result, the incidence rate and experimental time would be much shortened (Figure 8.17).

In addition to cancer transplantation, radiation or carcinogen chemicals also are used for the animal model preparation. This method may mimic the natural process of tumor generation, influenced by the environment. The resulting tumors would be syngeneic tumors. Usually, this procedure takes many months until the tumor is formed, and variations among individuals are large, which require statistical handling of the experimental design and results.

With the molecular level understanding of tumorigenesis, various oncogenes and tumor suppressor genes are discovered and studied. Therefore, by the genetic engineering, overexpression of oncogene and/or suppression of tumor suppressor genes could generate tumors in animals. If only one gene is mutated in an individual with all other genes are identical with its cohort, their relationships are congenic.

Mice are the most popular and common animal model for tumor research due to their small size, high reproduction rate, low cost of maintenance, availability of full genomic information, and so on. As a mammal, mice are genetically close to human and 99% of mouse genes have an equivalent in humans. Still, there are significant

Figure 8.18 Animal model for cancer study.

differences in tumor biology between mice and humans, bigger and closer species to humans are often required for better clinical relevance. Rats are bigger than mice, and closer to humans than mice. Mice and rats make up approximately 95% of all research animals. The next level would be guinea pigs, hamsters, rabbits, pigs, dogs, sheep, and monkeys. With the increased size of the higher-level animals, the difficulties of optical imaging also increase, similar to clinical situation. Therefore, smaller than mice and optically transparent animal models are also pursued for convenient application. Among nonmammalian vertebrates, zebrafish has emerged as an alternative cancer animal model, especially due to its optical transparency [49]. *Xenopus* (African frog) is another vertebrate model which is developed for cancer phenotype model [50]. With the convenient preparation from eggs, chicken embryos have also been developed and used for cancer models [51]. Even nonvertebrate organisms such as *Caenorhabditis elegans* [52], fruit flies (*Drosophila*) [53], and yeasts [54] have been used directly or indirectly for cancer studies combined with imaging (Figure 8.18).

8.2.8 Ex Vivo 3D Tumor Culture Model

Animal tumor models are important research tools, but with the ethical and animal welfare issues, surrogate models have been sought. Ex vivo 3D tumor culture is a prominently emerging platform as an intermediate model between in vitro 2D monolayer culture and in vivo solid tumor. There are several types of 3D culture of tumors.

Tumor sphere is a solid, spherical cell cluster developed in principle from a single cancer stem cell. As a clonal colony, the cells tightly fuse together forming a spherical structure in tumor sphere and are easily distinguishable from simply aggregated cells. For tumor sphere formation, the cells are cultures in serum-free, non-adherent condition in which only cancer stem cells can survive and proliferate [55]. The size

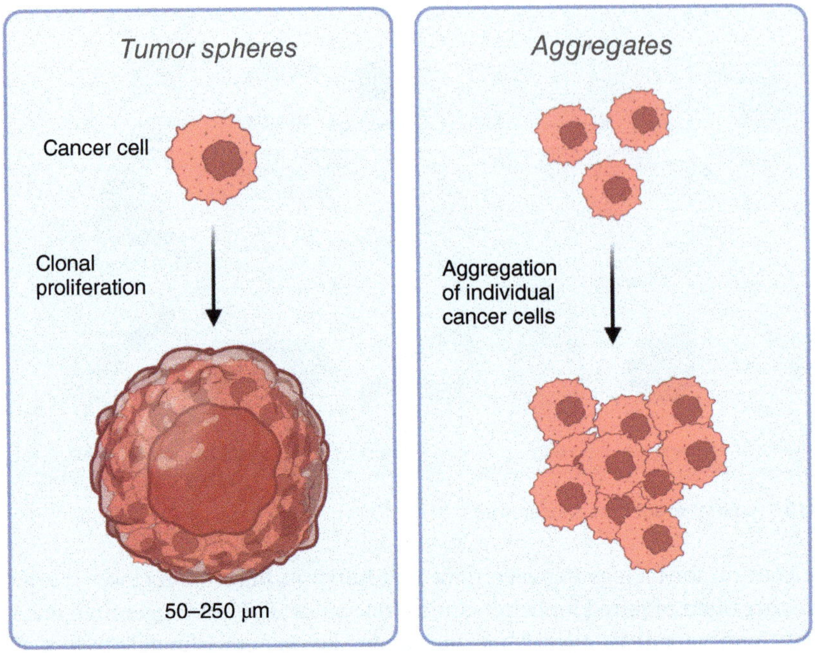

Figure 8.19 Tumor sphere and aggregated cells (from the reference or drawn figure).

of tumor sphere varies between 50 and 250 μm and the number of tumor sphere formed can be used as the index of aggressiveness for the given tumor sample. Tumor sphere is a special case of tumor structure, derived in artificial laboratory condition (Figure 8.19).

The real solid tumor is composed of mixed populations of cancer and also associated other cells including immune cells, which is called as tumor microenvironment. The complex 3D structure is called tumor spheroid. Spheroid, which means sphere-like, is mathematically defined as "rotational ellipsoid." Construction of tumor spheroid could be approached from two opposite directions: top-down and bottom-up approaches.

In top-down approach, the solid tumor is harvested and chopped into desirable size and cultured in proper media. For long-term culture and study, maintaining the original character of solid tumor is challenging. The relatively large size of spheroid makes it difficult to supply nutrients evenly to all the cells in and out, and also media change is more cumbersome compared to the conventional 2D culture. Various biomaterial and microfluidic culture systems are under development to circumvent the problems of 3D culture [56]. As a miniaturized solid tumor with all the cell components included, the tumor spheroid is an excellent platform for drug discovery and immune check point studies [57].

Bottom-up approach uses a mixture of cancer cells and associated cells to build up the solid tumor-like cell complex. The resulting tumor spheroid is often called cancer organoid [58]. The cancer cells are heterogenous, and they can be selected as cancer stem cells and/or differentiated cancer cells with different degrees and directions. The tumor-associated cells are available with much higher diversity and depending on the study purpose, the choice option is broad. For example, vascular cells can be introduced to form blood vessels in the tumor spheroid and be used for angiogenesis-related studies. Incorporation of various immune cells may provide the platform for immuno-oncology and immunotherapy studies. Fibroblasts are mesenchymal origin connected tissue cells, and the role of cancer-associated fibroblast (CAF) in tumor development has attracted more and more attention as the target for cancer treatment [59]. The multicellular tumor sphere allows the study of cell-to-cell and cell-to-extracellular matrix interactions. For example, co-culture of cancer and normal cells in the same tumor spheroid can provide a model how cancer cells interact with normal cells [60]. Spheroid formation is usually performed on a low-cell-adhesion plate with round bottom or by hanging drop method [61].

While tumor spheroids are self-assembled 3D structures, scaffolds can also be used to help the construction of 3D structure of the cells. As extracellular matrix is important for the cell survival and proliferation, various hydrogels mimicking the natural matrix are popularly used for spheroid formation [62]. Hydrogels are formed by interconnected polymers with high water contents, which allow efficient material transport in and out, such as nutrients and waste. Various types of hydrogel have been developed and used, including extracts from animal extracellular matrix, plant-based cellulose hydrogels, protein/peptide-based hydrogels, and synthetic polymer hydrogels. For predesigned shape of the organoid, more rigid solid scaffolds can be utilized (Figure 8.20).

As the size of the tumor spheroid gets bigger than the light penetration limit, a tissue clearing procedure may be necessary for optical imaging [63].

8.2.9 Clinical Imaging of Tumor for Diagnosis and Prognosis

Animal models provide more relevant information for the tumor compared to 2D cell cultures on the plate. Still, the eventual goal of tumor imaging would be the

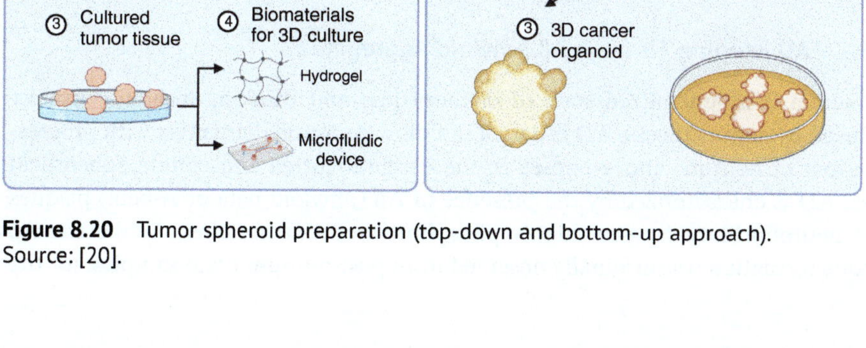

Figure 8.20 Tumor spheroid preparation (top-down and bottom-up approach). Source: [20].

clinical application. Optical imaging of clinical tumors requires several limitations and considerations. Most important consideration would be the safety of using imaging probes, which requires thorough toxicity test over several species before applying to human body. So far, there are five US-FDA-approved optical probes. They are ICG, methylene blue, 5-ALA, H-ALA, and fluorescein. Sodium salt of fluorescein has long been used for angiography of the retina and iris. With accumulated safety data, it has been recently applied to visualize intradural tumor [64] and ganglioglioma [65], expanding the scope to other tumors.

8.2.10 Intraoperative Imaging of Tumor

In cancer surgery, the complete resection of tumor tissue is critical for favorable prognosis, as a residual tumor may cause recurrence in due course. However, it is often difficult to define the boundary between a tumor and normal tissue. It would be a dilemma between minimum resection to save as much tissue as possible with a high chance of recurrence and maximum resection sacrificing more normal tissue. Intraoperative imaging probes, which can define the healthy and tumor tissue boundary clearly during the surgery, would be a great tool for the surgeon to make an optimum decision. This process is called fluorescence image-guided cancer surgery, and the development of clinically relevant tumor imaging probes is actively under way.

8.3 Neurodegenerative Disease Imaging

Neurodegenerative diseases are defined by the progressive loss of structure and/or function of neuronal system. Aging is the most common natural cause of neurodegeneration, and the disease includes Alzheimer's disease (AD), Parkinson's disease (PD), and Huntington's disease (HD). Each disease has different diagnostic biomarkers and progress pattern, but also shares common features including atypical protein aggregation and resulting cell death [66]. Other than aging and genetic mutations, the molecular causes are not fully understood, but oxidative stress [67] and chronic inflammation [68] seem to commonly contribute to the disease progression. Currently, there is no known treatment to reverse the progressive neurodegeneration, so early diagnosis and prevention would be a practical approach for managing the disease. Among various neurodegenerative diseases, most of the optical imaging studies have been focused on AD.

8.3.1 AD Imaging Through Aβ Amyloid Aggregates

Dementia is a generalized state of memory loss and thinking disability usually induced by aging process. AD is one of the most common dementias with progressive loss of neurons and synapses in the cerebral cortex and certain subcortical area. AD is characterized by the presence of Aβ (amyloid beta or A-beta) plaques and neurofibrillary tangles (highly phosphorylated tau proteins). While the Aβ plaque formation was originally observed from post-mortem brain samples, for the diagnostic purpose, a noninvasive imaging probe is necessary. With the definition of AD, Aβ amyloid was naturally the first target of AD imaging.

Aβ amyloids are commonly observed as 42 or 40 amino acid long peptide and they tend to form fibrous aggregation through oligomeric intermediate. The Aβ amyloids are originally folded into alpha helical soluble protein, but by exposing the hydrophobic area to the aqueous media, they change the conformation into beta sheet structure, and eventually form fibrous aggregates. Historically, ThT (Thioflavin T) [69] and Congo red [70] have been widely used as turn-on fluorescent probes for many protein aggregates and especially for Aβ amyloid aggregate sensing. Congo red is an excellent sensor for in vitro Aβ amyloid detection, but due to the large size and negative charges, the BBB penetration efficiency is poor. One of the early successful clinical PET probes for Aβ amyloid aggregate was PIB (Pittsburgh compound B) which is a ^{11}C derivative of ThT. By removing the positive charge of ThT, PIB has much higher affinity (μM to nM change of Kd) to Aβ and increased BBB penetration. With the improved property and as the first available clinical probe, ^{11}C-PIB has been used in PET imaging scanning in thousands of AD patients [71]. The conversion strategy of fluorescent probe into PET probe made the two-stage approach possible; Stage I is to carry out all the in vitro or ex vivo test using optical mode (fluorescence) and Stage II is in vivo or clinical test with PET mode (Figure 8.21).

With the demand of better clinically relevant Aβ amyloid probe, the discovery and development of novel probes have been extensively explored [72, 73]. From diversity oriented fluorescence library (DOFL), styryl-based Aβ amyloid probes with various colors were elucidated. Interestingly, these styryl-based probes showed higher sensitivity to Aβ-40 over Aβ-42, in contrast to ThT. When the best probes 2C40 (red) and 2E10 (green) were applied to AD model brain with ThT (blue), the styryl probes stained core amyloids and ThT stained peripheral amyloids with higher preference, implying the heterogeneous distribution of different Aβ amyloids in the plaque [74] (Figure 8.22).

While promising, the styryl probes are charged molecules and have limitation to penetrate BBB. To improve BBB penetration and binding affinity to Aβ amyloids, a series of 2C40 and 2E10 derivatives were generated and tested. Among them, STB-8 by removing the charge of 2E10 showed a clear BBB penetration upon IP injection and a selective staining of Aβ amyloid in the brain through in vivo two photon microscopy [75] (Figure 8.23).

The reduced charge and small rigid conformation of the probes are in general preferable to the BBB penetration. Using the idea, the probes were further trimmed and modified into a Stilbene compound 42 with an improved property [76] (Figure 8.24).

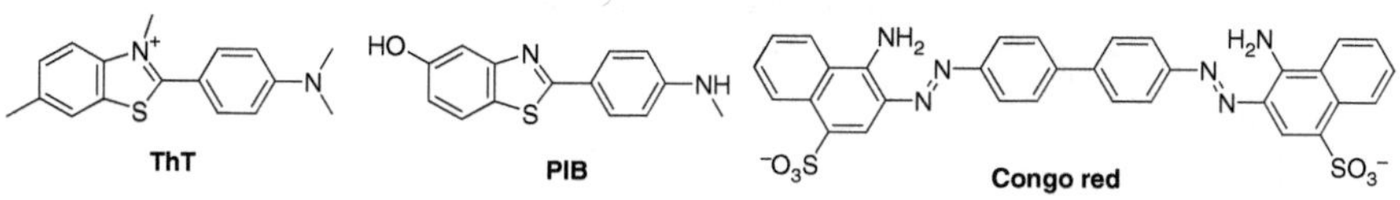

Figure 8.21 Structure of ThT/PIB and Congo red.

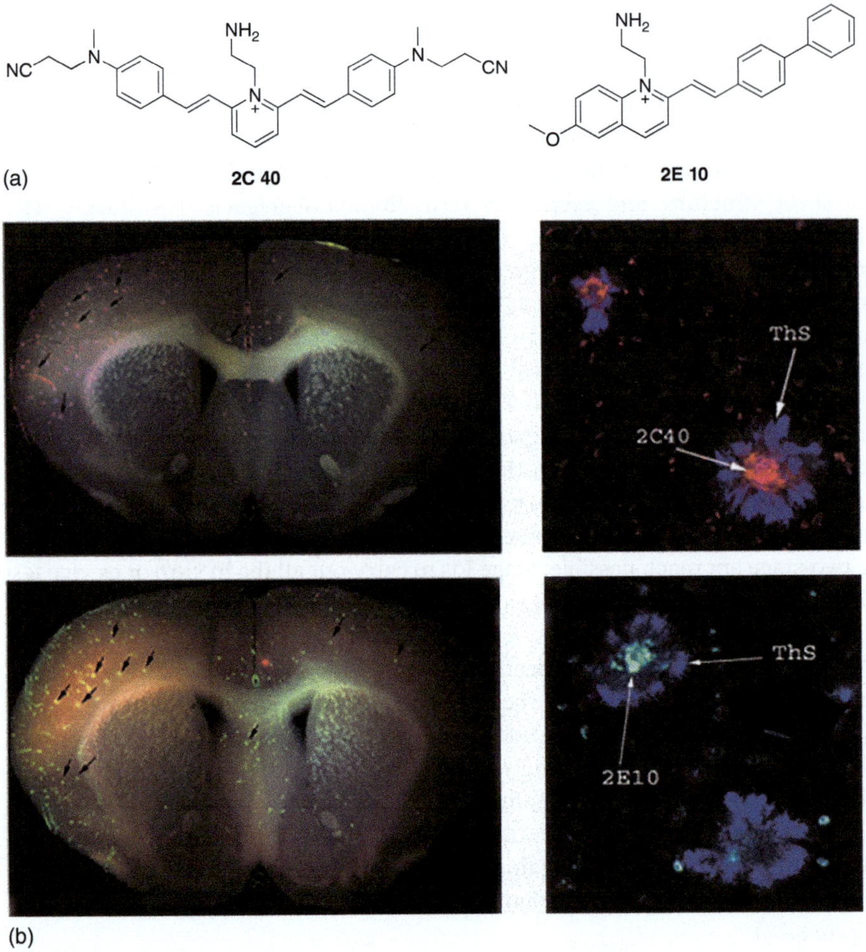

Figure 8.22 Structure of 2C40 and 2E10 (a) and brain images (b).

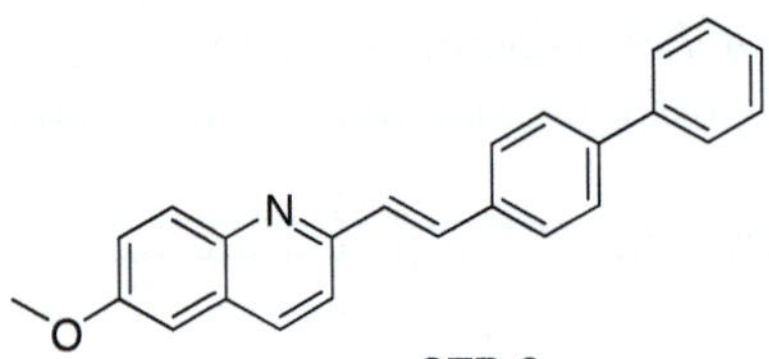

STB-8

Figure 8.23 Structure of STB-8 (conversion from 2E10).

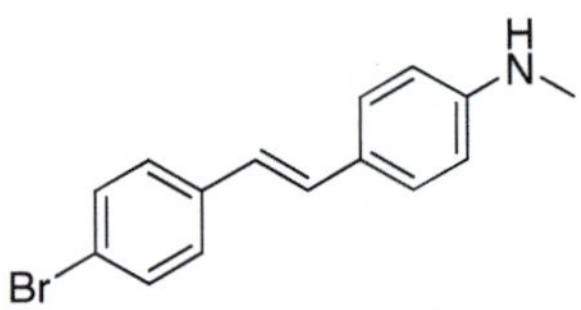

Stilbene 42

Figure 8.24 Structure of stilbene 42.

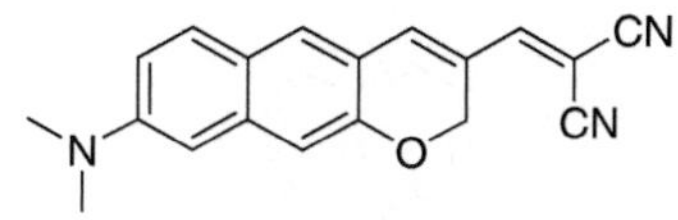

Figure 8.25 Structure of Aβ probe 5.

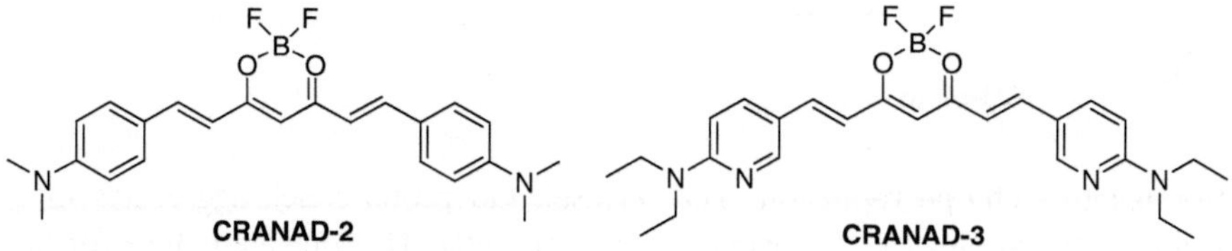

Figure 8.26 Structure of CRANAD.

Dicyanovinyl group is a well-known electron acceptor and molecular rotor, which can be used as a sensor motif for binding to rigid structures such as amyloid aggregates. Aβ probe 5 is an example, which showed a good sensing property for Aβ and deep tissue imaging capability using two-photon fluorescence mode [77] (Figure 8.25).

Curcumin, a component of curry, is known to have anti-inflammation and antioxidant activity. Curcumin has also been used for AD treatment with improving effect for cognitive function of AD patients [78]. The structure and hydrophobic property of curcumin suggested its high-affinity binding to Aβ, but the intrinsic fluorescence of curcumin is very low. By modifying the enolate moiety of curcumin with boron complex (similar to BODIPY dyes), CRANAD probes were developed for Aβ imaging. CRANAD-2 displayed the selective binding only for Aβ aggregates and CRANAD-3 showed response toward Aβ monomers, dimers, and oligomers [79] (Figure 8.26).

While various Aβ amyloid aggregate sensors and probes have been developed, the pathologically active component seems toxic oligomers, not the aggregates. Therefore, oligomer-selective probe would visualize the actively ongoing pathological area. The first oligomer-selective probe has been developed through DOFL screening with kinetic monitoring of monomer–oligomer–polymer transition over an extended time scale. The developed BD-oligo showed a spiked fluorescence signal at the oligomer stage and then the fluorescence was diminished during the further polymerization. In contrast, ThT reaches to maximum signal at the oligomer state and maintains its fluorescence. When injected, BD-oligo penetrates BBB in APP/PS1 mouse AD model and stains Aβ amyloid aggregates and oligomers [80] (Figure 8.27).

8.3.2 AD Imaging Through Tau

In human, there are six isoforms of tau protein with a range of 352–441 amino acids [81]. In healthy neuron, tau binds to microtubules and stabilizes the polymer structure of microtubules. Therefore, tau is called as neuron-specific microtubule-associated protein (MAP). The binding site of tau is positively charged to allow the binding to negatively charged microtubule. In pathological conditions such as AD, tau is highly phosphorylated and dissociates from microtubules due

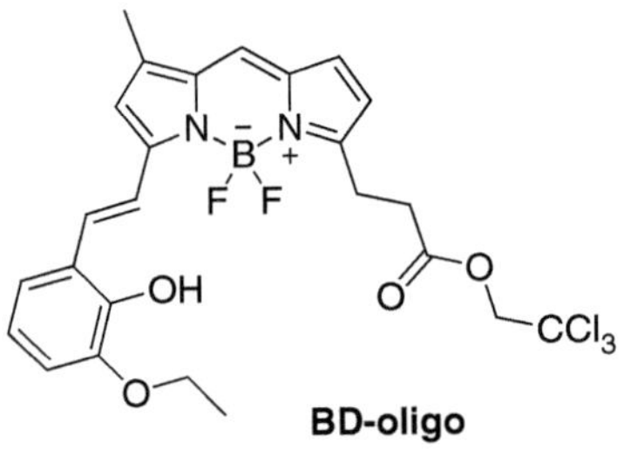

Figure 8.27 Structure of BD-oligo and response to Aβ polymerization.

to the negative charge repulsion. The released tau proteins self-aggregates into neurofibrillary tangles (NFT) inside the neuron cells. The prevalent hyperphosphorylation of tau in AD may be due to the imbalance of kinase and phosphatase function. The abnormal NFT damage the cell transport system and induce neuronal cell death [82]. Therefore, it seems NFT formation and Aβ amyloid aggregation has a correlation, but the exact relationship between them and which has more important or earlier clinical impact are not clear yet. Therefore, tau aggregate is an alternative potential biomarker of AD. However, the selective tau response over Aβ amyloid is challenging to achieve due to their apparently similar properties as aggregated proteins.

Extensive screening with more than 2000 molecules were performed to achieve the higher affinity for NFT over Aβ amyloid has been tried and a series of arylquinoline compounds are discovered as potential candidates [83]. One of the candidates, THK-523, was first used for tau aggregate imaging over Aβ amyloid plaque with high binding affinity and fast BBB penetration [84] (Figure 8.28).

For live-cell tau aggregation monitoring in optical mode, a cell line with GFP-BiFC (Bimolecular Fluorescence Complementation) was constructed [85]. In this genetic reporter design, GFP-fragment was attached to tau protein, and only when tau aggregates, the complementary GFP fragments bind to each other and form a fluorescent GFP giving out a signal. For tau-imaging probe development, DOFL was screened against pre-aggregates and aggregates of tau, and a new probe BD-tau was elucidated. BD-tau showed a specific response to tau aggregates in live cells, which could not be detected by other contrast agents such as thioflavin S or curcumin derivatives, demonstrating the unique potential of BD-tau as a tau-selective imaging probe [86] (Figure 8.29).

Through a molecular docking with a crystal structure of peptide (306VQIVYK311), the R3 binding region of p-tau protein, a selective tau probe, Tau-1, was developed [87]. Tau-1 showed a good BBB penetration and low toxicity with in vivo imaging capability in a transgenic mouse model (Figure 8.30).

Figure 8.28 Structure of THK-523.

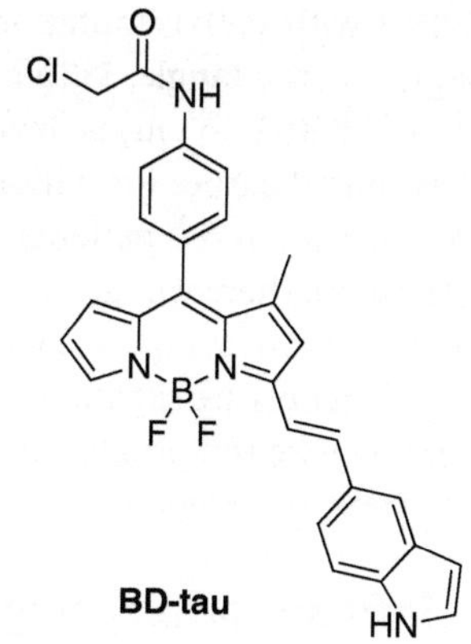

Figure 8.29 Structure of BD-tau.

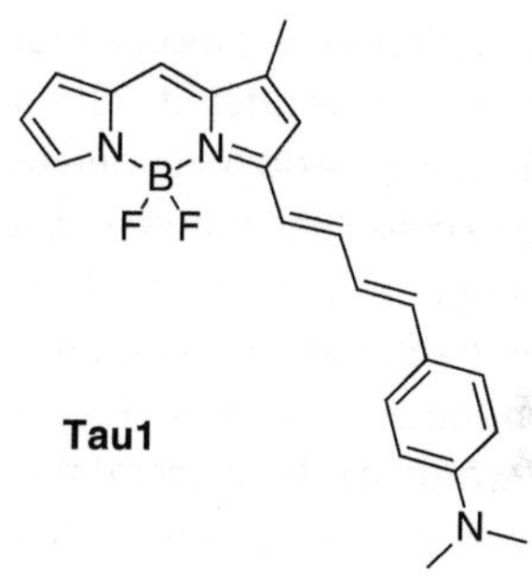

Figure 8.30 Structure of Tau1.

8.3.3 Animal Model for AD

AD is characterized by Aβ plaque and tau tangle, so animal model is also focused on the phenotypes. Aβ plaques are extracellular deposits formed by secreted Aβ peptides, with typically 40 and 42 amino acid residues. Amyloid precursor protein (APP) is a 100–140 kDa size transmembrane glycoprotein that exists as several isoforms [88]. APP is cleaved by beta and gamma secretase to release the problematic Aβ peptide into the extracellular matrix. Alpha secretase product is shorter than Aβ, and does not induce aggregation. The genetic analysis from familial Alzheimer's disease (FAD) cases elucidated mutations correlated to high AD occurrence, which are mainly on APP and PSEN 1 and 2 (components of gamma secretase) genes [89]. Using the genetic information, a large number of mutation or transgenic mouse model were developed and tested (Figure 4.40).

5XFAD transgenic mouse is one of the well-validated and widely used models that possesses total five mutations (three in APP and two in PSEN1). 5XFAD has been a popular model due to fast development of Aβ and progressive cognitive decline [90]. Originally, the model was assumed to express only Aβ and used for single target model, but the concurrent formation of tau aggregates was also observed to complicate the quantification of Aβ [91].

Interestingly, unlike human, wild-type mouse does not develop tau tangle with the aging [92]. For efficient tau tangle formation in the mouse model, mutated human tau gene seemed to be essential. The first mouse model for tau tangle formation was JNPL3 with P301L mutation of human tau protein [93]. For expressing both Aβ plaque and tau tangle, 3xTg mouse model (Swedish mutation in APP, M146V in PSEN1, and P301L in tau) is developed and widely used. Interestingly, 3xTg mouse model exhibited earlier formation of Aβ plaque formation before tau tangle [94], as observed in human AD patients.

While useful, there was always concerns that if mouse model is fully relevant with human AD patient symptoms with various discrepancy observations. To fill the gap, the animal models are expanded to other species, including rats, cats, dogs, and monkeys with genetic mutations and/or natural aging model [95].

8.4 Inflammation Imaging

Inflammation is a complex biological response to harmful stimuli, such as invading pathogens or altered cells such as cancer. As part of innate immunity, inflammation is a generic body defense mechanism to recover from general infection or damaged cells. The antigen-specific process is driven by adaptive immunity, which takes longer time to respond to the damages. The typical symptom of inflammation is fever, pain, redness, swelling, and loss of function [96]. However, over-activation of immune system can induce life-threatening side effects; cytokine storm or septic shock would be the most fatal acute immune responses. If the immune system does not clear out the causes of inflammation during the initial stage, chronic inflammation persists damaging the tissue or whole-body system for a prolonged period. Many diseases such as neurodegeneration or even tumors are suspected to be seriously affected by the chronic inflammation. If the immune cells start to recognize part of the body as a self-antigen, autoimmune diseases also can occur, making the situation even worse.

8.4.1 Inflammation Imaging by Environmental Changes

With the physical and biological symptoms of inflammation, the microenvironment would be drifted from normal status. For example, inflammation often induces physical swelling of the tissue, which may result in the local viscosity increase. A viscosity sensor, HBTD-V, was developed with a molecular rotor motif, showing a linear fluorescence response to viscosity changes. When an inflammation mouse model was monitored by HBTD-V, a higher viscosity in the lung could be observed than that of normal mouse [97]. Another probe VO was developed as a dual sensor for peroxynitrite and viscosity with two different fluorescence responses (fluorescence increases at 770 nm to viscosity and at 635 nm to peroxynitrite). VO was used to monitor the liver viscosity changes by acetaminophen-induced hepatotoxicity [98] (Figure 8.31).

Most of the pH sensors have been focused on the acidic pH range, targeting lysosome or other acidic organelles [35]. For sensitive monitoring of the change of pH in inflammation, sensors working near neutral or slightly basic condition would complement to the available sensors. In chronic inflammation, high-level proteases are activated to reorganize the extracellular matrix as part of the wound healing. The skin surface of a healthy human is slightly acidic (pH 4.2–5.9) [99], but in acute

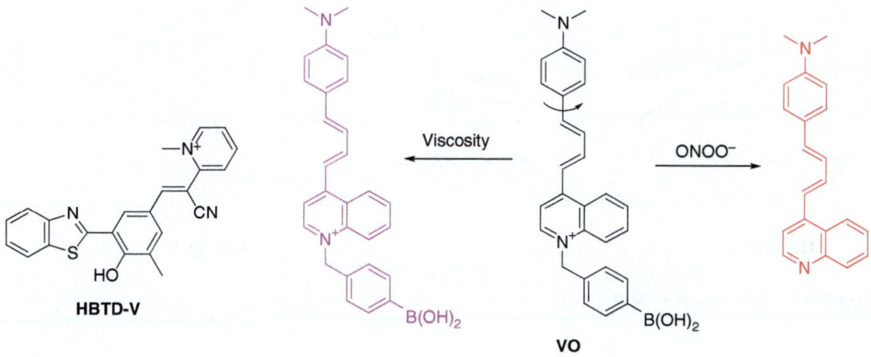

Figure 8.31 Structure of HBTD-V and VO (with two responses).

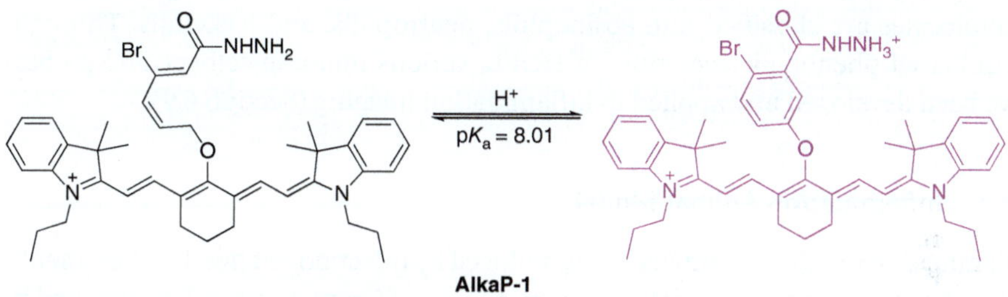

Figure 8.32 Structure of Alka-P1.

wounds, pH is neutralized by body fluid leakage. In chronic wounds, it could be even basic (pH 7.1–8.9) under certain pathological circumstances. A series of pH sensors were designed and tested to match the optimum pH change of the wound, and AlkaP-1 was selected as the best candidate with pK_a of 8.01 (turn-on in acidic condition) [100]. Alka-P1 successfully visualized the alkalization in the chronic wound developed from an acute wound to chronic stage in a transgenic diabetic mouse model (Figure 8.32).

The pH change is not always one direction to the alkalization in inflammation. A highly sensitive NIR range pH sensor NIR-OH-1 with pK_a 7.77 was developed to monitor the slightly alkaline area pH change in the biological systems [101]. NIR-OH-1 turned on in basic conditions sharply responding to the pH change, and was demonstrated in the subtle pH change in mitochondria. NIR-OH-1 was also demonstrated in the acidification of abdominal acidification by lipopolysaccharide (LPS)-injected mouse model (Figure 8.33).

In inflammation tissue, as part of immune cell action, various ROS are generated and accumulated. Naturally, the ROS sensors are widely used for inflammation imaging, which is covered in Section 7.6 and in an excellent review paper [102].

8.4.2 Inflammation Imaging Through Immune Cells

The main players of the inflammation process are immune cells using signaling molecules across the blood vessels to target tissues. Therefore, rather than targeting

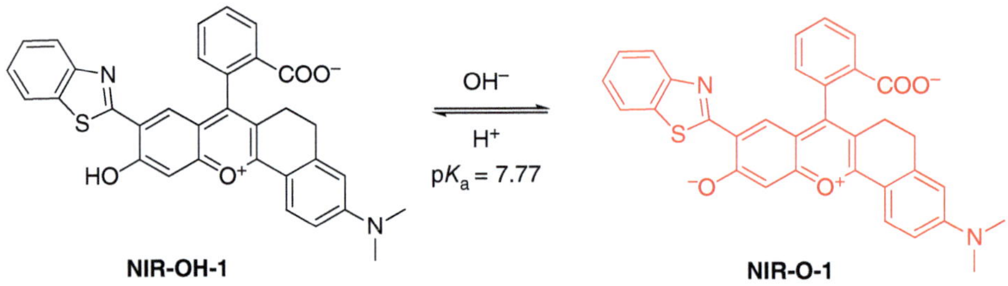

Figure 8.33 Structure of NIR-OH-1.

each environmental factor, monitoring the immune cells would be an alternative approach for the inflammation imaging. Immune cells are composed of B and T lymphocytes, monocytes (precursor cells for macrophage), and granulocytes. Granulocytes are classified into eosinophils, neutrophils, and basophils. Through an unbiased phenotypic screening of DOFL, various immune-cell-specific probes have been developed and applied to inflammation imaging (Section 6.9).

8.4.3 Inflammation Animal Model

Inflammation in cells or animals can be induced by infection or chemical treatment. LPS is the main component of the outer membrane of Gram-negative bacteria, and it mimics the bacterial infection by stimulating the immune response in animals. LPS is composed of lipid and polysaccharide composed of O-antigen, and often called endotoxin [103]. Injection of LPS into mouse paw or intraperitoneal cavity comprises the most common acute inflammation animal model.

Rheumatism or rheumatoid arthritis is an autoimmune disease occurring mainly in joints with severe chronic inflammation. Type II collagen, the main constituent protein of articular cartilage has been used for inducing autoimmune response to make rheumatism model animal. Intravenous injection of type II collagen into susceptible strains of rodents or nonhuman primates induces experimental autoimmune disease. For faster acute model, collagen antibody can be used to directly attack collagen of the host.

Among transgenic models, Apolipoprotein E (ApoE) is widely used for atherosclerosis model. ApoE is a plasma protein and involved in lipid transport and lipoprotein metabolism. Knock out of ApoE coupled with a high-fat diet induces inflammation and vulnerable tissue in the artery along with accumulation of foam cells, activated macrophages.

8.5 Diabetes Imaging

Diabetes is induced by uncontrolled high concentration of glucose in the blood stream. Glucose is the first metabolic product of photosynthesis and the very basic energy source of the body both for animals and plants. Therefore, low concentration of glucose is a direct threat to the body, especially for the brain. At the same time, however, glucose is a chemically reactive and thus biologically toxic material, especially at high concentration. Glucose contains reactive reducing aldehyde group, and it readily reacts with amines to form the imine bond. This imine formation is a reversible reaction through hydrolysis, but the high concentration of glucose can push the reaction to further conjugation through glycation, which is irreversible. The result is crosslinking of proteins (through their multiple amine groups) and caramelization, accompanied by ROS accumulation. It is paradoxical situation that the essential energy material glucose is toxic to the body at high concentration.

Insulin is the hormone to control the blood concentration, by activating the glucose uptake into the muscle or adipose tissues from the bloodstream. As the result, insulin reduces the circulating free glucose from the bloodstream. Therefore, low insulin in the blood would generate diabetic symptoms. Pancreatic β-cells are the production site of insulin and acute destruction of the β-cells is the cause of type I diabetes, also known as juvenile diabetes. As insulin treatment cure the type I diabetes, it is also called as insulin-dependent diabetes. Insulin injection or a slow insulin secretion device would be a possible treatment. Alternatively, pancreatic islet transplantation may provide a long-term solution as far as the transplanted islets survive in the patient's body.

For type I model, streptozotocin (STZ) has been used to selectively destroy β-cells to induce an acute diabetic model animal. STZ is a toxic glucose analogue that hijacks the glucose recognition machineries in β-cell. The selective uptake of STZ induces the β-cell death and subsequently abolishes the insulin production, inducing diabetes symptoms.

Type II diabetes is a more chronic and complicated disease, induced by high glucose concentration. The sustained high glucose in the blood reduces the sensitivity of insulin response to the receptors, and demands higher insulin production for the compensation. So, in the early stage of diabetes, the function of β-cells for insulin production is increased, but at the later stage, the overloaded β-cells collapse to lead to critical end stage of diabetes. Due to the lowered sensitivity, insulin treatment also has limited effect for the therapy in type II diabetes.

Currently, the clinical monitoring of diabetes is mainly focused on blood glucose concentration measurement. Glycation level of hemoglobin is another ancillary parameter to reflect the accumulated exposure to the glucose. If there are imaging probes for α- or β-cell status, they would provide comprehensive view for the disease progress. Several α- or β-cell-selective fluorescent probes were developed and used for pancreatic tissue imaging (Section 5.3.1), but for the whole-body imaging, there is a critical tissue penetration issue of light. For whole-body imaging of pancreas, PET is a possible solution and several peptide-based PET probes have been developed [104]. While the peptide motifs are selected from the hormone structures, without optical reporter, they are limited only to the PET application. PiF was one of the fluorescent β-cell probes and demonstrated favorable tissue distribution and pharmacokinetic property in the body. After all the characterization is achieved by fluorescent property, by simple replacing of the built-in ^{19}F into ^{18}F, the conversion of modality from fluorescence to PET was demonstrated (Section 6.7). These types of multimodality probes may be ideal for covering in vitro, ex vivo, and eventually in vivo imaging applications.

8.6 Liver Disease Imaging

Liver is the most complex chemical reaction center in the body. Also, with the high regenerative potential from resection or various damages, the monitoring of the dynamic liver status would be of great interest. The usual liver disease progresses from inflammation, through fibrosis and cirrhosis, to cancer. The causes of liver inflammation are viral infection (hepatitis B virus [HBV] and hepatitis C virus [HCV]), fatty liver, alcohols, and so on. HCVhas been one of the major threats to liver health with high rate of cirrhosis and cancer occurrence. With the recent novel nucleotide prodrug development, HCV becomes almost completely curable disease. HBV has an older history than HCV, and is still waiting for the magic bullet treatment.

With the modern diet and life style, nonalcoholic fatty liver disease (NAFLD) is currently one of the most common liver diseases induced by chronic metabolic problem, including obesity and diabetes. The progress of NAFLD starts from nonalcoholic fatty liver (NAFL, simple steatosis) and advances to nonalcoholic steatohepatitis (NASH). NAFLD with advanced fibrosis increases the risk of liver cirrhosis and cancer. The current gold standard for NAFLD diagnosis is liver biopsy, so noninvasive diagnostic method development is in great demand. As the underlying mechanism of NAFLD is not fully understood, the development of a high-quality imaging technique and proper animal model is important for research.

Conventionally, the chemical liver damage model was made by hepatotoxic chemicals such as CCl_4, which is also quite toxic to humans. In addition, CCl_4 is one of the powerful ozone-depletion materials, and a more sophisticated model for liver diseases has been developed. Methionine–choline deficient (MCD) diet is a common dietary model of NASH. MCD is composed of high sucrose (45%) and fat (10%), but lacks methionine and choline, which are essential for hepatic mitochondrial beta-oxidation and very low-density lipoprotein (VLDL) synthesis. With MCD diet, an acute steatohepatitis accompanied by pericellular fibrosis is induced, resembling human NASH [105].

A further advanced fibrosis model could be made by STZ treatment followed by a long-term high-fat diet. The hyperglycemia induced by STZ treatment and long-term metabolic stress by high-fat diet results in a chronic condition of hepatic steatosis with advanced fibrosis. The model animal shows the accumulation of lipid droplets in hepatocytes, collagen fibers in perisinusoidal region, and the disruption of microvasculature. Similar to pancreatic imaging case, liver is also located at a deeper location than the optical signal can reach. While NIR probes have been developed for various liver imaging, the obtainable signals are limited to the near skin area with low spatial resolution. For higher-resolution imaging, intravital technique using surgery and two-photon probes seems the practically accessible optical modality [106].

Alternative noninvasive imaging of liver was mainly relied on ultrasound technique. Although cheap and convenient, the ultrasound images are relatively low resolution and provides information on physical changes only for the surface part of the liver. A modified ultrasound technique, transient elastography (FibroScan), measures liver stiffness and elasticity for the assessment of the degree of hepatic fibrosis, complementing the conventional ultrasound imaging. A fully noninvasive liver imaging with high 3D resolution and molecular information may require a breakthrough of technology and probes.

8.7 Aging

It is controversial to classify aging as a disease. Aging is a process of getting old, after growth and development. The concept can be applied to individuals, organs, tissues, and cells. Quoting the Hayflick limit, human cells were considered to have the ability of approximately 50 times cell division. After the limitation, the cells may stop division by entering senescence state or commit suicide through apoptosis. This division number of 50 has been considered the natural biological limit for cell growth and known as "Hayflick limit" following the discoverer's name.

Cell division is regulated by the coordinated procedure of cell cycle (Section 10.12). When the cells stop dividing, cells are considered to enter G0 state or resting state. G0 states are categorized as either reversible (quiescent) or irreversible (senescent). Quiescent cells can return to the cell cycle after activation in response to environmental stimuli. They are usually characterized by low RNA content, lack of cell proliferation markers, and increased label retention due to low cell turnover [107]. Senescence is an irreversible state in response to DNA damage or aging-related degradation. Therefore, imaging senescent cells may serve as the reasonable visualization method of the aged tissues and organs.

Currently, senescence-associated β-Gal is the gold standard biomarker of senescent cells. However, the standard procedure requires fixing of cells, hampering the study of live-cell function. Although dozens of fluorescent probes have been developed to overcome the limit, still a practical and universal probe and method are yet to be validated.

Compared to cancer or inflammation study with rich biomarkers developed over decades, aging related biomarkers are rare. Development of novel probes for aged cells would be straightforward and the probes will cast the clues to elucidate novel biomarkers. In addition, the functional manipulation of the stained cells may provide the control power of aging process. Therefore, the whole aging field seems to be fully open to accommodate new wave of research.

8.8 Theranostics

The term "theragnosis" is the blend of therapy and diagnosis, meaning the combination of diagnostic tests and therapy with one molecule. As isotopes share almost same chemical property with only mass difference, isotope-exchanged molecule can be considered almost "same molecule." Therefore, the term has been used for a molecule with exchangeable isotopes of imaging and therapy application. ^{123}I and ^{131}I are one of the examples. ^{123}I is a gamma-ray-emitting isotope and can be used for probe for gamma camera or SPECT. ^{131}I irradiate high-energy beta ray, which can kill cells nearby. If ^{123}I probe provides an image of excellent targeting to cancer area, one ideal approach is the exchange of ^{123}I with ^{131}I, then the new isotope may be localized in the cancer tissue and kills cancer cells. A similar approach could be applied to metal isotopes with same chelators. The emerging field is called "theranostics" and increases its application especially in precision oncology [108].

The same concept could be applied to fluorescent probes. For example, TiY is a tumor-initiating cell probe with vimentin as the binding target. At low concentration, TiY visualizes the position of tumor-initiating cell and facilitate the isolation of the cells. At high concentration, interestingly, TiY kills selectively the target cells. Vimentin is an important intermediate filament for cell migration, and TiY inhibits its function. At low concentration, the partial or minimal inhibition may not induce a significant effect to the cells. At high concentration, significant amount of vimentin function is blocked and the cells may have gone to death (Section 4.1.3.1).

TiNIR is another tumor-initiating cell probe with the target of HMOX2. In a similar manner with TiY, TiNIR also exhibits theragnostic performance showing another example of theranostics (Section 4.1.3.1). These are the cases when the binding target of imaging probe is a functionally important protein.

Even in case that the target is not functionally important, the fluorescent probes have unique option of inducing cell toxicity through photodynamic activation. Fluorescent molecules have the ability of light absorption, and a part of the energy is used for emission. With a stronger light irradiation, another pathway of photodynamic effect becomes important and the resulting ROS can kill the cells. If the fluorescent probe is selectively bind to a specific cell population, only stained cells will be killed. If the target cell is a tumor, this approach would be similar to precision oncology in nuclear medicine. 5-ALA forms porphyrin structure in tumors and has been used for photodynamic therapy (Section 8.2.2).

Another example is demonstrated with CDy1, an embryonic stem probe. Embryonic stem cells are important source for cell therapy in tissue regeneration, but carries the possibility of teratoma formation in the receiver's body. Therefore, partially differentiated cells into adult stem cells are commonly used for transplantation. Still, small amount of leftover embryonic stem cells could cause a problem. Utilizing the ability of stem-selective localization of CDy1, the remaining stem cells are labeled and eliminated by sequential irradiation to suppress the teratoma formation (Section 4.1.1).

References

1 Maeda, H., Nakamura, H., and Fang, J. (2013). *Adv. Drug Delivery Rev.* 65: 71–79. https://doi.org/10.1016/j.addr.2012.10.002.

2 Unsoy, G. and Gunduz, U. (2018). *Curr. Drug Targets* 19: 202–212. https://doi.org/10.2174/1389450117666160401124624.

3 van Manen, L., Handgraaf, H.J.M., Diana, M. et al. (2018). *J. Surg. Oncol.* 118: 283–300. https://doi.org/10.1002/jso.25105.

4 An, F., Yang, Z., Zheng, M. et al. (2020). *J. Nanobiotechnol.* 18: 49. https://doi.org/10.1186/s12951-020-00603-8.

5 Cwalinski, T., Polom, W., Marano, L. et al. (2020). *J. Clin. Med. Res.* 9: 3538. https://doi.org/10.3390/jcm9113538.

6 Chau, C.H., Steeg, P.S., and Figg, W.D. (2019). *Lancet* 394: 793–804. https://doi.org/10.1016/s0140-6736(19)31774-x.

7 Yang, L., Hu, Z., Luo, J. et al. (2017). *Bioorg. Med. Chem.* 25: 3531–3539. https://doi.org/10.1016/j.bmc.2017.05.002.

8 Liu, T., Wu, L.Y., Kazak, M., and Berkman, C.E. (2008). *Prostate* 68: 955–964. https://doi.org/10.1002/pros.20753.

9 van Dam, G.M., Themelis, G., Crane, L.M.A. et al. (2011). *Nat. Med.* 17: 1315–1319. https://doi.org/10.1038/nm.2472.

10 Uddin, M.J., Crews, B.C., Blobaum, A.L. et al. (2010). *Cancer Res.* 70: 3618–3627. https://doi.org/10.1158/0008-5472.CAN-09-2664.

11 Krall, N., Pretto, F., Decurtins, W. et al. (2014). *Angew. Chem. Int. Ed.* 53: 4231–4235. https://doi.org/10.1002/anie.201310709.

12 DeBin, J.A. and Strichartz, G.R. (1991). *Toxicon* 29: 1403–1408. https://doi.org/10.1016/0041-0101(91)90128-E.

13 Yamada, M., Miller, D.M., Lowe, M. et al. (2021). *Contemp. Clin. Trials Commun.* 23: 100830. https://doi.org/10.1016/j.conctc.2021.100830.

14 Tang, C., Du, Y., Liang, Q. et al. (2020). *Mol. Imag. Biol.* 22: 476–485. https://doi.org/10.1007/s11307-019-01389-4.

15 Wu, X., Wang, R., Kwon, N. et al. (2022). *Chem. Soc. Rev.* 51: 450–463. https://doi.org/10.1039/D1CS00543J.

16 Miampamba, M., Liu, J., Harootunian, A. et al. (2017). *Theranostics* 7: 3369–3386. https://doi.org/10.7150/thno.20678.

17 Whitley, M.J., Cardona, D.M., Lazarides, A.L. et al. (2016). *Sci. Transl. Med.* 8: 320ra4. https://doi.org/10.1126/scitranslmed.aad0293.

18 Ofori, L.O., Withana, N.P., Prestwood, T.R. et al. (2015). *ACS Chem. Biol.* 10: 1977–1988. https://doi.org/10.1021/acschembio.5b00205.

19 Tholen, M., Yim, J.J., Groborz, K. et al. (2020). *Angew. Chem. Int. Ed.* 59: 19143–19152. https://doi.org/10.1002/anie.202006719.

20 Widen, J.C., Tholen, M., Yim, J.J. et al. (2021). *Nat. Biomed. Eng.* 5: 264–277. https://doi.org/10.1038/s41551-020-00616-6.

21 Li, H., Yao, Q., Sun, W. et al. (2020). *J. Am. Chem. Soc.* 142: 6381–6389. https://doi.org/10.1021/jacs.0c01365.

22 Li, H., Yao, Q., Xu, F. et al. (2018). *Biomaterials* 179: 1–14. https://doi.org/10.1016/j.biomaterials.2018.06.028.

23 Usama, S.M., Inagaki, F., Kobayashi, H., and Schnermann, M.J. (2021). *J. Am. Chem. Soc.* 143: 5674–5679. https://doi.org/10.1021/jacs.1c02112.

24 Kitagawa, Y., Tanaka, S., Kamiya, M. et al. (2021). *Clin. Cancer Res.* 27: 3936–3947. https://doi.org/10.1158/1078-0432.CCR-20-4518.

25 Doura, T., Kamiya, M., Obata, F. et al. (2016). *Angew. Chem. Int. Ed.* 55: 9620–9624. https://doi.org/10.1002/anie.201603328.

26 Gardner, L.C. and Cox, T.M. (1988). *J. Biol. Chem.* 263: 6676–6682. https://doi.org/10.1016/S0021-9258(18)68695-8.

27 Von Wettstein, D., Gough, S., and Kannangara, C.G. (1995). *Plant Cell* 7: 1039–1057. https://doi.org/10.1105/tpc.7.7.1039.

28 Yamamichi, G., Nakata, W., Yoshimura, A. et al. (2020). *Int. J. Urol.* 27: 213–218. https://doi.org/10.1111/iju.14170.

29 Matsumoto, T., Murayama, Y., Matsuo, H. et al. (2020). *Gastric Cancer* 23: 725–733. https://doi.org/10.1007/s10120-020-01044-w.

30 Tetard, M.-C., Vermandel, M., Mordon, S. et al. (2014). *Photodiagn. Photodyn. Ther.* 11: 319–330. https://doi.org/10.1016/j.pdpdt.2014.04.004.

31 Miyake, M., Nakai, Y., Nishimura, N. et al. (2021). *BJU Int.* 128: 244–253. https://doi.org/10.1111/bju.15368.

32 Shirazi, H.S., Chan, K.M., Rouget, J. et al. (2020). *Biointerphases* 15: 031002. https://doi.org/10.1116/6.0000047.

33 Chen, L.Q. and Pagel, M.D. (2015). *Adv. Radiol.* 2015: 206405. https://doi.org/10.1155/2015/206405.

34 Gillies, R.J., Raghunand, N., Karczmar, G.S., and Bhujwalla, Z.M. (2002). *J. Magn. Reson. Imaging* 16: 430–450. https://doi.org/10.1002/jmri.10181.

35 Wang, S., Ren, W.X., Hou, J.-T. et al. (2021). *Chem. Soc. Rev.* 50: 8887–8902. https://doi.org/10.1039/D1CS00083G.

36 Koide, Y., Kojima, R., Hanaoka, K. et al. (2019). *Commun. Chem.* 2: 94. https://doi.org/10.1038/s42004-019-0194-4.

37 Song, Y., Zhang, H., Wang, X. et al. (2021). *Anal. Chem.* 93 (3): 1786–1791. https://doi.org/10.1021/acs.analchem.0c04644.

38 Yu, L., Zhang, J.F., Li, M. et al. (2020). *Chem. Commun.* 56: 6684–6687. https://doi.org/10.1039/D0CC02943B.

39 Hayflick, L. and Moorhead, P.S. (1961). *Exp. Cell. Res.* 25: 585–621. https://doi.org/10.1016/0014-4827(61)90192-6.

40 Scherer, W.F., Syverton, J.T., and Gey, G.O. (1953). *J. Exp. Med.* 97: 695–710. https://doi.org/10.1084/jem.97.5.695.

41 Earle, W.R., Schilling, E.L., Stark, T.H. et al. (1943). *J. Natl. Cancer Inst.* 4: 165–212. https://doi.org/10.1093/jnci/4.2.165.

42 Gey, G.O. (1952). *Cancer Res.* 12: 264–265.

43 Skloot, R. (2011). *The Immortal Life of Henrietta Lacks Paperback*. Jenson Books Inc.

44 Masters, J.R. (2002). *Nat. Rev. Cancer* 2: 315–319. https://doi.org/10.1038/nrc775.

45 Petschel, R., Irion, G., Naujok, A. et al. (1988). *Histochemistry* 90: 219–232. https://doi.org/10.1007/bf00492511.

46 Macville, M., Schröck, E., Padilla-Nash, H. et al. (1999). *Cancer Res.* 59: 141–150. PMID: 9892199.

47 Heng, H.H. (2013). *Nature* 501: 167. https://doi.org/10.1038/501167d.

48 Arlett, C.F. (2001). *Lancet Oncol.* 2: 467. https://doi.org/10.1016/S1470-2045(01)00450-8.

49 Hason, M. and Bartůněk, P. (2019). *Genes (Basel)* 10: 935. https://doi.org/10.3390/genes10110935.

50 Hardwick, L.J.A. and Philpott, A. (2018). *Front. Physiol.* 9: 1660. https://doi.org/10.3389/fphys.2018.01660.

51 Jarvis, S., Koumadoraki, E., Madouros, N. et al. (2021). *Cancer Treat. Res. Commun.* 27: 100307. https://doi.org/10.1016/j.ctarc.2021.100307.

52 Kirienko, N.V., Mani, K., and Fay, D.S. (2010). *Dev. Dyn.* 239: 1413–1448. https://doi.org/10.1002/dvdy.22247.

53 Halder, G. and Mills, G. (2011). *Oncogene* 30: 4063–4066. https://doi.org/10.1038/onc.2011.128.

54 Pray, L. (2008). *Nat. Educ.* 1: 183. https://www.nature.com/scitable/topicpage/l-h-hartwell-s-yeast-a-model-808/Resolved.

55 Johnson, S., Chen, H., and Lo, P.-K. (2013). *Bio-Protocol* 3: e325. https://doi.org/10.21769/BioProtoc.325.

56 Carvalho, M.R., Lima, D., Reis, R.L. et al. (2015). *Trends Biotechnol.* 33: 667–678. https://doi.org/10.1016/j.tibtech.2015.09.009.

57 Aref, A.R., Campisi, M., Ivanova, E. et al. (2018). *Lab Chip* 18: 3129–3143. https://doi.org/10.1039/C8LC00322J.

58 LeSavage, B.L., Suhar, R.A., Broguiere, N. et al. (2022). *Nat. Mater.* 21: 143–159. https://doi.org/10.1038/s41563-021-01057-5.

59 Biffi, G. and Tuveson, D.A. (2021). *Physiol. Rev.* 101: 147–176. https://doi.org/10.1152/physrev.00048.2019.

60 Kunz-Schughart, L.A., Heyder, P., Schroeder, J., and Knuechel, R. (2001). *Exp. Cell Res.* 266: 74–86. https://doi.org/10.1006/excr.2001.5210.

61 Tung, Y.-C., Hsiao, A.Y., Allen, S.G. et al. (2011). *Analyst* 136: 473–478. https://doi.org/10.1039/C0AN00609B.

62 Tibbitt, M.W. and Anseth, K.S. (2009). *Biotechnol. Bioeng.* 103: 655–663. https://doi.org/10.1002/bit.22361.

63 Ueda, H.R., Ertürk, A., Chung, K. et al. (2020). *Nat. Rev. Neurosci.* 21: 61–79. https://doi.org/10.1038/s41583-019-0250-1.

64 Olguner, S.K., Arslan, A., Açık, V. et al. (2020). *Front. Oncol.* 10: 618579. https://doi.org/10.3389/fonc.2020.618579.

65 Höhne, J., Acerbi, F., Falco, J. et al. (2020). *J. Clin. Med.* 9: 2405. https://doi.org/10.3390/jcm9082405.

66 Bredesen, D., Rao, R., and Mehlen, P. (2006). *Nature* 443: 796–802. https://doi.org/10.1038/nature05293.

67 Pereira, T.M.C., Côco, L.Z., Ton, A.M.M. et al. (2021). *Antioxidants (Basel)* 10: 1845. https://doi.org/10.3390/antiox10111845.

68 Stephenson, J., Nutma, E., van der Valk, P., and Amor, S. (2018). *Immunology* 154: 204–219. https://doi.org/10.1111/imm.12922.

69 Naiki, H., Higuchi, K., Hosokawa, M., and Takeda, T. (1989). *Anal. Biochem.* 177: 244–249. https://doi.org/10.1016/0003-2697(89)90046-8.

70 Lorenzo, A. and Yankner, B.A. (1994). *Proc. Natl. Acad. Sci. U.S.A.* 91: 12243–12247. https://doi.org/10.1073/pnas.91.25.12243.

71 Klunk, W.E., Engler, H., Nordberg, A. et al. (2004). *Ann. Neurol.* 55: 306–319. https://doi.org/10.1002/ana.20009.

72 Jun, Y.W., Cho, S.W., Jung, J. et al. (2019). *ACS Cent. Sci.* 5: 209–217. https://doi.org/10.1021/acscentsci.8b00951.

73 Lee, D., Kim, S.M., Kim, H.Y., and Kim, Y.S. (2019). *ACS Chem. Neurosci.* 10: 2647–2657. https://doi.org/10.1021/acschemneuro.9b00199.

74 Li, Q., Lee, J.-S., Ha, C. et al. (2004). *Angew. Chem. Int. Ed.* 43: 6331–6335. https://doi.org/10.1002/anie.200461600.

75 Li, Q., Min, J., Ahn, Y.-H. et al. (2007). *ChemBioChem* 8: 1679–1687. https://doi.org/10.1002/cbic.200700154.

76 Hong, M.C., Kim, Y.K., Choi, J.Y. et al. (2010). *Bioorg. Med. Chem.* 18: 7724–7730. https://doi.org/10.1016/j.bmc.2010.06.044.

77 Kim, D., Moon, H., Baik, S.H. et al. (2015). *J. Am. Chem. Soc.* 137: 6781–6789. https://doi.org/10.1021/jacs.5b03548.

78 Lalatsa, A. and Barbu, E. (2016). *Int. Rev. Neurobiol.* 130: 115–153. https://doi.org/10.1016/bs.irn.2016.05.004.

79 Zhang, X., Tian, Y., Zhang, C. et al. (2015). *Proc. Natl. Acad. Sci. U.S.A.* 112: 9734–9739. https://doi.org/10.1073/pnas.1505420112.

80 Teoh, C.L., Su, D., Sahu, S. et al. (2015). *J. Am. Chem. Soc.* 137: 13503–13509. https://doi.org/10.1021/jacs.5b06190.

81 Sergeant, N., Delacourte, A., and Buée, L. (2005). *Biochim. Biophys. Acta* 1739: 179–197. https://doi.org/10.1016/j.bbadis.2004.06.020.

82 Lippens, G., Sillen, A., Landrieu, I. et al. (2007). *Prion* 1: 21–25. https://doi.org/10.4161/pri.1.1.4055.

83 Okamura, N., Furumoto, S., Harada, R. et al. (2013). *J. Nucl. Med.* 54: 1420–1427. https://doi.org/10.2967/jnumed.112.117341.

84 Fodero-Tavoletti, M.T., Furumoto, S., Taylor, L. et al. (2014). *Alzheimer's Res. Ther.* 6: 11. https://doi.org/10.1186/alzrt240.

85 Tak, H.J., Haque, M., Kim, M.J. et al. (2013). *PLoS One* 8: e81682. https://doi.org/10.1371/journal.pone.0081682.

86 Lim, S., Haque, M.M., Su, D. et al. (2017). *Chem. Commun.* 53: 1607–1610. https://doi.org/10.1039/C6CC08826K.

87 Verwilst, P., Kim, H.-R., Seo, J. et al. (2017). *J. Am. Chem. Soc.* 139: 13393–13403. https://doi.org/10.1021/jacs.7b05878.

88 Selkoe, D.J. (1996). *J. Biol. Chem.* 271: 18295–18298. https://doi.org/10.1074/jbc.271.31.18295.

89 Glenner, G.G. and Wong, C.W. (1984). *Biochem. Biophys. Res. Commun.* 122: 1131–1135. https://doi.org/10.1016/0006-291x(84)91209-9.

90 Oakley, H., Cole, S.L., Logan, S. et al. (2006). *J. Neurosci.* 26: 10129–10140. https://doi.org/10.1523/JNEUROSCI.1202-06.2006.

91 Shin, J., Park, S., Lee, H.Y., and Kim, Y.S. (2021). *Sci. Rep.* 11: 1617. https://doi.org/10.1038/s41598-021-81304-6.

92 Chen, F., David, D., Ferrari, A., and Götz, J. (2004). *Curr. Drug Targets* 5: 503–515. https://doi.org/10.2174/1389450043345236.

93 Lewis, J., McGowan, E., Rockwood, J. et al. (2000). *Nat. Genet.* 25: 402–405. https://doi.org/10.1038/78078.

94 Eriksen, J.L. and Janus, C.G. (2007). *Behav. Genet.* 37: 79–100. https://doi.org/10.1007/s10519-006-9118-z.

95 Mckean, N.E., Handley, R.R., and Snell, R.G. (2021). *Int. J. Mol. Sci.* 22: 13168. https://doi.org/10.3390/ijms222313168.

96 Ferrero-Miliani, L., Nielsen, O.H., Andersen, P.S., and Girardin, S.E. (2007). *Clin. Exp. Immunol.* 147: 227–235. https://doi.org/10.1111/j.1365-2249.2006.03261.x.

97 Tang, L., Zhou, L., Yan, X. et al. (2020). *Dyes Pigm.* 182: 108644. https://doi.org/10.1016/j.dyepig.2020.108644.

98 Deng, Y. and Feng, G. (2020). *Anal. Chem.* 92: 14667–14675. https://doi.org/10.1021/acs.analchem.0c03199.

99 Korting, H.C. and Braun-Falco, O. (1995). *Clin. Dermatol.* 14: 23–27. https://doi.org/10.1016/0738-081x(95)00104-nx.

100 Mai, H., Wang, Y., Li, S. et al. (2019). *Chem. Commun.* 55: 7374–7377. https://doi.org/10.1039/C9CC02289A.

101 Yuan, J., Peng, R., Cheng, D. et al. (2021). *Chem. Asian J.* 16: 342–347. https://doi.org/10.1002/asia.202001350.

102 Bai, X., Ng, K.K.-H., Hu, J.J. et al. (2019). *Annu. Rev. Biochem.* 88: 605–633. https://doi.org/10.1146/annurev-biochem-013118-111754.

103 Rietschel, E.T., Kirikae, T., Schade, F.U. et al. (1994). *FASEB J.* 8: 217–225. https://doi.org/10.1096/fasebj.8.2.8119492.

104 Keliher, E.J., Reiner, T., Thurber, G.M. et al. (2012). *ChemistryOpen* 1: 177–183. https://doi.org/10.1002/open.201200014.

105 Conn, P.M. (2017). Animal model for liver disease. In: *Animal Models for the Study of Human Disease*, Academic Press, 2e. Hardcover ISBN: 9780128094686; eBook ISBN: 9780128096994.

106 Jieun Moon, Jehwi Jeon, Eunji Kong, Sujung Hong, Jingu Lee, Eun Kyung Lee, Pilhan Kim *Biomed. Opt. Express* 2021, 12, 7918–7927. https://doi.org/10.1364/BOE.442608

107 Fukada, S.-i., Uezumi, A., Ikemoto, M. et al. (2007). *Stem Cells* 25: 2448–2459. https://doi.org/10.1634/stemcells.2007-0019.

108 Langbein, T., Weber, W.A., and Eiber, M. (2019). *J. Nucl. Med.* 60: 13S–19S. https://doi.org/10.2967/jnumed.118.220566.

9

Non-optical Imaging Probes

Fluorescent probes have many excellent advantages, but also have critical limitations for whole-body imaging. The tissue penetration depth of input and emission light is limited to less than 1 mm, even with long-wavelength or two-photon probes. To avoid the penetration limit of optical modalities, there are several other options for whole-body imaging by molecular probes.

9.1 Ultrasound Imaging Probes

Ultrasound has been widely used for noninvasive clinical imaging with a long history usage [1]. Ultrasound imaging is safe and the response is in real time, typically 50 frames per second [2]. The most popular application of ultrasound is fetal scanning. As the amniotic fluid contains very few scattering particles, the image of the fetus stands out in the black background. Combined together with computed tomography (CT) technique, even a three-dimensional (3D) image of the fetus could be constructed to provide realistic facial features. With the real-time motions, time dimension added movie is called 4D ultrasound imaging (Figure 9.1).

As the largest internal organ, the liver is a popular target of ultrasound imaging. The general ultrasound provides the information for the morphological changes and scars on the liver. Although the penetration of ultrasound is deeper than lights, still ultrasound images are mainly for the surface of liver. Using transient elastography, FibroScan provides the information on the stiffness of the liver, and the technique is useful for the fibrosis or cirrhosis of liver.

When ultrasound imaging is applied to blood flow or perfusion measurement, the poor backscatter from the blood pool does not generate a clear contrast. Ultrasound contrast agent (UCA) can enhance the scattering of the blood pool. Usually, a UCA consists of stabilized air microbubbles, decorated by targeting motifs such as antibodies on the surface. When UCAs are injected intravenously, they can traverse the capillaries, as they are smaller than red blood cells. Even though their size is smaller than red blood cells, their scattering cross-section is up to nine orders of magnitude higher. With the high scattering, UCA enables the quantification of the perfusion of various organs [3]. It is interesting to use air as the contrast agent for more vivid imaging (Figure 9.2).

Sensors and Probes for Bioimaging, First Edition. Young-Tae Chang and Nam-Young Kang.

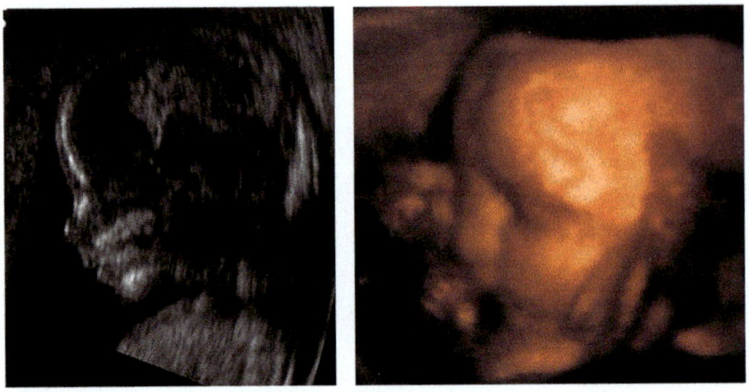

Figure 9.1 (a) 2D and (b) 3D images of fetus by ultrasound imaging. Source: (a) X.Compagnion / Wikimedia Commons, Public Domain and (b) Unknown author / Wikimedia Commons, Public Domain.

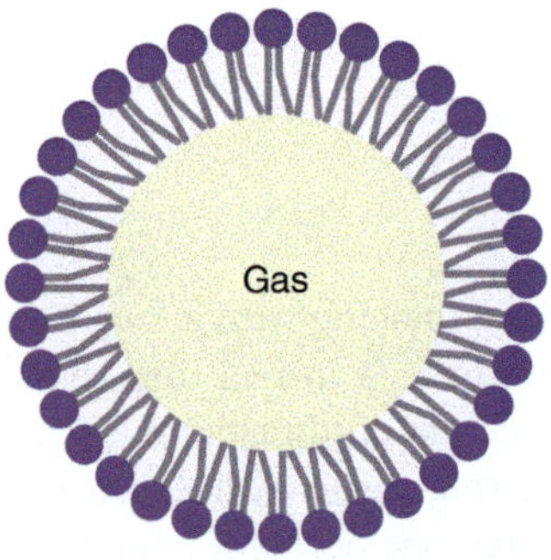

Figure 9.2 Structure of UCA (single layer of phospholipid).

9.2 X-Ray Contrast Agents

X-ray imaging has been widely used in radiology. With the penetration difference among tissues, X-ray imaging is an excellent noninvasive tool to look at the inside of the body, especially for the electron-rich bone structure. Combined with a CT algorithm, X-ray CT can provide the image for 3D structure of the body. Without a physical surgery, sectional images of the body can be reconstructed by digital tomography. While powerful and popular, X-ray CT still has low contrast for the soft tissues such as blood vessels.

For the better contrast in X-ray imaging, electron-rich heavy elements are desired. For the gastrointestinal imaging by ingestion, barium sulfate has been widely used [4]. Barium sulfate is relatively nontoxic and provides the disease information from esophagus, through stomach, down to intestine (Figure 9.3).

For the systemic application, iodine has been the most popular choice for X-ray contrast agent. Simple sodium and lithium iodide were the first tried contrast agents, but due to the toxicity, they are not being used for clinical trials. As a result, covalently bound iodine-containing organic molecules are in real use. To increase the water solubility of the contrast agents, the iodinated molecules are

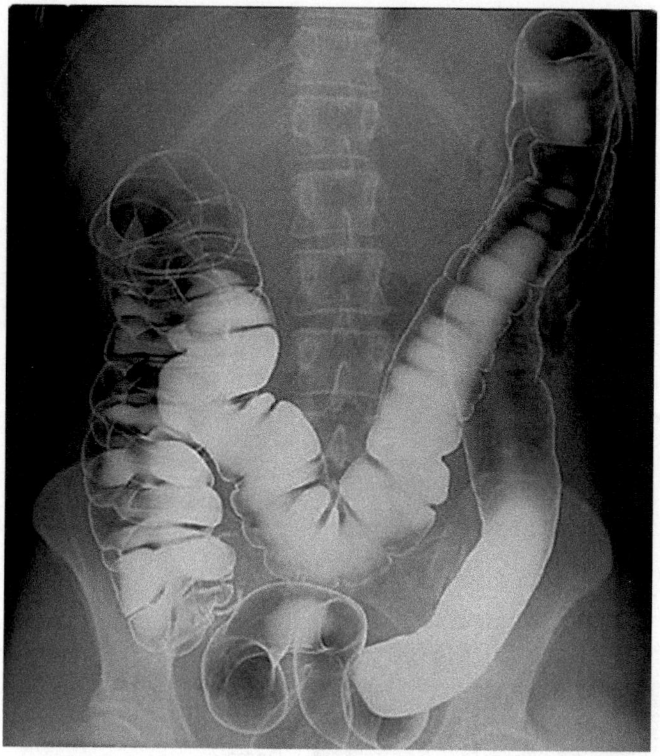

Figure 9.3 X-ray image by barium sulfate. Source: Glitzy queen00 / Wikimedia Commons, Public Domain.

often negatively charged or decorated with hydrogen-bond-forming functional groups such as hydroxyl or amide. To increase the iodine loading per molecule, a dimerization of two units is also commonly used. The representative X-ray contrast agents are described in Figure 9.4.

For the biologically friendly formulation, the contrast agents need to be readily soluble in aqueous solution with low viscosity, otherwise can cause a renal toxicity. To achieve enough contrast, the agents need to increase the contrast of the target tissue by twofold, and their retention time in the target tissue should be long enough to complete the CT scan of two to four hours. Eventually, they need to be cleared from the body in a reasonably short time, usually within 24 hours [5].

For clear contrast in angiography (image of the blood vessel), the concentration of the X-ray contrast agents needs to be very high, near the molar concentration range. Therefore, the intravenously injected agents should be extremely safe and nontoxic even at a very high dose. This limits the application of soluble iodinated agents to blood vessel imaging rather than molecular targeting. This is the disadvantage of X-ray contrast agents compared to other modality of imaging technique, such as magnetic resonance imaging (MRI) or nuclear imaging techniques, positron emission tomography (PET) or single-photon emission computed tomography (SPECT) (Figure 9.5).

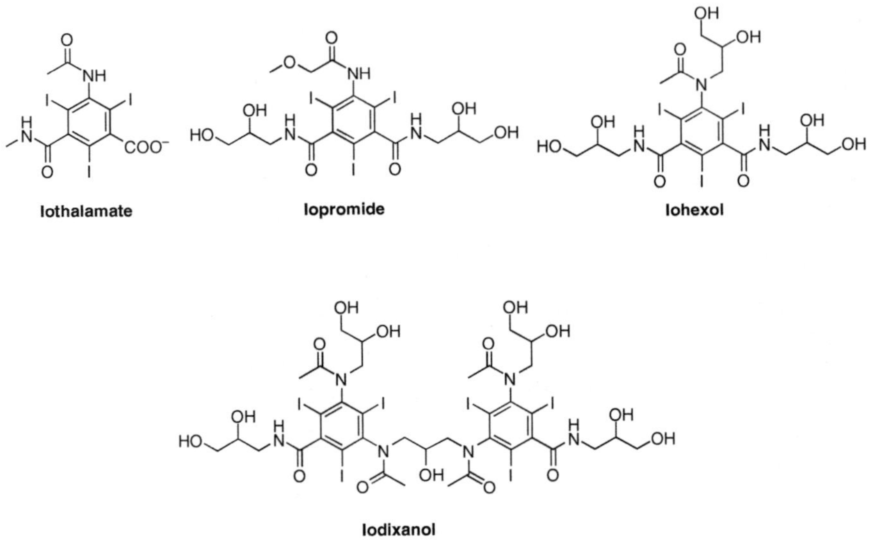

Figure 9.4 Structure of X-ray contrast agents.

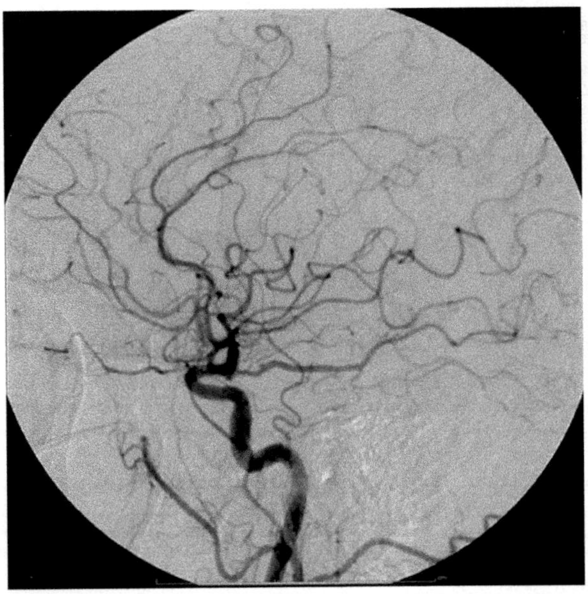

Figure 9.5 Angiogram by X-ray contrast agents. Source: Glitzy queen00 / Wikimedia Commons, Public Domain.

Contrast agents with a long blood circulation period are called "blood-pool" agents. Liposomes have served as a vehicle for water-soluble iodinated contrast agents, but have suffered from phagocytic clearing in the blood. To increase the blood circulation lifetime, so-called stealth liposomes, decorated with polyethylene glycol (PEG), have been developed by avoiding the phagocytic clearing. Stealth liposomes containing iohexol were an example of an effective blood-pool contrast agent [6]. Compared to the free iohexol, the ~100 nm size liposome-based contrast

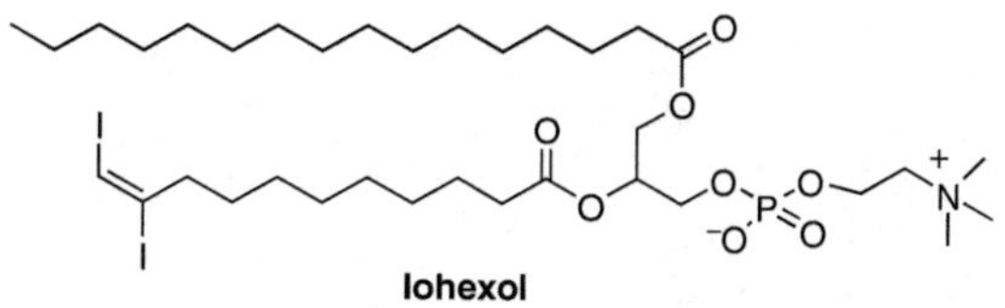

Figure 9.6 Stealth liposome with iohexol cargo.

agent provided good visualization of vascular and cardiac structures for a long-term imaging (Figure 9.6).

Alternative approach to the cargo method is forming the liposome with covalently labeled phospholipids. The empty cavity of the liposome could be used for drug delivery or enhanced contrast with soluble agents such as iohexol [7] (Figure 9.7).

Gold or bismuth nanoparticles are emerging new source for X-ray contrast agents, but they have not been reported to be applied to clinical usage yet.

9.3 MRI Contrast Agents

MRI is an excellent imaging platform for whole-body tomography. MRI utilizes the protons in water and lipids as the intrinsic reporters and is proper for imaging soft tissues, which is complementary to X-ray CT. MRI is based on nuclear magnetic resonance(NMR) principle, but the term "nuclear" was intentionally removed to avoid the negative impression connected to radioactivity. NMR has nothing with nuclear reaction or radioactivity.

The nucleus of hydrogen is proton, and proton exists spinning or with spinning effect. As a proton is a positively charged material, the spinning of a proton induces a magnetic field. In other words, a spinning proton is like a small magnet. The orientation of each magnet (made of proton) is random, but they are aligned in a high magnetic field along the field axis. Depending on the direction of the proton spinning, two energy states are generated, parallel and antiparallel. The energy gap increases proportionally to the external magnetic field strength. In the lower energy state, the parallel proton absorbs electromagnetic wave matching the energy gap to move to the antiparallel state. In the excited state, proton releases the absorbed energy, getting back to the parallel state, and the process is called relaxation. The relaxation is recorded to gain information on the protons. Combined with CT technique, proton position is reconstructed in a 3D map with the notion of different relaxation properties. High relaxation may be correlated with high density of protons, which may be imaged bright. The dark empty space implies there is no proton in the area.

The density of proton is not all. Relaxation process is complicated and contains a lot of information. Protons in different environments have unique relaxation patterns and speeds. For example, the speed of relaxation can be imaged in a different gray scale, and each proton is expressed in different brightness. As a result, MRI provides highly sophisticated 3D images of the body, distinguishing water (proton attached to oxygen) rich and lipid (proton attached to carbon) rich area. The higher the magnetic field, the higher resolution and quality images could be obtained. However, due to the potential unknown danger of high magnetic field to human body, clinical application of high-field MRI is constricted.

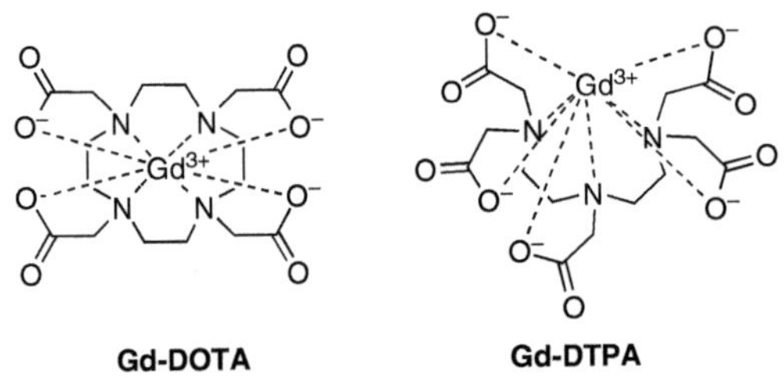

Figure 9.7 Representative example of MRI contrast agents.

If the contrast of the target area is not strong enough, MRI contrast agents could be employed. MRI contrast agents are usually metal complexes that accelerate the relaxation process. By the way, the relaxation is not a one-directional event, but divides into T1 and T2 relaxation processes. T1 relaxation measures the z-component recovery, and T2 relaxation measures the loss of x-y component. Depending on which information is counted more, T1 weighted vs. T2 weighted images can provide different images, reflecting different aspects of the system. Accordingly, there are separate T1 and T2 contrast agents. T1 contrast agents are paramagnetic complexes of Gd^{3+} or Mn^{2+}. T2 contrast agents include super paramagnetic materials such as iron oxide.

For clinical application, Gd^{3+} is the dominant component of T1 contrast agents. As T1 relaxation measures the recovery of z-component, the faster relaxation induced by contrast agents provides the target area a brighter contrast. The Gd^{3+} agents are mainly used for angiography of enhanced blood vessel imaging or brain tumor detection by the leakage of blood–brain barrier (BBB). The representative Gd^{3+} agents are described in Figure 9.7 [8].

For imaging of large blood vessels such as aorta, the dose can be as low as 0.1 mmol kg^{-1} (sub mM concentration), but for finer vascular structures, higher concentrations are often used for better contrast [9]. The high concentration requirement casts concerns about safety of the contrast agents. The main cause of toxicity is due the released Gd^{3+} ion by the stability issue of the metal complex agents. The free Gd^{3+} ion is toxic with LD_{50} = 0.34 mmol kg^{-1} in mouse, and the LD_{50} is increased 50 times in the chelated complex [10]. The clinical application of Gd^{3+} agents can cause acute or chronic kidney problem. Therefore, the efforts to increase the stability of the complex and possible low dose usage by MRI technical improvement are under progress.

Mn^{2+} ions have the property of T1 contrast agents and are usually used for research purposes. Intrinsic concentration of Mn^{2+} in biological system is very low, so the background of Mn^{2+} in MRI would be negligible. Mn^{2+} can mimic Ca^{2+} and can enter the cells by Ca^{2+} transporter. Considering the signaling molecule role of Ca^{2+} in the neuronal system, Mn^{2+} has been used for functional brain imaging [11].

The superparamagnetic iron oxide is a strong T2 contrast agent. For clinical application, the iron oxide is made in nanoparticle size (USPIO: ultra-small superparamagnetic iron oxide) and formulated as suspended colloids. When injected, USPIO accumulates in the absorbing target tissues and induces the accelerated T2 relaxation [12]. As T2 relaxation is a decrease of x-y components, the area of T2 contrast agent accumulation is darker than the surrounding tissues.

Hemoglobin contains iron ion and has two states: oxygenated and deoxygenated forms. Oxygenated hemoglobin is diamagnetic (no unpaired electron), and deoxyhemoglobin is paramagnetic (unpaired election number is 4), which is a natural T2 contrast agent. Therefore, it is possible to distinguish the high oxygen area from low oxygen area using MRI. This MRI technique is called BOLD (Blood-Oxygen Level Dependent). As the high oxygen area is interpreted as the activated area with high oxygen consumption, the BOLD images can provide the information on activation of specific areas especially in brain function. Therefore, this technique is also called as functional MRI (fMRI) imaging.

9.4 SPECT Probes

Radioactivity is the ability to emit high-energy radiation by nuclear reaction of atoms. The identity of an atom is defined by the number of protons in the nucleus, which is the same as the number of electrons of the atom in the neutral state. Atoms with the same proton number, but with different neutron number are called isotopes. Some isotopes are stable, and others are unstable with excess nuclear energy. The unstable atom undergoes autonomous nuclear reaction, releasing high-energy radiation. The nuclear reaction can be either fission or fusion, with the change of nuclear composition, to form more stable atoms. The representative radiations from radioactive isotopes are alpha, beta, and gamma rays. Alpha ray is a fragmented heavy particle composed of two protons and two neutrons, which is the nucleus of helium. Beta ray is high-energy electrons. Gamma ray is an electromagnetic wave with shorter wavelength than X-ray.

If a gamma-ray-generating radioisotope is introduced in the body, the emitting gamma ray can be recorded by a gamma camera, imaging the position and density of the radioisotopes. The camera and images may look similar to X-ray techniques, but the source of high-energy electromagnetic wave is different. In an X-ray image, the X-ray was provided by an external light source, and only penetrated light through the body sample reaches camera, generating images. As the hard tissue absorbs more X-ray, the bone area may be darker and soft tissue area may be brighter. If you see an X-ray image with bone white, it should be an arbitrarily inverted image as in film versus real picture. In gamma camera, the origin of light is the radioactive isotope in the body, and there is no external light source.

If multiple gamma camera images from different angles are processed by CT algorithm, 3D images of gamma radiator can be constructed. For the directional information of the radioisotope, a collimator is placed in front of the gamma camera, and the collimator defines the angle of incoming gamma ray. This tomographic imaging technique is called SPECT. Therefore, SPECT probes are gamma-radiating isotopes, and popular choices are ^{99m}Tc ($t_{1/2}$ = 6 hours), ^{111}In ($t_{1/2}$ = 67 hours), ^{67}Ga ($t_{1/2}$ = 78 hours), ^{201}Tl ($t_{1/2}$ = 73 hours), ^{123}I ($t_{1/2}$ = 13 hours), and ^{125}I ($t_{1/2}$ = 59 day). Another iodine isotope, ^{131}I ($t_{1/2}$ = 8 day), radiates both beta and gamma rays and can be used for radiotherapy using the cytotoxic beta ray (high-energy electron). As a reference, the stable and most abundant isotope of iodine is ^{127}I.

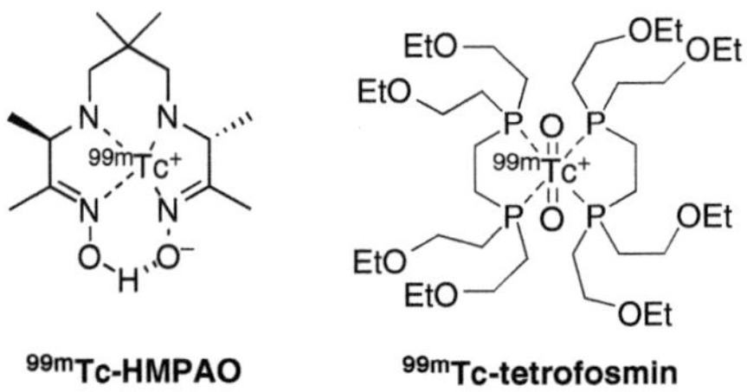

Figure 9.8 Structure of ^{99m}Tc-HMPAO and ^{99m}Tc-tetrofosmin.

Among SPECT probes, ^{99m}Tc is the most common isotope used in clinical diagnosis, representing up to 80% of all nuclear medicine practice [13]. While most of the radioisotopes are produced by a heavy facility cyclotron, ^{99m}Tc can be produced by a portable generator from a longer-lived parent nuclide ^{99}Mo ($t_{1/2}$ = 67 hours). The simple procedure (often called as "milking") of ^{99m}Tc production made it the most widely used SPECT nuclide. The metal ion ^{99m}Tc needs to be complexed with chelators before injection into the biological system. The representative formulations are ^{99m}Tc-HMPAO (hexamethylpropylene amine oxime, also called as exametazime) and ^{99m}Tc-tetrofosmin (Figure 9.8).

Utilizing its access to the brain blood flow, ^{99m}Tc-HMPAO is commonly used in functional brain imaging, such as cerebral perfusion in stroke [14]. The supply of blood is proportional to the metabolism rate of the brain region, and in Alzheimer dementia, the uptake of ^{99m}Tc-HMPAO is reduced compared to the normal brain [15].

^{99m}Tc-HMPAO was also used for leukocyte labeling to monitor the inflammation or infection site in the body [16]. For this practice, the patient blood was taken and ex vivo labeled with ^{99m}Tc-HMPAO, and then reinjected into the patient. Considering the potential contamination of the sample during the procedure, an in situ reagent, which can be directly injected and label leukocytes, would be ideal.

^{99m}Tc-tetrofosmin has been mainly used for cardiac blood flow monitoring, so-called myocardial perfusion imaging. By injection, ^{99m}Tc-tetrofosmin is rapidly taken up by myocardial tissue and reaches its maximum level in about five minutes. The blood supply to the heart is through the coronary artery, and the probe is used to monitor a reversible ischemia and infarcted area of the heart [17] (Figure 9.9).

While metal ions are required to make a complex with chelators, iodine can be easily introduced to small organic molecule through a covalent bond formation. One disadvantage, however, is that iodine is larger than hydrogen (van der Waal's radius of iodine is 2.15 Å), and the introduction of iodine into the probe molecule may induce a great steric perturbation and introduce extra hydrophobicity. One advantage of iodine is the availability of various radioactive isotopes. While ^{123}I is the most common SPECT isotope as efficient gamma ray radiation among iodine, ^{125}I is a longer lifetime, low-energy probe and could be used for preclinical studies. By simple replacement of the isotope, ^{131}I isotope can be used for radiotherapy through its beta emission [18]. As the isotopes exhibits the almost the same chemical properties other than the molecular weight, the isotope-exchanged molecules are considered as chemically or biologically equivalent. With the flexibility of the chemical modification of iodine, any kind of high-affinity small molecule against target protein

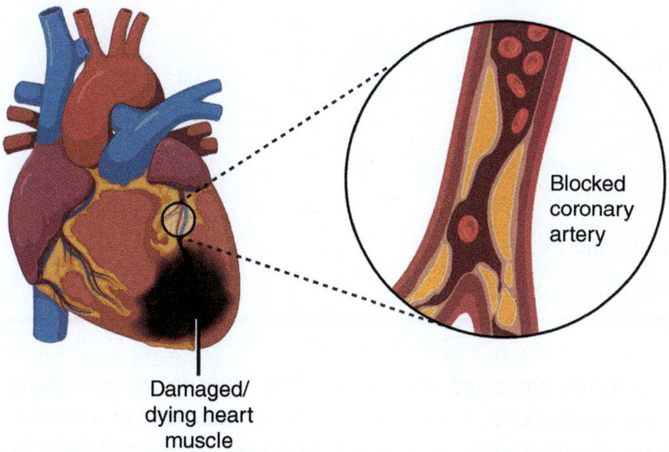

Figure 9.9 Heart and coronary artery with damage.

Figure 9.10 Design of SPECT probe from PK11195.

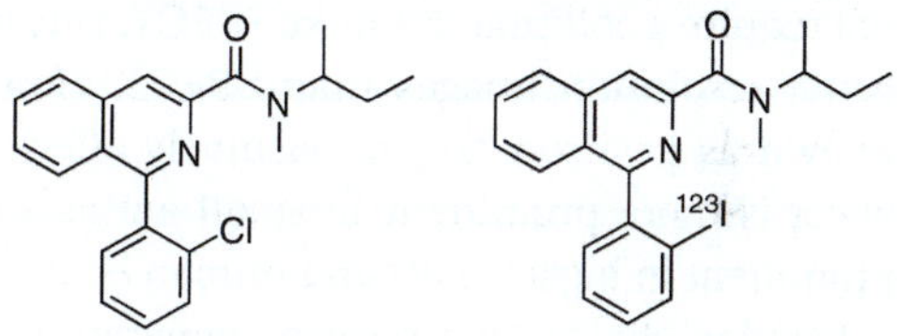

could be a candidate for SPECT probe. It includes agonist or antagonist of specific receptor and inhibitor of enzyme overexpressed in the target area.

For example, the translocator protein TSPO (18 kDa peripheral benzodiazepine receptor) is a famous target protein, whose brain expression is increased in chronic neurodegenerative diseases such as Alzheimer dementia [19]. The well-known ligand PK11195 has nM range affinity for TSPO and was used as the parent molecule for iodinated probe design. As an example, the chlorine of PK11195 was replaced with ^{123}I as a SPECT probe [20] (Figure 9.10).

Interestingly, iodide ion itself is used as a SPECT probe. Thyroid is the gland which makes and secretes thyroid hormones. The representative thyroid hormones are T3 and T4 (Figure 9.11), and they contain multiple iodide in the structure. The thyroid hormones control the metabolic speed of our body. For the synthesis of the hormones, thyroid cells express NIS (sodium/iodide symporter, SLC5A5) for facilitated iodide transport into the cells. In hyperthyroidism, the heart rate and metabolic speed increase, with body weight loss albeit a good appetite. Graves' disease is a representative hyperthyroidism with exophthalmos symptoms. In the thyroid cells in Graves' disease patients, NIS is overexpressed and accumulates iodide compared to other neighboring tissues. $^{123}I^-$ is used for Graves' disease and thyroid tumor diagnosis.

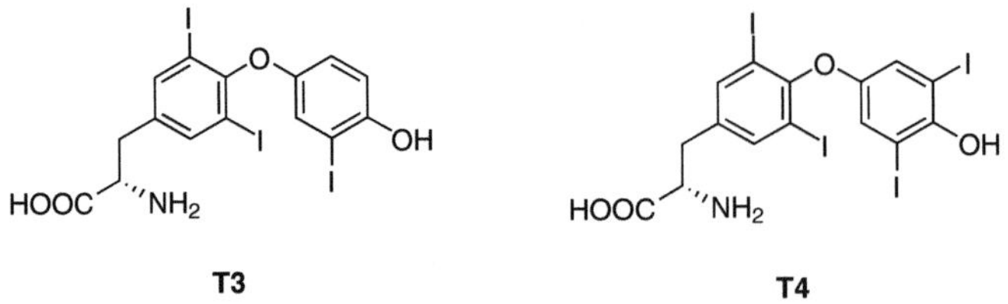

Figure 9.11 Structure of T3 and T4.

9.5 PET Probes

PET is another imaging technique based on radionuclides that decay, generating positron, a positively charged electron. Positron is an anti-particle of electron. When released positron travels around and meets electron, they annihilate by releasing two gamma ray photons at 180° to each other. The pair of gamma rays is (almost) simultaneously detected by the surrounding gamma camera, providing the information on the angle to annihilation location. Due to the pair signal, PET does not require a collimator, unlike SPECT. Due to this advantage, PET usually provides better resolution images than SPECT. However, the distance of positron travel, known as positron range, negatively affects the spatial resolution. The diffusion effect is more prominent in small animals due to the size of the object and less prominent in big animals and human bodies.

Popular choices for positron-generating isotopes are ^{11}C ($t_{1/2}$ = 20 minutes), ^{13}N ($t_{1/2}$ = 10 minutes), ^{15}O ($t_{1/2}$ = 2 minutes), ^{18}F ($t_{1/2}$ = 110 minutes), ^{64}Cu ($t_{1/2}$ = 13 hours), ^{68}Ga ($t_{1/2}$ = 68 minutes), ^{76}Br ($t_{1/2}$ = 16 hours), and ^{124}I ($t_{1/2}$ = 4.2 day). Due to the extremely short half-life, ^{13}N and ^{15}O are mainly used in the form of ammonia and water.

Among iodine isotopes, ^{124}I emits positron and can be used for PET imaging. It would be interesting to test side by side with ^{123}I-labeled probe for the comparison of the technical performance of SPECT and PET, as the two probes are chemically and biologically almost identical to each other. ^{76}Br would be sometimes a better choice than ^{124}I due to the less steric disturbance and more ideal half-life of 16 hours compared to four days of ^{124}I, as too long half-life may result in too low positron signal.

The metal ion such as ^{64}Cu or ^{68}Ga needs a complex formation similar to ^{99m}Tc in SPECT. DOTA has been popularly used for the chelating agent for the metals, and recently introduced NOTA provides stronger binding with higher stability [21] (Figure 9.12).

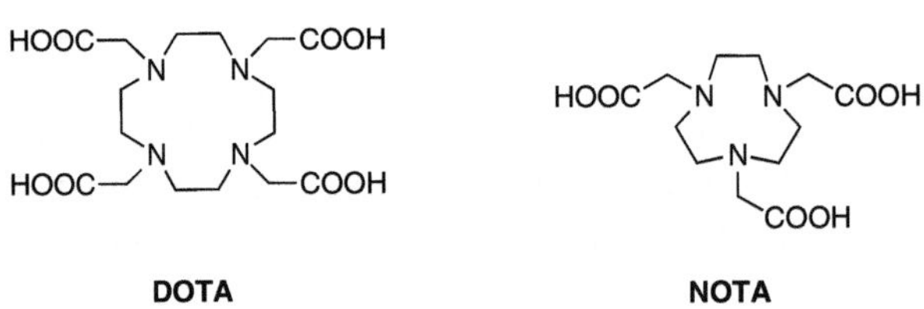

Figure 9.12 Structure of DOTA and NOTA.

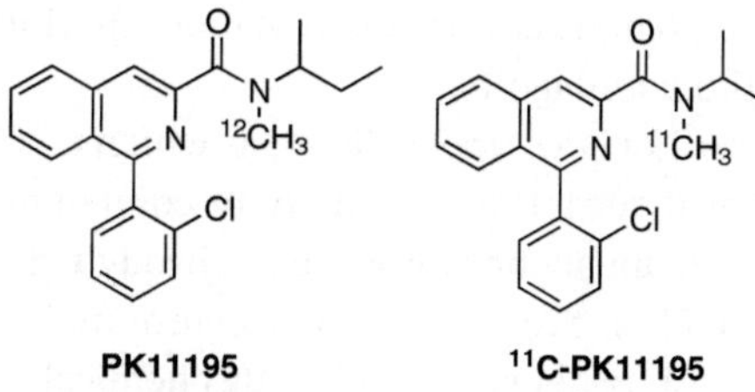

Figure 9.13 11C labeled PK11195.

For organic small-molecule labeling, ^{11}C is an ideal choice, because the replacement of stable isotope ^{12}C with ^{11}C may generate negligible isotope effect. The example is shown in ^{11}C labeled PK11195 (Figure 9.13) [22]. The problem is the relatively short half-life of ^{11}C, which is 20 minutes. The synthesis and purification should be completed within this time frame, which is technically quite challenging (Figure 9.13).

That is why an alternative popular option is ^{18}F, which has much longer half-life of about two hours. Fluorine is the most electronegative atom, but the atomic size is similar to that of hydrogen (van der Waal's radii of hydrogen and fluorine are 1.20 and 1.35 Å, respectively). So, fluorine is considered as a reasonable bioisosteric replacement of hydrogen, which may minimize the steric disturbance as seen in iodine replacement in SPECT. Due to the high electronegativity of fluorine, it is also common to replace the hydroxyl group with fluorine, considering the electronic property. With these various advantages, ^{18}F has been the most popular isotope for the PET probe design. As a result, a broad range of ^{18}F-based PET probes were developed by modifying known high-affinity ligands and used in clinical applications.

9.5.1 PET Probes for Tumor

Among PET probes, ^{18}F-FDG (2-fluoro-2-deoxyglucose) is by far the most used probe [23]. As a glucose analogue, ^{18}F-FDG enters the cells through the glucose transporter, despite its hydrophilic property. In the cell, 6-position of ^{18}F-FDG is phosphorylated as the first step of glycolysis, and due to the 2-position fluorine, further metabolic modification is hampered. As a result, the phosphorylated ^{18}F-FDG is captured in the cell and accumulated. With the metabolic reasons, ^{18}F-FDG can label cells and tissues with high glucose uptake or demand. Cancer is the typical case due to increased aerobic glycolysis, called Warburg effect. Therefore, ^{18}F-FDG has been used for various tumor imaging techniques (Figure 9.14).

One problem of using ^{18}F-FDG for oncology is its natural high uptake in normal tissues such as brain and bladder, which makes it difficult to use ^{18}F-FDG for glioma (a brain tumor) or bladder or prostate cancers. Also, inflammation can increase the local uptake of ^{18}F-FDG, leading to false-positive imaging results. These limitations

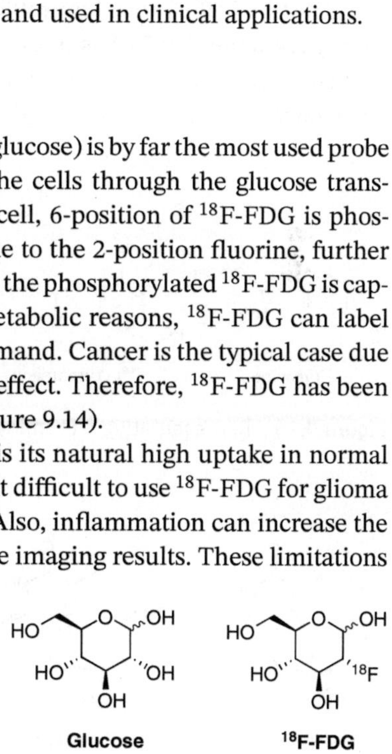

Figure 9.14 ^{18}F-FDG structure and tumor imaging.

prompted the development of more specific cancer probes and inflammation probes without ambiguity.

Inositol is a carbocyclic sugar which is abundant in the brain and other tissues. It is an important component in the cell membrane to form an anchor for proteins through glycosylphosphatidylinositol (GPI) or as a precursor of calcium-mobilizing second messenger, inositol trisphosphate (IP3). By chemical composition, inositol has the same chemical formula with glucose ($C_6H_{12}O_6$), that is it is a structural isomer of glucose. Therefore, ^{18}F-inositol is an isomer of ^{18}F-FDG with the same chemical formula. ^{18}F-inositol had a very low brain uptake, but kept the uptake to breast cancer, demonstrating the potential for more tumor specificity [24] (Figure 9.15).

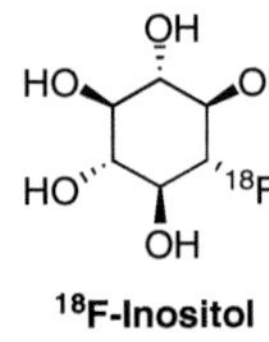

Figure 9.15 Structure of ^{18}F-inositol.

Amino acids are essential nutrients for cells, and especially for proliferation of cells through enhanced protein synthesis. Naturally, many tumor cells have high expression of amino acid transporters. Various ^{18}F-labeled amino acids are developed for tumor imaging including ^{18}F-tyrosine, ^{18}F-FET (fluoroethyltyrosine), and ^{18}F-L-DOPA (dihydroxy phenylalanine), which accumulated in tumors, therefore potentially complementary to ^{18}F-FDG [25] (Figure 9.16).

Another target for tumor detection would be enhanced DNA synthesis. Highly proliferating cells may need to aggressively up take the building blocks for DNA synthesis. A nucleoside analogue, fluorothymidine (FLT), can be taken up into cells and metabolized by thymidine kinase 1, which is overexpressed in tumor cells. For PET, ^{18}F-FLT is the probe for proliferation and tumor imaging [26]. But, in a sense, strictly speaking, ^{18}F-FLT is not a real proliferation probe as it is not incorporated into DNA. In that context, ^{18}F-FMAU is directly incorporated into DNA synthesis and showed selective tumor imaging in various tissues [27] (Figure 9.17).

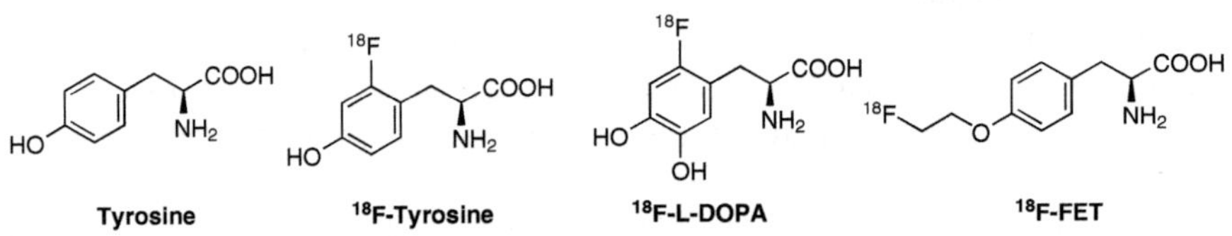

Figure 9.16 Representative ^{18}F-amino acids.

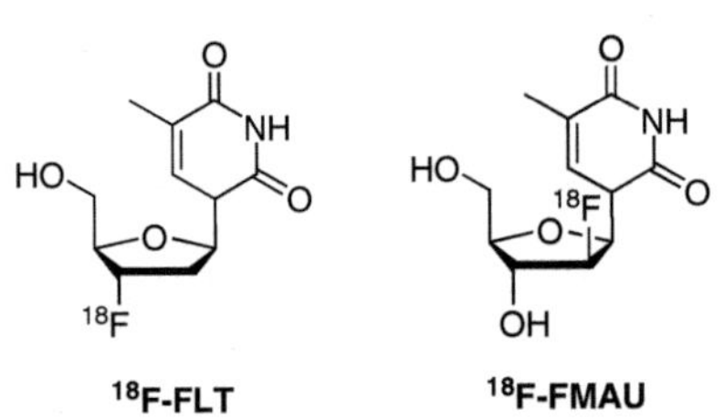

Figure 9.17 Structure of ^{18}F-FLT and ^{18}F-FMAU.

9.5.2 PET Probes for Brain Function

Brain is a high-energy consumption organ and actively takes up glucose. That is why ^{18}F-FDG signal is strong in the brain, interfering the tumor detection. From the different view of point, ^{18}F-FDG signal can be used for measuring the degree of brain function in terms of glucose metabolism rate. Among different regions of the brain, metabolically more active area may take up higher amounts of ^{18}F-FDG. This principle was applied to define the sensor processing region of the brain. For example, when a light stimulation is given to eyes, the area with increased ^{18}F-FDG signal may be vision-processing area in the brain.

For disease diagnosis, the generally decreased ^{18}F-FDG signal could be interpreted with deteriorated brain function as in dementia. In addition to cognitive test, ^{18}F-FDG would be a functional brain test for the diagnosis. Alzheimer's disease (AD) is one of the aging-related dementias and characterized with accumulated Aβ-amyloid aggregates and tau tangles in the brain tissue. Significant efforts have been poured to develop specific sensors and probes for Aβ and tau (Section 8.2) and some of them are converted into PET probes. Starting from ^{11}C-PIB, many PET probes were developed for Aβ for clinical trial and tau probes are following. With the more advanced stage of Aβ study, the combined information of low ^{18}F-FDG signal and high Aβ probe signal is the current molecular signature for the diagnosis for AD (Figure 9.18) [28].

Aβ was also pursued as a therapeutic target for AD treatment with small-molecule drugs and antibodies. Unfortunately, most of the trials do not seem to work as expected. After the failure of numerous trials of Aβ drugs, tau is an emerging alternative biomarker for imaging and target for therapy. Due to the similar property as aggregated proteins of Aβ and tau, the selectivity for tau over Aβ is challenging to achieve. However, based on screening and molecular design, the efforts for developing a tau-specific probe is still in progress.

Parkinson's disease (PD) is a kind of neurodegenerative disease, which begins with a movement disorder, typically followed by dementia. PD is characterized by loss of dopamine neurons in the brain, damaging the neural transmission system. L-DOPA, a precursor of dopamine, is a common treatment for PD to improve the mobility problem. L-DOPA is converted into dopamine by aromatic amino acid decarboxylase (AADC) in the brain to replenish the dopamine supply. ^{18}F-DOPA, an analogue of L-DOPA, was developed and used as a false substrate for AADC. ^{18}F-DOPA can be converted into ^{18}F-dopamine in the brain and provides the information for the neuronal activity. High PET signal implies a healthy status of dopamine neuron system in the brain (Figure 9.19).

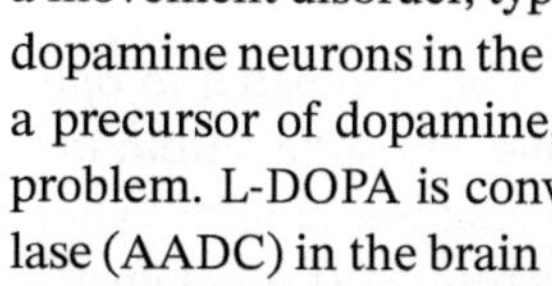

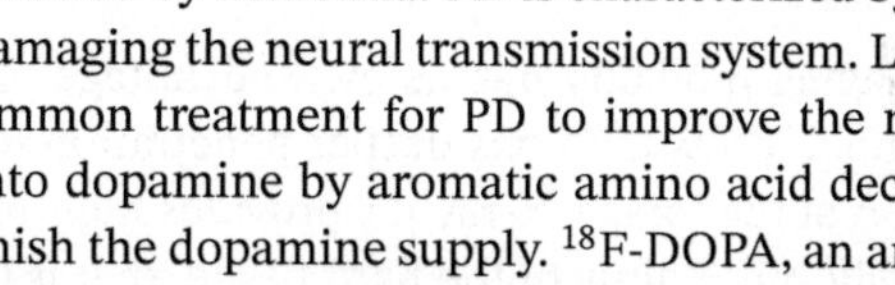

Figure 9.18 Representative of Aβ-targeting PET probe.

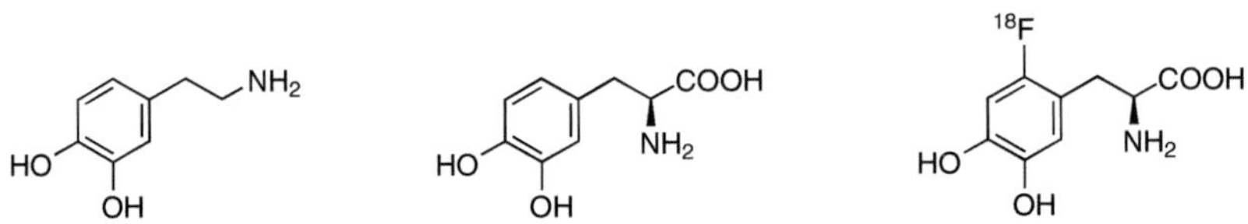

Dopamine L-DOPA ^{18}F-L-DOPA

Figure 9.19 Dopamine, L- DOPA, and ^{18}F-L-DOPA.

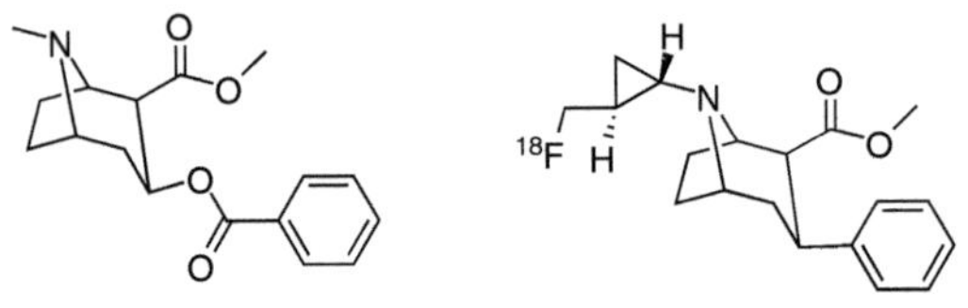

Cocaine ^{18}F-PR17.MZ

Figure 9.20 Structure of cocaine and ^{18}F-cocaine derivative.

Dopaminergic neurons contain vesicles packed with high concentration of dopamine similar to the loaded guns near the end of axonal synapse. When the neuron is activated, dopamine is released by exocytosis of the vesicles from the presynaptic neuron into the synaptic gap. Some of the dopamine reaches the postsynaptic receptors to transfer the signal, but plenty of unutilized dopamine remains in the gap space. Dopamine transporter (DAT) is localized at the presynaptic neuron and take up the remaining dopamine from the synaptic cleft. Vesicular monoamine transporter 2 (VMAT2) is responsible for the repacking of the dopamine into vesicles. Cocaine has a high binding affinity to DAT, and ^{18}F-cocaine derivatives are PET probes for brain function (Figure 9.20) [29].

9.6 Multimodality

Each imaging modality has its own advantage and limitation. Compared to optical imaging, ultrasound has deeper penetration to the tissue, but suffers from low resolution of the image. X-ray CT provides high-resolution anatomical information of the body, but mainly of the hard tissue, that is bone structure. MRI supplements X-ray CT by covering the soft tissues with high-resolution tomographic images. Both X-ray CT and MRI are great for the anatomical imaging, but they lack the molecular information on the body. SPECT or PET provides the missing molecular linkage by presenting the location of receptors or transporters. However, the spatial resolution of SPECT and PET is much poorer than X-ray CT or MRI. Therefore, multimodal imaging is common to supplement the weak points of each modality.

As a result, it is common to see the combined technique, including SPECT-CT or PET-CT in clinical application. MRI was slower to be introduced to the coupled machine with SPECT or PET, partially due to the requirement of high magnetic field. However, the recent technical solutions seem to overcome the difficulties and SPECT-MRI or PET-MRI is emerging as the state-of-the-art technique [30].

Photoacoustic mode is another combination of optical input with ultrasound output. Fluorescent probes absorb light and reemit (usually) redshifted light, but photoacoustic probes utilize the nonradiative portion of the absorbed light. The photoacoustic probes represent the molecular positioning, and the absorbed light energy generates ultrasound signal for the detection. The better tissue penetration of ultrasound makes more sensitive and higher resolution of tomographic images.

Single probe could be designed not only for technique combinations, but also for multimodal imaging. For example, iron nanoparticle coated with fluorescent motif could acts as optical and MRI probes. Some of the metal isotopes irradiate both beta and gamma rays, then gamma ray can be used for imaging, and beta ray can be used for therapy. Iodide has multiple isotopes, which can be used for SPECT or PET or even for radiotherapy. If a fluorescent molecule contains iodide in its structure, the optical probe could be converted into another modality probe with minimal change to the chemical property. In a similar way, a fluorine-containing fluorescent molecule could be easily converted to a PET probe [31].

References

1 Howry, D.H., Stott, D.A., and Bliss, W.R. (1954). *Cancer* 7: 354–358. https://doi.org/10.1002/1097-0142(195403)7:2%3C354::aid-cncr2820070220%3E3.0.co;2-9.

2 Versluis, M., Stride, E., Lajoinie, G. et al. (2020). *Ultrasound Med. Biol.* 46: 2117–2144. https://doi.org/10.1016/j.ultrasmedbio.2020.04.014.

3 Quaia, E. (ed.) (2005). *Contrast Media in Ultrasonography*. Berlin: Springer.

4 Yang, X., Lovell, J.F., Zhang, Y. et al. (2019). *ChemBioChem* 20: 462–473. https://doi.org/10.1002/cbic.201800589.

5 Lusic, H. and Grinstaff, M.W. (2013). *Chem. Rev.* 113: 1641–1666. https://doi.org/10.1021/cr200358s.

6 Kao, C.Y., Hoffman, E.A., Beck, K.C. et al. (2003). *Acad. Radiol.* 10: 475–483. https://doi.org/10.1016/S1076-6332(03)80055-7.

7 Elrod, D.B., Partha, R., Danila, D. et al. (2009). *Nanomed. Nanotechnol. Biol. Med.* 5: 42–45. https://doi.org/10.1016/j.nano.2008.06.007.

8 Xiao, Y.D., Paudel, R., Liu, J. et al. (2016). *Int. J. Mol. Med.* 38: 1319–1326. https://doi.org/10.3892/ijmm.2016.2744.

9 Lentschig, M.G., Reimer, P., Rausch-Lentschig, U.L. et al. (1998). *Radiology* 208: 353–357. https://doi.org/10.1148/radiology.208.2.9680558.

10 Penfield, J.G. and Reilly, R.F. Jr., (2007). *Nat. Clin. Pract. Nephrol.* 3: 654–668. https://doi.org/10.1038/ncpneph0660.

11 Lin, Y.J. and Koretsky, A.P. (1997). *Magn. Reson. Med.* 38: 378–388. https://doi.org/10.1002/mrm.1910380305.

12 Gkagkanasiou, M., Ploussi, A., Gazouli, M. et al. (2016). *J. Neuroimaging* 26: 161–168. https://doi.org/10.1111/jon.12318.

13 Vaz, S.C., Oliveira, F., Herrmann, K., and Veit-Haibach, P. (2020). *Br. J. Radiol.* 93: 20200095. https://doi.org/10.1259/bjr.20200095.

14 Moretti, J.L., Defer, G., Cinotti, L., and Cesaro, P. (1990). *Eur. J. Nucl. Med.* 16: 17–22. https://doi.org/10.1007/BF01566007.

15 Bonte, F.J., Harris, T.S., Hynan, L.S., and Bigio, E.H. (2006). *Clin. Nucl. Med.* 31: 376–378. https://doi.org/10.1097/01.rlu.0000222736.81365.63.

16 Weldon, M.J., Joseph, A.E., French, A., and Saverymuttu, S.H. (1995). *Gut* 37: 557–564. https://doi.org/10.1136/gut.37.4.557.

17 Elhendy, A., Schinkel, A.F., van Domburg, R.T., and Bax, J.J. (2005). *J. Nucl. Med.* 46: 2003–2008. PMID: 16330563.

18 Pimlott, S.L. and Sutherland, A. (2011). *Chem. Soc. Rev.* 40: 149–162. https://doi.org/10.1039/B922628C.

19 Chauveau, F., Boutin, H., Van Camp, N., and Dollé, F. (2008). *Eur. J. Nucl. Med. Mol. Imaging* 35: 2304–2319. https://doi.org/10.1007/s00259-008-0908-9.

20 Pimlott, S.L., Stevenson, L., Wyper, D.J., and Sutherland, A. (2008). *Nucl. Med. Biol.* 35: 537–532. https://doi.org/10.1016/j.nucmedbio.2008.02.013.

21 Guleria, M., Das, T., Amirdhanayagam, J., and Sarma, H.D. (2018). *Cancer Biother. Radiopharm.* 33: 8–16. https://doi.org/10.1089/cbr.2017.2337.

22 Camsonne, R., Crouzel, C., Comar, D., and Maziere, M. (1984). *J. Labelled Compd. Radiopharm.* 21: 985–991. https://doi.org/10.1002/jlcr.2580211012.

23 Couturier, O., Luxen, A., Chatal, J.F. et al. (2004). *Eur. J. Nucl. Med. Mol. Imaging* 31: 1182–1206. https://doi.org/10.1007/s00259-004-1607-9.

24 Vasdev, N., Chio, J., van Oosten, E.M. et al. (2009). *Chem. Commun.* 5527–5529. https://doi.org/10.1039/B913317H.

25 McConathy, J. and Goodman, M.M. (2008). *Cancer Metastasis Rev.* 27: 555–573. https://doi.org/10.1007/s10555-008-9154-7.

26 Mier, W., Haberkorn, U., and Eisenhut, M. (2002). *Eur. J. Nucl. Med. Mol. Imaging.* 29: 165–169. https://doi.org/10.1007/s00259-001-0675-3.

27 Fei, X., Wang, J.Q., Miller, K.D., and Sledge, G.W. (2005). *Eur. J. Nucl. Med. Mol. Imaging* 32: 15–22. https://doi.org/10.1007/s00259-004-1713-8.

28 Zhu, L., Ploessl, K., and Kung, H.F. (2014). *Chem. Soc. Rev.* 43: 6683–6691. https://doi.org/10.1039/C3CS60430F.

29 Riss, P.J., Piel, M., Bockhart, V. et al. (2010). *ChemMedChem* 5: 1686–1688. https://doi.org/10.1002/cmdc.201000237.

30 Cassidy, P.J. and Radda, G.K. (2005). *J. Royal Soc. Interface* 2: 133–144. https://doi.org/10.1098/rsif.2005.0040.

31 Kang, N.Y., Lee, J.Y., Lee, S.H., and Song, I.H. (2020). *J. Am. Chem. Soc.* 142: 3430–3439. https://doi.org/10.1021/jacs.9b11173.

10

Fluorescence Imaging Techniques and Analysis Methods

Basic light microscopes usually have a light source, sample, and detector in linear alignment. The resulting images are the results of the light absorption or scattering by the sample. In comparison, the basic fluorescence microscope requires a light source and detector located usually at 90° to each other with the sample at the pivot point. Fluorescence light is generated by the excitation of a dye molecule and the subsequent emission through the relaxation of the excited electron. As the input light is usually much stronger than the emission light, this 90° optical design is to minimize the interference by the input light (Figure 10.1). The basic fluorescence microscopes are with a single excitation with a single emission, and they measure the fluorescence intensity difference of the location in a two-dimensional (2D) sample. If the probe is a Ca^{2+} sensor, a stronger intensity in the endoplasmic reticulum (ER) lumen than that in the cytosol may be interpreted as a higher concentration of Ca^{2+} in the ER lumen. However, there is a possibility that the sensor concentration is different in each organelle, and the above observation could be the result of the preferred localization of the sensor in ER compared to cytosol. Time-course measurement of the fluorescence intensity change would serve as a self-normalized approach, and the Ca^{2+} secretion from ER into cytosol could be imaged by an increased signal in cytosol and a decreased signal in ER. Still, the absolute amount of Ca^{2+} in organelles requires more references. Ratiometric sensor is one solution for this problem.

10.1 Multicolor Imaging

Our eyes have three color sensors. Digital color camera or color display screen usually has three or four-color sensors (RGB or RGBC – R: red, G: Green, B: blue, C: cyan). If multiple probes with different colors are used in the staining of a biological sample, colorful images could be obtained. Fluorescence microscope could be equipped with a three- or four-color camera or a single photo multiplier tube (PMT) for high sensitivity. As PMT does not distinguish different colors, each color should be defined by a different dye itself or by proper filters. If the dyes have sharp excitation and emission spectra, many probes could be simultaneously used,

Sensors and Probes for Bioimaging, First Edition. Young-Tae Chang and Nam-Young Kang.

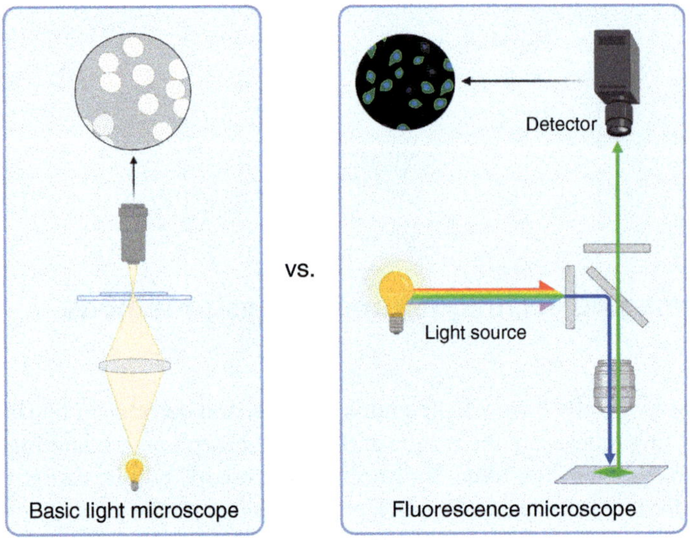

Figure 10.1 Microscope design.

but unfortunately most of the dyes have a broad spectrum. Therefore, using the standard fluorescent dyes, the practical maximum number of multiple colors is 3 or 4 (RGB + Y: yellow). If pushed to a near-infrared (NIR) range, the number could be 5. NIR cannot be detected by human eyes, but the signal could be converted into false color for human perception.

In typical cell imaging, the nucleus is stained blue, due to the excellent nuclear probes: DAPI and Hoechst. Then, other organelles or proteins are labeled with the combination of the colors. Each dye may have its own optimum excitation and emission wavelengths, usually fixed excitation and emission filters are assembled in a cube, including a dichromic mirror in the middle with a 45° angle. If four dyes are used for imaging, four matching cubes are used. For flexibility of dye choice, the cubes are usually replaceable, but the installed cube slots are usually not too many due to the intrinsic limit of the spectrum overlap. For acquiring the multicolor images, the cubes should slide in to the light path one by one. The cube movement can be automated by a motor, but apparently it takes time and generates mechanical noise. If the measurement is high-speed kinetics, the cube change time could be the practical limit for the time resolution.

One alternative option to overcome the limit is filter wheels. Multiple numbers of excitation filters are assembled in one wheel, and emission filters are assembled in another wheel. By rotating the filter wheels, the combinations of excitation and emission could be made for each dye imaging. As the rotating motion could be faster than sliding, automated filter wheel can improve the kinetic performance of multicolor imaging. Also, the choice of excitation and emission combination is much higher than filter cubes. If there are 10 filters for excitation and emission each, the possible combination would be about 50 (half of 10×10, as the emission is in longer wavelength than the excitation in most cases) (Figure 10.2).

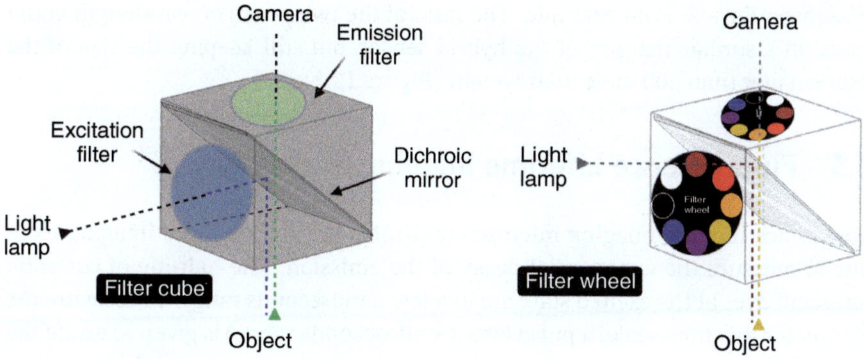

Figure 10.2 Filter cube and filter wheel.

10.2 Ratiometric Measurement

If a Ca^{2+} sensor contains two color moieties (one Ca^{2+} sensitive and the other Ca^{2+} insensitive with a different color), the ratio of the fluorescence signal from the two moieties can provide the Ca^{2+} concentration without the influence of sensor concentration. In this case, the insensitive dye serves as the reference. The unit of single fluorescence measurement is "emission photon #/probe concentration," but ratiometric measurement has no unit as all the factors are cancelled out. So, in principle, 2 sensors in one molecule may generate the same result. With the advantage, ratiometric sensor is popular especially for biological applications due to the complex situations of dynamic events in life. One potential problem for this hybrid sensor is its size. For the smooth penetration to the cell, usually a small molecular weight (<500) is preferred, but the hybrid molecule easily goes over the criteria, hampering the cell permeability. Alternative option is a single, but color-changing sensor, usually through the electronic structure change. The famous

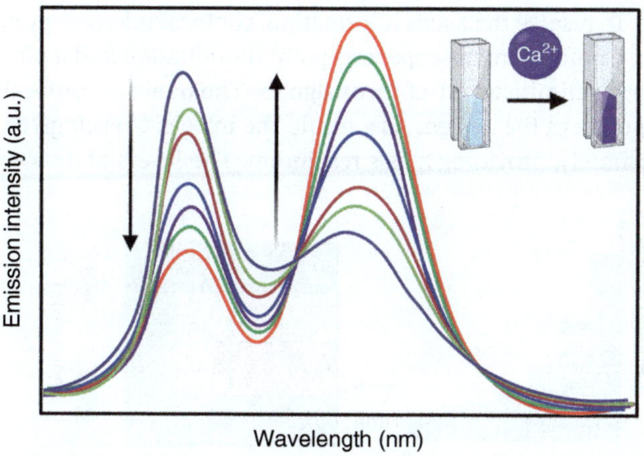

Figure 10.3 Spectrum with two peaks for ratiometric measurement.

Ca^{2+} sensor, Fura-2, is an example. The ratio of the two peaks of wavelength could be used in a similar manner of the hybrid sensor, but still keeping the size of the sensor smaller than 500 molecular weight (Figure 10.3).

10.3 Fluorescence Lifetime Imaging Microscopy

Fluorescence lifetime imaging microscopy (FLIM) is a fluorescence imaging technique to measure the exponential decay of the emission. The half-life of common fluorescent dyes at the excited state are in a few nanoseconds range. To measure the emission for this time scale, a pulse laser (femtoseconds range) is given to excite the dyes, and after a few picoseconds time delay, the emission is measured to calculate the lifetime of the dye. The information of lifetime for each localization can be reconstructed as 2D images, providing the analyte or environmental information. FLIM data are also concentration-independent values and are becoming a more popular technique, despite the high specifications of the hardware such as the pulse laser and time-resolved detector. The sensitivity of the detector should be much higher in FLIM compared to common fluorescence imaging, as the time-resolved measurement monitors only a small fraction of total fluorescence along the time course. FLIM could be either time domain or frequency domain, and can generate more detailed information for the probes and their environment (Figure 10.4).

10.4 Confocal Fluorescence Microscopy

The basic fluorescence microscope takes images from up or downside of the samples. If the sample is transparent, the images are accumulated probe images through the whole thickness of the sample. If the sample is not transparent, the image would be mainly surface area images in tens to hundreds of micrometers depth. This type of images is called wide-field or epi-fluorescence image, where epi- means above or on top of. The epi-fluorescence image contains the overlapped information of dyes from different depths. To resolve the z-axis information, confocal microscopy is the right choice technique. Confocal microscope uses point illumination and a pinhole in front of the detector to eliminate out-of-focus signals. The name of confocal stems from the two focal points in the design. As a result, the images of each given depth can be obtained separately, providing *z*-axis resolution. The layers of images

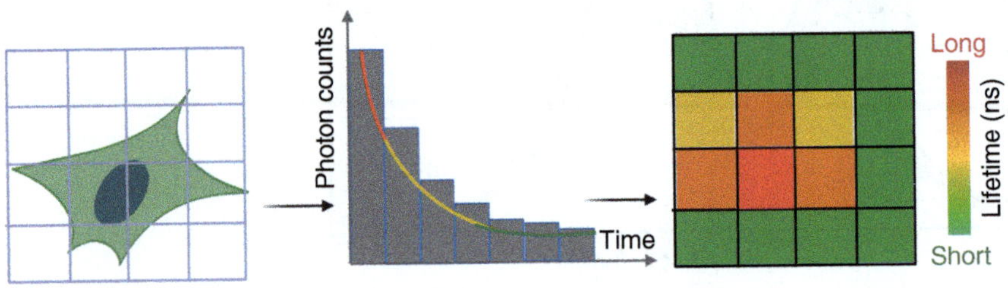

Figure 10.4 FLIM spectrum and reconstructed images.

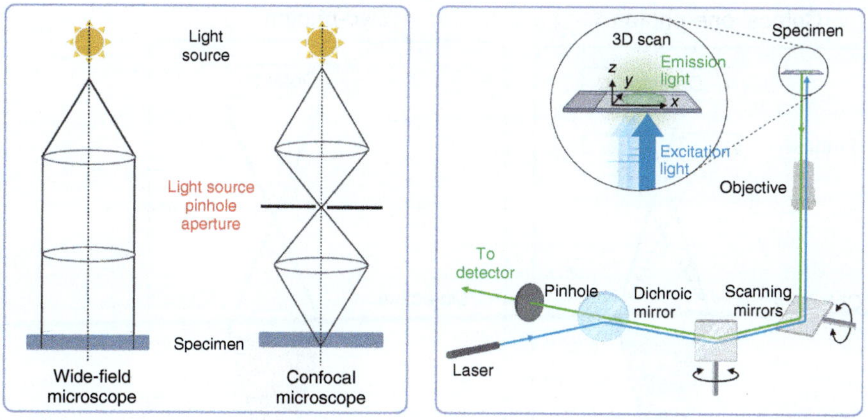

Figure 10.5 Design of confocal microscopy.

can be stacked to reconstruct a tomographic 3D image of the sample. In confocal imaging, the x–y spatial resolution (~200 nm) is usually higher than z-axis resolution (~500 nm) (Figure 10.5).

Although confocal fluorescence is a powerful imaging technique, the fluorescence signal is much weaker than wide-field images, as only fractional signal from the defined depth is obtained and also significant out-of-focus signal is wasted. That is why higher concentration of probe staining and high-power laser are often required to get high-quality confocal images. Although the input light is focused on the defined depth, the light will activate all the probes on the path above the focal depth. Therefore, the high-power laser can cause bleaching of the stained dye in the samples and can also damage the tissue with overheating.

10.5 Two-Photon Excitation Fluorescence Imaging and Harmonic Generation

Common fluorescence imaging uses a shorter wavelength of excitation light and records a longer wavelength emission light. The gap of this energy difference is known as Stokes shift. Two-photon excitation microscopy is based on a simultaneous excitation by two photons with longer wavelength than the emission light. The result is that two low-energy photons are combined to make a high-energy photon. This nonlinear optical phenomenon requires ultra-high-power laser and focusing in depth. As a result, the two-photon excitation technique provides spatially resolved images similar to confocal imaging. Even better, the excitation can occur only at the focal point where enough power of light is concentrated, unlike confocal imaging. Therefore, the photo bleaching is also much less in two-photon excitation imaging than in single photon method. Due to the requirement of long-wavelength light, most likely NIR light, the light penetration depth could be much deeper than that in confocal imaging, allowing up to a millimeter-deep tomographic imaging (Figure 10.6).

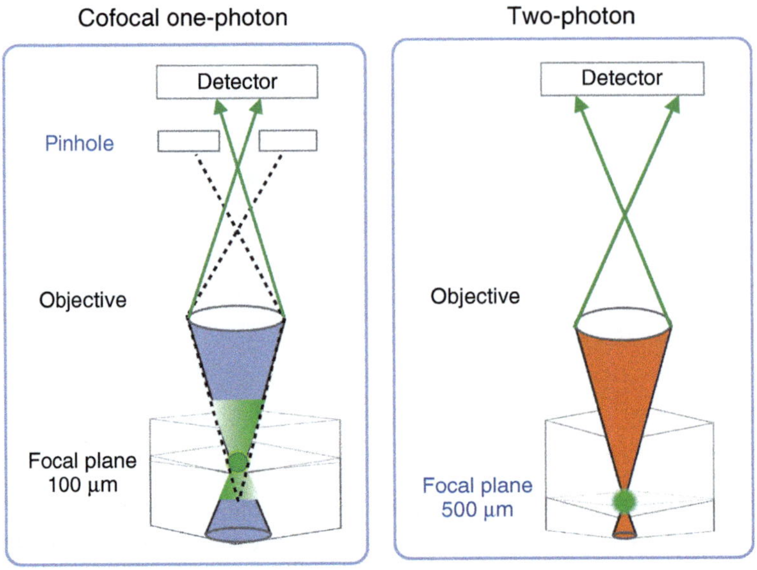

Figure 10.6 Comparison between confocal and two-photon imaging.

Second-harmonic generation (SHG), also called frequency doubling, is another nonlinear optical phenomenon in which two photons interact with a nonlinear material and generate a frequency-doubled photon. SHG, like other even-order nonlinear optical phenomena, does not occur in the material with an inversion symmetry. In some cases, almost 100% of the input light energy can be converted to the second-harmonic frequency. SHG shares a common aspect of the nonlinear optical phenomenon with two-photon excitation fluorescence imaging, and requires a high-power pulse laser. As a result, almost same optical setup could be shared for the two methods. One unique advantage of SHG is that it does not require any labeling with fluorescent dye, making it a label-free technique if only the material is susceptible for SHG. In biological applications, SHG is mainly used for high-resolution optical imaging of collagen. Using a femtosecond pulse laser and proper filters, the excitation light can be easily separated from the emitted, frequency-doubled SHG signal. This technique allows for high axial and lateral resolution, which is comparable to that of confocal microscopy without having to use pinholes. As a special case, third-harmonic generation (THG), with three photons involved, is also available for biological application and mainly used for lipid imaging.

10.6 Selective Plane Illumination Microscopy

Both confocal and two-photon excitation imaging require point scanning. For one depth plane, the point should scan the 2D space in a zig-zag motion, then move

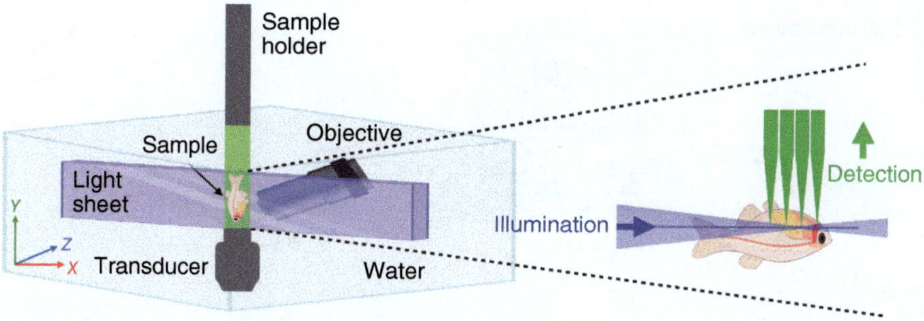

Figure 10.7 Selective plane illumination microscopy.

to the next depth layer. Naturally, the scanning time is longer than single-snapshot imaging. Selective plane illumination microscopy (SPIM) is a technique to solve the point scanning limit. To improve the scanning speed, a plane light or light sheet is generated, and irradiated from the side of the specimen in a x–y plane direction at a fixed z depth. As the probes in the given depth are all excited, a single-snapshot shooting completes the imaging of the layer. Then, the light plane moves to the next layer on the z-axis. The speed of imaging in SPIM is higher by two-dimension compared to confocal or two-photon excitation method. Therefore, SPIM would be an ideal technique for large samples and for dynamic change monitoring due to the high-speed imaging. For the large samples, a clearing procedure is necessary similar to other optical imaging (Figure 10.7).

10.7 Total Internal Reflection Fluorescence Microscopy

Another interesting technique for the defined z-axis (depth) imaging is total internal reflection fluorescence (TIRF) microscopy. TIRF microscopy can be applied to specimens attached on a glass slide with inverted microscope configuration, where the objective is located under the slide (the opposite configuration is called upright configuration). By applying the excitation light from a tilted angle, the total internal reflection is induced between the sample and the solid glass. As a result, only a very thin layer (usually less than 200 nm) of the sample near the glass surface can be selectively excited, breaking the resolution limit by the Abbe's principle (half of the wavelength of light) in the z-axis direction. Cell membrane is about 4 nm thick and TIRF microscopy can provide the information on the events on the membrane or right beneath the membrane, such as movement of membrane receptors or endocytosis/exocytosis (Figure 10.8).

The efficiency of internal reflection is dependent on the angle of the input light. By adjusting the tilted angle of the excitation light, the amount of light entering the sample can be controlled, activating the fluorescent probes at different thicknesses. The results can be converted into 3D tomographic images of areas near the cell membrane.

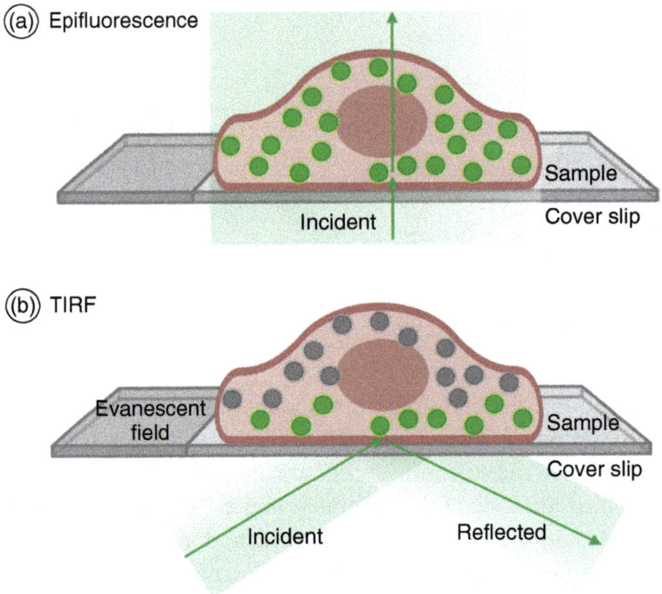

Figure 10.8 Total internal reflection fluorescence.

10.8 Super-Resolution Imaging

The diffraction limit by Abbe's rule has been considered as an unbreakable barrier in the resolution of optical imaging (hundreds of nanometers at best). Considering the size of bacteria or mitochondria as about a micrometer, it is difficult to image the details of the organelle structure. Most of the intracellular vesicles are even smaller (tens to hundreds of nanometers), and they could be images as just a spot, even worse, a diffused spot at best. TIRF or near-field imaging [1] tricks were developed to partially improve the resolution, but they require special design of the microscopes, limiting the imaging field and general applications. Structured illumination microscopy (SIM) is a digital deconvolution approach, which can improve the resolution severalfold [2]. Various new techniques for a more general super-resolution microscopy have been developed to make the real breakthrough, with stimulated emission depletion (STED) and stochastic optical reconstruction microscopy (STORM) as the representative techniques.

The optical resolution is defined as the minimum distance to distinguish two spots. Even single-molecule emission is observed blurry with a broad peak of hundreds of nanometers width due to the diffraction. STED uses two laser pulses, the excitation light of the fluorophore at the center and de-excitation light surrounding the center point to remove the diffused emission. The mechanism of de-excitation is based on stimulated emission depletion. As a result, the distribution of target spots becomes sharper than the original spot image. The imaging would be point scanning similar to confocal imaging, which makes the imaging slow. Another potential problem is that the strong de-excitation light may photobleach the fluorescent probes

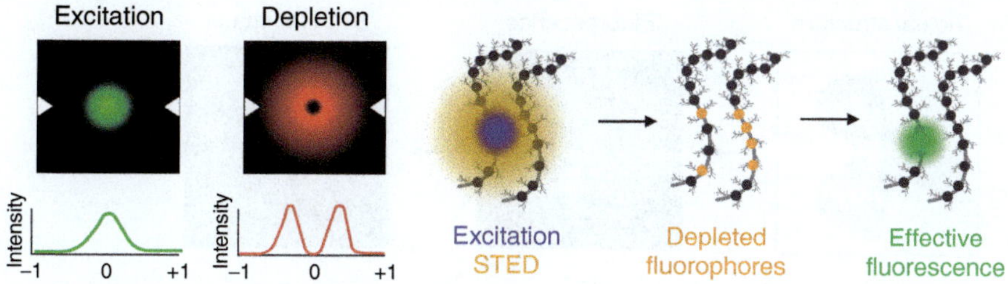

Figure 10.9 Stimulated emission depletion.

during the imaging process, which may limit the choice of probes to photostable dyes (Figure 10.9).

STORM or photo-activated localization microscopy (PALM) is based on another principle using photo-switchable (commonly called blinking) dyes. When two fluorescent molecules are close to each other within the Abbe's limit, they cannot be distinguished, that is not resolved. The higher the concentration of probes, the higher the possibility of the probe molecules are in the Abbe's limit. If the probe concentration is low enough, the spots are broad, but distinguishable. So, the central positions of each diffused spot can be mathematically calculated and imaged as a pointage. But, due to the low concentration of probes, the information is not plenty enough to show the whole structure. If the dye molecule is switchable stochastically (randomly), for example, by turning on 0.1% of dyes, one frame of the image can be obtained. Then, turn off all the dyes, turn on another random 0.1% of dyes, image the second frame, and repeat the process 1000 times. By overlaying the 1000 images (or more) a high-resolution of image by pointage method could be reconstructed. The key requirement for this technique is photoswitchable dyes and switching methods. The switching can be accelerated by strong laser or chemical methods, and each makes up different names of the improved technique such as dSTROM (direct STORM). While widely used, the mechanism of switching of dyes, conversion between bright and dark status before photobleaching, is not fully understood for all the dye cases (Figure 10.10).

Binding-activated localization microscopy (BALM) is another pointage painting method. Instead of photoswitching of dyes, BALM utilizes the fluoresce turn-on of probe by binding to the target. For example, a dye has low fluorescence in the media, but after binding to the DNA, and it turns on the fluorescence property. Important point is that the binding affinity of the probe to the target should be low with the equilibrium shifted more to off-state (to keep the on-state at low population). Also, the on–off exchange should be fast enough.

With the super-resolution techniques maturing, tens of nanometers resolution are within the possible range of imaging even in the live cells. The breaks in the resolution limit have uncovered much more details of the small world, such as chromosomes, organelles, and cytoskeletal structure. Once the resolution reaches a single-digit nanometer, the imaging should be called nanoscopy rather than microscopy (Figure 10.11).

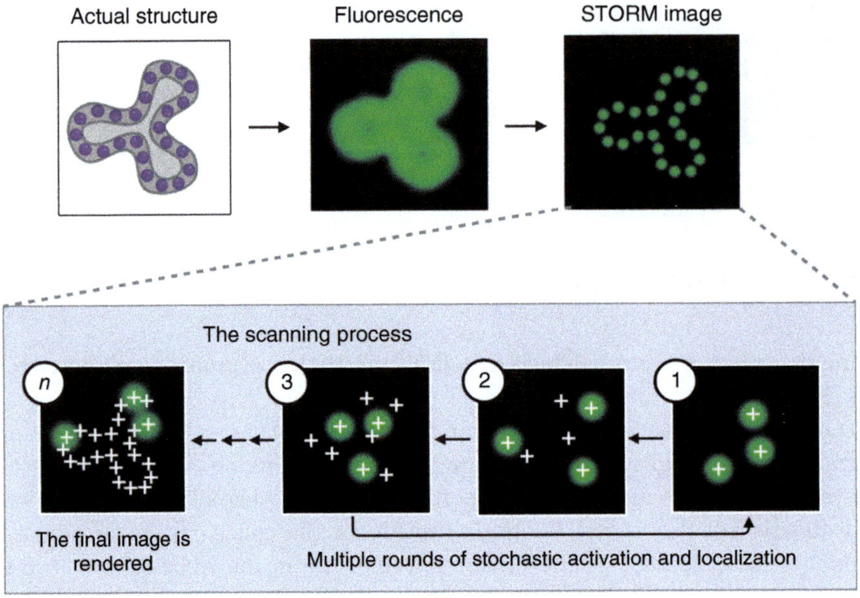

Figure 10.10 Stochastic optical reconstruction microscopy.

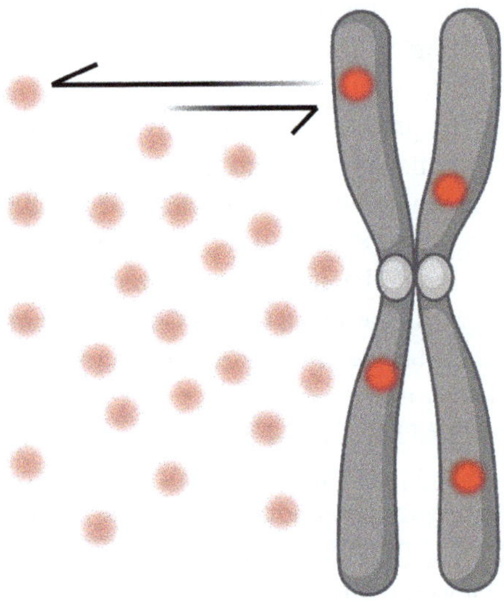

Figure 10.11 Binding-activated localization microscopy.

10.9 Single-Molecule Imaging

For molecular events such as receptor–ligand interaction or receptor dimerization, single-molecule imaging capability would be critical. For single-molecule imaging, a sensitive detector and high magnification optical setup are required. Photostability of the dye is another important factor. Single-molecule imaging may not necessarily mean single-photon detection. A single molecule can be excited

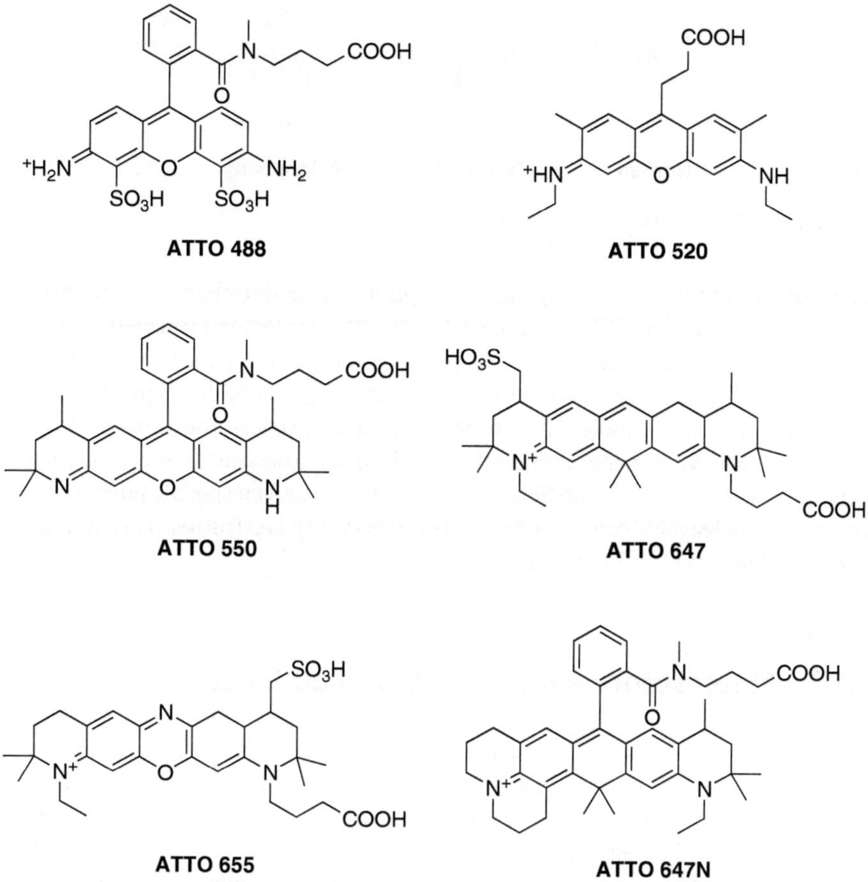

Figure 10.12 Structure of representative Atto dyes.

and emit fluorescence, and go for repetitive cycles of excitation and emission until photobleaching. Photobleaching usually means physical damage and an irreversible change of the dye structure, which is different from the dark status of photoswitching. The stability of the dye determines the possible tracking time of the single-molecule event. Most of the dyes have limited cycle number and several seconds of maximum lifetime for the single-molecule imaging. Among the commercially available dyes, Atto dyes are claimed to be the most photostable dyes, but their hydrophobic character is not ideal for biological application. Super stable fluorescent dye with good physiologically friendly properties are yet to be developed (Figure 10.12).

10.10 Photoactivation of Caged Molecule

For single-molecule imaging, a low concentration of probe treatment on the sample and a search for the proper imaging site are typical approaches. Alternative option

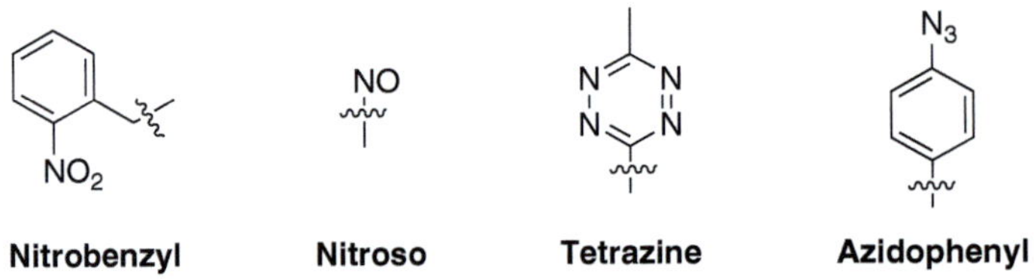

Figure 10.13 Photocleavable cage groups.

is activation of the fluorogenic by light. The photoactivatable chemical structure is called the caging group. When the caged molecule (molecule protected with the caging group) get irradiated, the caging group is removed to generate the original molecule. If a biologically active molecule is caged, by irradiating light in a given specific area, the active molecule can be released in a space- and time-controlled manner (also known as spatiotemporal control). If a fluorescent molecule is caged, photoactivation can turn on the fluorescence only in the given specific position. The common photocleavable cage groups are nitrobenzyl [3], NO (nitroso) [4], tetrazine [5], and azidophenyl [6] (Figure 10.13).

10.11 Fluorescence Recovery After Photobleaching

For measuring the kinetics of diffusion, fluorescence recovery after photobleaching (FRAP) is an option. For example, by irradiating strong light to the labeled membrane, fluorescent dyes in the defined area can be permanently quenched or photobleached. By measuring the recovery of the fluorescence by the diffusion of nearby fluorescent probes, the diffusion rate or membrane mobility can be calculated. The FRAP technique has been used to distinguish more ordered lipid raft domain and less ordered area of the cell membrane [7]. Spectroscopic method to achieve such a diffusion kinetics is called fluorescence correlation spectroscopy (FCS), which also can provide the information about the number of fluorescent dyes in the given area. If multiple dyes are used for relative diffusion kinetics, it is called fluorescence cross-correlation spectroscopy (FCCS) (Figure 10.14).

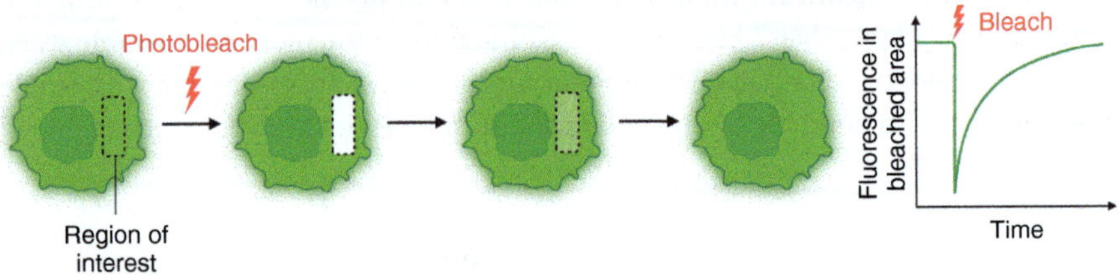

Figure 10.14 Fluorescence recovery after photobleaching.

10.12 Flow Cytometry Technique

Flow cytometry is a powerful technique for cell analysis using fluorescence labeling. Flow cytometry means measurement (metry) of cells (cyto) through flow. Flow cytometry itself is not an imaging technique, but provides quantitative information for the probe number in each cell. By flowing the diluted cells in a tube, each "single" cell can be analyzed by light irradiation and detection. When the input light hits cells, scattered light could be detected from different angles. Forward scatter (FSC) increases with the increasing size of the cells, and side scatter (SSC) increases when the cells contain more irregular organelles such as granules. The scattering information can define the blood cells and classify them into red blood cells, lymphocytes, monocytes, and granulocytes.

If the cells are labeled with fluorescence, the cells can be discriminated according to the fluorescence intensity in addition to the scattering information. Multiple color labeling is also possible, as far as the signals can get resolved. The same input light can be used for both scattering and fluorescence, by positioning different detectors with proper dichromic mirrors (Figure 10.15).

One of the basic and popular application of flow cytometry is cell cycle analysis by nuclear staining probes. DNA probes are almost nonfluorescent in the media, and enhance their fluorescence dramatically upon binding to DNA. Therefore, the fluorescence intensity of a DNA dye in a cell reflects the total amount of the cellular DNA. The human cells in their resting state (G0) have 2N amount of DNA, where one N is from the father and the other N is from the mother. Those cells are called diploid

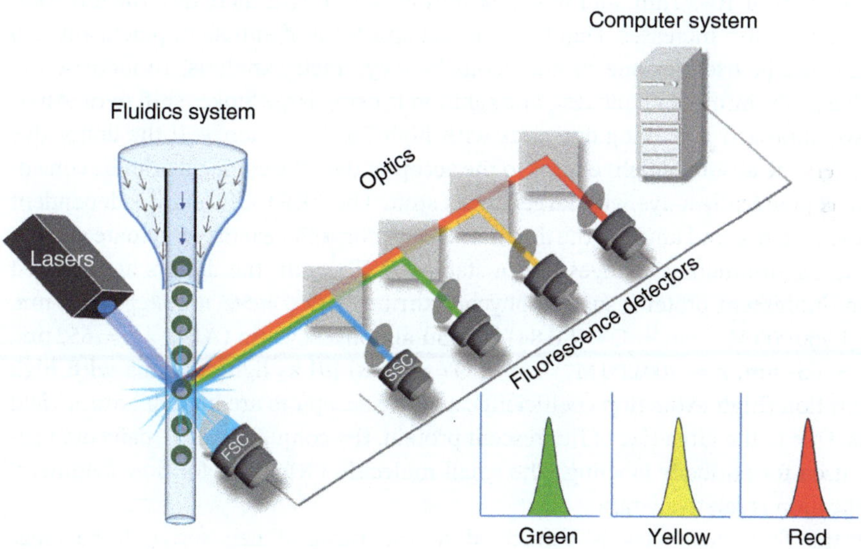

Figure 10.15 Flow cytometry.

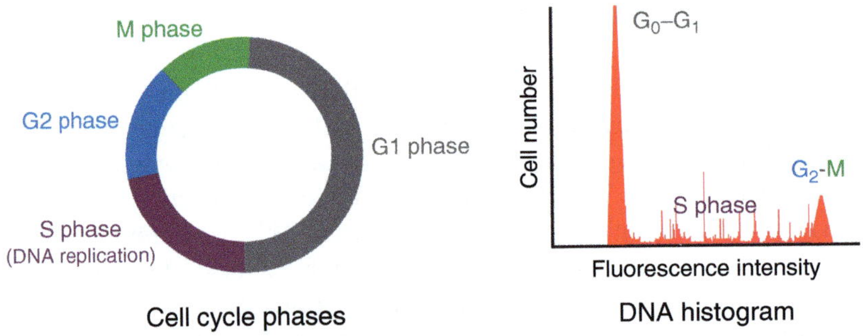

Figure 10.16 Cell cycle and flow cytometry data with DNA probe.

(2N) cells. When the cells go into cell cycle (from G0 to G1) by environmental stimuli, the cells collect materials for cell division, which requires doubling the cell body and duplication of genetic information DNA. Once enough materials are secured, the cells start DNA synthesis (S means synthesis in S phase). In S phase, the DNA duplication starts, and the amount of DNA increases from 2N to 4N. In G2, the cells check if everything is OK for cell division, and the cells are physically divided into two during the M phase. The process of the whole cell cycle could be monitored by DNA probes in live or dead cell status (Figure 10.16).

Usually the typical DNA probes are blue color dyes such as DAPI or Hoechst, which can be excited by UV light. Longer-wavelength channels are usually assigned for labeled antibodies or small-molecule probes. Most of the common dyes have Stokes shift of 10–30 nm, and when excitation wavelength increases, the emission wavelength also increases. Due to the broad spectrum of emission, practically 3–5 dyes could be used for one simultaneous flow cytometry analysis. To increase the usable probe number in one test, one solution is using large Stokes shift dyes. Alternative approach is making dye pairs with high FRET efficiency. If the donor dye transfers the absorbed light energy to the acceptor dye, the conjugate can be considered as practically a dye with large Stokes shift. The FRET efficiency is dependent on the distance and angle of the donor and acceptor, and also may fluctuate according to environmental changes. For a stable FRET result, the donors are selected from fluorescent proteins such as phycoerythrin (PE, $\lambda_{ex} = 565$ nm, $\lambda_{em} = 573$ nm, $\varepsilon = 1\,960\,000\,M^{-1}\,cm^{-1}$, QY = 0.84) [8] and allophycocyanin (APC, $\lambda_{ex} = 652$ nm, $\lambda_{em} = 658$ nm, $\varepsilon = 700\,000\,M^{-1}\,cm^{-1}$, QY = 0.68) [9] as light antenna with high absorption (high extinction coefficient), and the acceptors are high quantum yield dyes. Due to the large size of fluorescent protein, the conjugate FRET pairs are usually used for antibody labeling. The small-molecule FRET pair for flow cytometry application is relatively rare.

If the flow cytometry is connected to the physical cell sorter, it becomes (fluorescence-activated cell sorter) (FACS, Figure 10.17). By the difference of the fluorescence or scattering signal, each cell can be physically isolated through differential charging. As the separation condition is mild enough, the isolated cells can be cultured for further study or analysis.

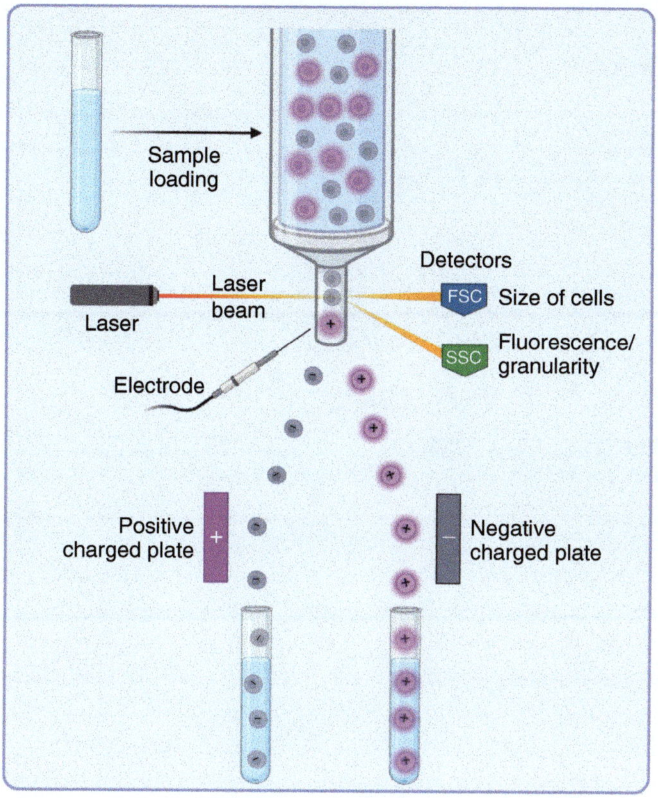

Figure 10.17 Fluorescence-activated cell sorter.

References

1 Fish, K.N. (2022). *Curr. Protoc.* 2: e517. https://doi.org/10.1002/cpz1.517.

2 Guerra, J.M. (1995). *Appl. Phys. Lett.* 66: 3555. https://doi.org/10.1063/1.113814.

3 Grimm, J.B., Klein, T., Kopek, B.G. et al. (2016). *Angew. Chem, Int. Ed.* 55: 1723–1727. https://doi.org/10.1002/anie.201509649.

4 Zhang, Z., Wu, J., Shang, Z. et al. (2016). *Anal. Chem.* 88: 7274–7280. https://doi.org/10.1021/acs.analchem.6b01603.

5 Loredo, A., Tang, J., Wang, L. et al. (2020). *Chem. Sci.* 11: 4410–4415. https://doi.org/10.1039/D0SC01009J.

6 Lord, S.J., Conley, N.R., Lee, H.L.D. et al. (2008). *J. Am. Chem. Soc.* 130: 9204–9205. https://doi.org/10.1021/ja802883k.

7 van der Weij-De Wit, C.D., Doust, A.B., Van Stokkum, I.H. et al. (2014). *Pharm. Res.* 31: 255–270. https://doi.org/10.1007/s11095-013-1146-9.

8 van der Weij-De Wit, C.D., Doust, A.B., van Stokkum, I.H.M. et al. (2006). *J. Phys. Chem B.* 110: 25066–25073. https://doi.org/10.1021/jp061546w.

9 MacColl, R. (2004). *Biochim. Biophys. Acta.* 1657: 73–81. https://doi.org/10.1016/j.bbabio.2004.04.005.

11
Perspectives for Future Probe Development

For the application of imaging probes, the most important factors would be the selectivity to the target and high signal-to-noise ratio of the probe. There are two mainly different approaches: rational design and diversity-based screening. The decision would be affected by the available information for the target. If a clear molecular target is already elucidated and the information is available, the design be more straightforward. This approach would be called "hypothesis-driven approach." If there is little information known about the target, it would be difficult to establish the molecular hypothesis. Then, phenotype screening with a diversity-based tool would be a more practical approach. This approach would be called "discovery-driven approach."

11.1 Design of Selective Probes

If the target for the imaging is known, the design of a selective probe would be doable. The molecular target is exclusively found in the target cell or tissue, and it be a "biomarker for the target." First approach may be elucidating the binding ligand to the biomarker. Most of the binding targets are proteins, but they also could be carbohydrates, lipids, and nucleic acids. If there is a known natural product or traditional treatment, it would be a good starting point. If the structure of the biomarker is known by X-ray crystallography or cryo-EM image, a computational docking could be applied. Once a promising binding ligand is elucidated, a reporter should be employed by conjugating the binding ligand and the reporter. The reporter could be an always-bright fluorescent molecule, and the amount of ligand binding to the target is proportional to the signal. This type of conjugated ligand is a typical "probe." If the reporter is a smart material which is turned on only by binding to the target or near the target area, it could be called "sensor." Sensor is a part of the probe, but may generate a higher signal-to-noise ratio, due to the low background in the non-target area. For the binding dependent turn-on can be either electronic changes, such as positron emission tomography (PET) or intramolecular charge transfer (ICT). Proton or metal ion coordination to the ligand would be a good example. Motion-induced changes of emission (MICE) is another popular

Sensors and Probes for Bioimaging, First Edition. Young-Tae Chang and Nam-Young Kang.

mechanism for the fluorescence change in sensors [1]. By binding to the target molecule, the motion of the sensor is affected, and so does the fluorescence. Other than the molecular interaction, the related environmental changes for MICE are viscosity and temperature.

Selective transporter is another type molecular target. Once identified, the substrate for the transporter could be designed. For example, tumor cells express high levels of glucose transporters to import large amounts of energy and material resources. Then, well-designed glucose derivatives would be a good candidate for the selective tumor imaging. Fluorescent false neurotransmitter is another example. By mimicking a natural neurotransmitter such as dopamine, a fluorescent derivative could selectively be taken up by dopamine neurons.

Utilizing the unique metabolism of the target cell is also a good strategy. One possible environmental factor is specific enzymes in the extracellular area of the target cell. In that case, a selective substrate development for the enzyme is a popular approach. Various proteases have been selected as the molecular target and fluorescence resonance energy transfer (FRET)-based substrate development was widely used. In principle, the approach can be applied to different intracellular enzymes, such as hydrolases, kinases, and phosphatases, if an efficient substrate can be designed. For example, imported glucose can be phosphorylated at 6-position as the first step of glycolysis, and can be trapped in the target cells, facilitating the accumulation of the probes. Esterase-selective turn-on probes for live-cell labeling is another example.

Small biological molecules or signaling molecules can be the target for the selective probes. Reactive oxygen species (ROS) for the activated immune cells or cancer cells have been widely studied for the selective cell staining. Various metal ions were also important target for distinguishing cells in disease status. pH can also be considered one of them, as pH is the scale of proton ion concentration. Biological battery molecule adenosine tri-phosphate (ATP) could be differentially produced in different cells, and ATP sensor could be used for cell-selective imaging purpose. Oxido-reduction metabolite NAD(P)H is another example for the small-molecule targeting for the cell selectivity (Figure 11.1).

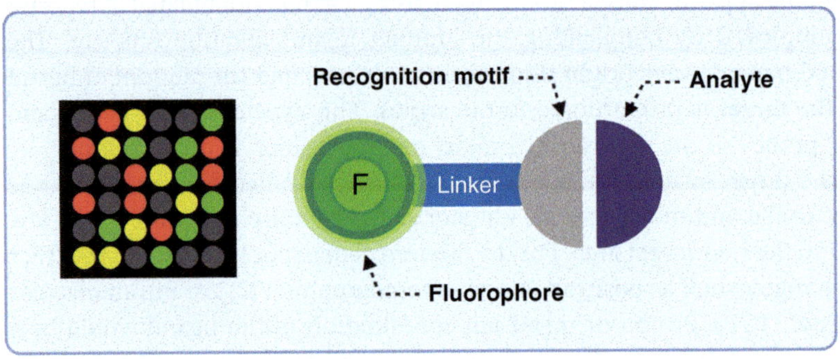

Figure 11.1 Hypothesis-driven approach. F, fluorescence.

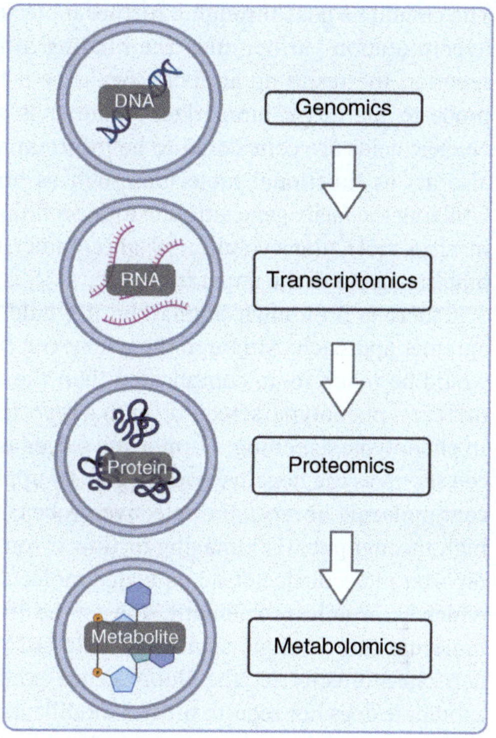

Figure 11.2 Genomics and proteomics data.

If the molecular target is not clearly known yet, but still desirable to find one, information-based search would be an option. Through various -omics data, such as genomic or proteomic differences, according to the disease or different tissue, the candidate of the molecular target could be selected. Among the identified potential candidates, the really relevant target should be elucidated by careful validation. In many cases, the practical problem is that the potential candidates are too many and the -omics data have usually a high error range. Familial study for the correlation of disease and mutation could be the supporting data for the selected candidates. To validate a potential target from the scratch, a genetic knockout (deletion of target gene) or overexpression is commonly used. By knocking out of the candidate gene, if the phenotype of the target tissue is changed into a background type, the gene could be a validated target, as the role of the gene is important for the phenotype. By overexpression of the target gene, an exaggerated or more sensitive disease phenotype may appear to support the role of the candidate (Figure 11.2).

11.2 Discovery of Selective Probes by Screening

Let's take the situation that biomarkers are known, but there is no available binding ligand. One traditional approach is to develop a specific antibody for the biomarker. By injection of the biomarker into different species body, the first antibody (germline antibody) is selected by a screening of the immune library.

The candidate goes through a further molecular evolution process called "somatic hypermutation" to optimize the binding affinity. By conjugating the fluorescent reporter, the resulting antibody could be a high-affinity, high-selectivity binding probe to the target biomarker. Aptamer is a nucleic-acid-based macromolecule. Nucleic acids are considered to be information storage and carriers, but they may also act as functional molecules such as binding ligands, similar to antibodies. Utilizing the facile gene amplification technique, polymerase chain reaction (PCR), in vitro molecular evolution of an aptamer is possible to develop a high-affinity binding ligand to the target molecule.

If there is no known biomarker, it would be difficult to go for an antibody or aptamer approach. Although the whole-cell-based screening is also tried, the result would be much more complicated than the case of a purified target molecule. In this case, phenotypic screening with a reporter library could be an excellent choice. In phenotypic screening, a tumor cell serves as the positive control cell and a normal cell serves as the negative control cell. By applying the reporter library (e.g. fluorescent molecule library), the selective probe candidate can be directly elucidated by high throughput (HT) imaging or flow cytometry. Important point here is that the reporter molecule is not a conjugated molecule (binding motif + fluorescent motif), which is usually problematic due to the increased size. The fluorescent library molecules are a single small molecule (<MW 500) with small diversity moieties decorated directly on the fluorescence scaffold. Therefore, the identified probe candidate does not require further modification, such as reporter conjugation.

The practical issue for this screening-based approach is the later stage of target identification of the probe. In a hypothesis-driven approach, the target information was used for the design of the probe. In a discovery-driven approach, the probe provides the clue for the target. If the biomarker is the binding target, affinity matrix is a possible approach. The conventional affinity matrix is made by conjugating the probe molecule to the solid phase matrix, such as hydrophilic carbohydrate polymers. Assuming that the binding target is a protein, proteins are extracted from cells by adding detergents. Chemically speaking, detergents have almost the same characteristics as biomembranes and destroy all the compartmentalization of the cell, losing the cellular localization information. Therefore, the protein mixture is collected in one pool and physical fishing out of the target protein is performed by the fishing rod (affinity matrix). Proteins that are optimized for their original environment may be denatured during the harsh preparation procedure and lose their original binding affinity to the probe. Some of the proteins may form complexes with other partners in the cell, but the majority of them may be dissociated by detergent treatment. When the affinity-matrix-bound proteins are recovered, there may be nonspecific binding proteins due to denaturing or their own sticky nature such as serum albumin. Such a nonspecific binding property is called "avidity," in comparison to specific interaction as "affinity." The uneven copy number of the proteins may also generate false-positive candidates. If the copy number of a protein is 100 times higher than the real target protein, even though the binding affinity is 100 times lower, the bound amount to the affinity matrix would be the same. Also, the recovery yield of the proteins from the affinity matrix depends on the washing procedure. Harsh and repeated washing may lose the target, and mild washing may allow too many nonspecific binders. Despite the many potential problems, the affinity matrix method has been widely used for target identification of unlabeled drug molecules.

A live-cell version of the fluorescent probe has a unique option. Due to the smooth live cell permeability, they can act as a harpoon without breaking the cells by detergents. The probe can be modified to make a covalent bond by the introduction of an electrophile such as chloroacetyl, and it may go to the live cells and form a covalent bond with the cysteine residue of the target protein. Considering the harsh changes that occurred during the cell extract preparation, the live-cell target labeling would be much more specific than the conventional affinity matrix approach. It is noteworthy that all the proteins are in their original compartment and environment, when the labeling occurs. The subsequent cell extract preparation does not bother the probe–target interaction, as it is already connected by stable covalent bonds. As the result, the labeled target protein can be visualized on the protein gel through the fluorescent labeling by the probe. Therefore, it is easy to cut out the protein band and analyze the protein sequences by mass spectrometry to identify the target protein. If the binding target is protein (POLD), the sequencing of protein is well established, but if the target is carbohydrate (COLD), the analysis is more complicated and often relies on an indirect approach (Figure 11.3). Relatively rare, but DNA also acts as the binding target (DOLD) for the probes, such as biofilm of bacteria or neutrophil extracellular trap (NET).

Even the lipid composition (LOLD) of the cell membrane could be used for cell discrimination. In LOLD, the flexibility of the membrane plays a role in kinetically differentiating the updated speed of probes into the cell membrane. For example, B cells have a more flexible membrane than T cells, and lipophilic aggregates of

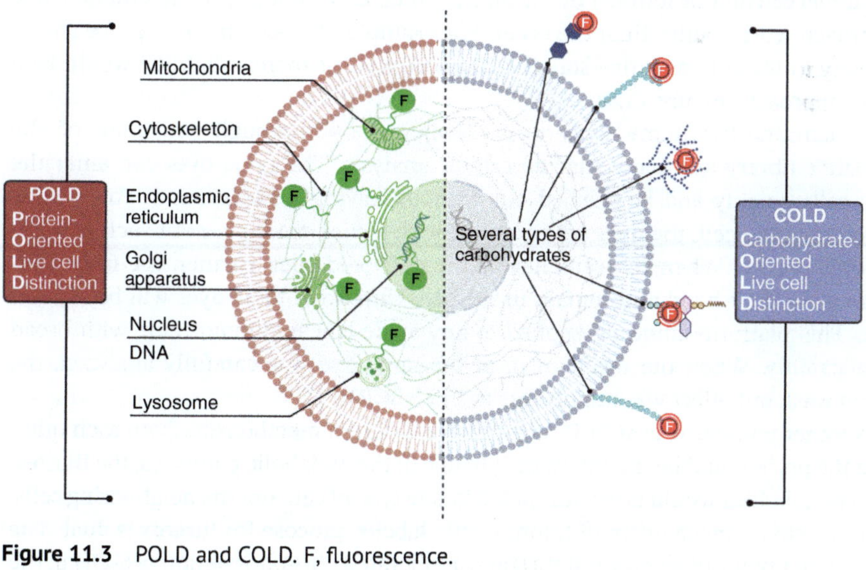

Figure 11.3 POLD and COLD. F, fluorescence.

probes tend to fuse with B-cell membranes faster than with T cells. As a result, B cells are stained by CDgB much stronger than T cells through the kinetic discrimination.

By the way, even if there is a selectively labeled protein band, a much larger amount of free probe may be found in the cell. If only 10% of the probe is in protein-bound form, and 90% of it in free-floating form in the cell, which portion dominates the cell staining? The binding targets may play a role, but their contribution may not be a major factor for the cell selectivity. There should be other mechanisms to explain the situation.

One possibility is GOLD. If the target protein is a transporter on the membrane, the probe would be a substrate or inhibitor for the target. Human and mouse cells has about 400 solute carrier (SLC) and 50 adenosine tri-phosphate-binding cassette (ABC) transporters on the plasma and organelle membranes. SLC transporters usually facilitate the intake of nutrients into the cell or organelles without energy consumption but by using the concentration gradient of substrate or other ions. ABC, as the name implies, consumes ATP for pumping out usually toxic materials out of cells. In the situation that GOLD is suspected as the potential mechanism, SLC- and/or ABC-expressing cell library could be used for the target identification. By genetic editing, the cells are engineered to carry one overexpressed transporter each. When a target SLC is overexpressed, the cells may be brightly labeled by the probe. If a target ABC is overexpressed, the probe staining will be decreased. The small population of cells either brighter or dimmer is collected and re-cultured for the next round of selection. By repeating the biospanning, the target transporter containing cells are enriched, and the identity of the target is elucidated by DNA sequencing. If CRISPR is used for the gene editing, both direction of activation (CRISPRa) and inhibition (CRISPRi) of the target expression are possible. For example, in the CRISPRi ABC library, the target ABC-suppressed cells may show brighter staining than the control by reduced pumping out of the probe. In the CRISPRi SLC library, the target cell may be dimmer by the reduced intake. Depending on the situation, one approach works better than the other. For example, the accumulated probe shows toxicity to the cell, negative staining (reduced SLC or increased ABC) would be a safer approach (Figure 11.4).

Background-free tame dyes were developed by thorough screening of the BODIPY library and molecular descriptor analysis. The tame dyes can enter the live cells quickly and be washed out without leaving a background. For specific labeling in the cell, the dyes are equipped with a click reaction motif such as azide or cyclooctyne. When the cells have a pre-delivered click partner, the tame dyes form a covalent bond and remain in the cell. Unreacted tame dyes will be washed out. This platform could be applied to any azide-labeled biomolecule with broad applicability. When the mechanism of the tame dye was carefully analyzed, the clean wash out effect was mainly due to the ABCB1.

Another possibility is MOLD. If cells have different metabolisms from each other and the probe could be incorporated into one of the metabolic pathways, the fluorescent metabolites would be accumulated in one type of cell, not the neighboring cells. The selectivity mechanism of fluorescently labeled glucose for tumors is dual. The increased uptake of glucose (GOLD) and accelerated phosphorylation of 6-OH in the

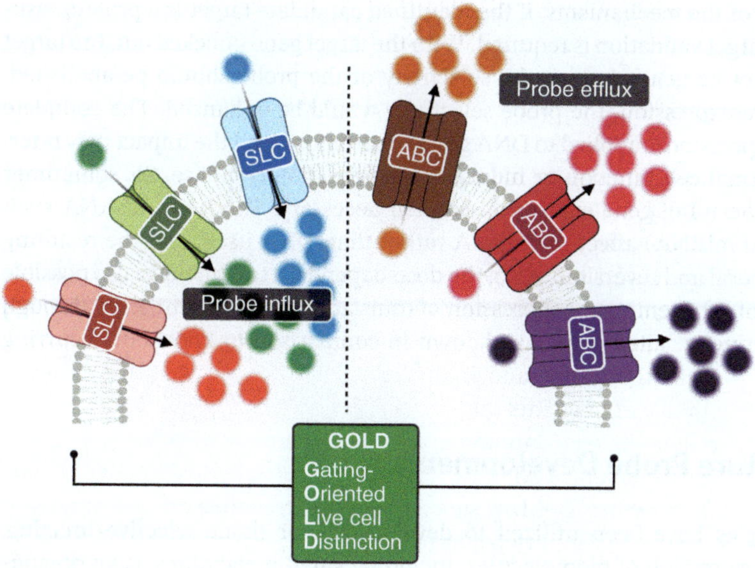

Figure 11.4 GOLD.

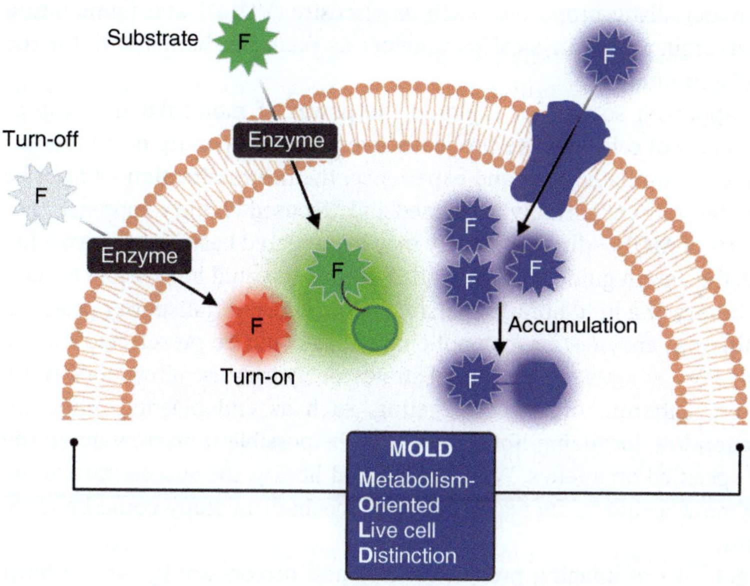

Figure 11.5 MOLD. F, fluorescence.

enhanced glycolysis (MOLD) facilitate the accumulation of the probe in the target tissue. Microglia probe CDr20 showed glucuronic acid addition as the mechanism of fluorescence turn-on, and NeutropG utilized the unique lipid droplet biosynthesis pathway for the selective staining of the neutrophil, competing with other granulocytes, such as eosinophils and basophils (Figure 11.5).

Regardless of the mechanisms, if the identified candidate target is a protein, usually genetic target validation is required. With the target gene knocked out, the target protein cannot be generated, so the selectivity of the probe should be abolished. In case of overexpression, the probe selectivity would be enhanced. The complete KO or overexpression is applied to DNA genome structure, and the impact may interfere with normal cell function or induce a compensation response. So, sometimes a temporal control of gene expression is often necessary. The RNAi (dsRNA such as siRNA and miRNA) affect the mRNA rather than DNA itself, and the resulting effect is temporal and reversible. Also, the dose-dependent treatment is also possible to some extent. The temporal suppression of translation (protein synthesis) through mRNA targeting is called gene knockdown in comparison to knockout occurring to DNA.

11.3 Future Probe Development

Many strategies have been utilized to develop cell- or tissue-selective imaging. Differential expression of biomolecules, including small metabolites, is an obvious biomarker for the probe development. General environmental changes, such as pH, polarity, or membrane potential, could be another target for biological discrimination. Non-materialistic properties, such as viscosity (VOLD) and temperature (TOLD), are emerging new physical parameters as phantom biomarkers for the different cell discrimination.

The library approach so far has explored the power of molecular diversity to increase the success of cell-selective probe discovery and to open up novel mechanisms. With the accumulated data and experience, the future direction will not be totally random but rather "partially" designed and "focused" library approach. For example, GOLD- or MOLD-directed library may be prepared based on the working mechanism as the design guide. For the GOLD library, it would be possible to have carbohydrate library or a lipid library, focusing on the relevant transporter proteins. For MOLD, different enzymes or metabolic reactions could be preselected for the library design. For example, potential substrates of hydrolysis, phosphorylation, or glycosylation. Different organelle targeting, such as endoplasmic reticulum (ER) or mitochondria, localizing library may also be possible to narrow down the targets to the specified organelles. With the focused library, the success rate of the probe development would be increased, and the mechanism study could be more straightforward.

For the target cells of imaging probes, cancers and tumors are by far the most extensively studied area. Despite the enormous efforts, each cancer seems unique and even continuously changing the identity by dynamic and heterogenous development. With the general trend of personalized medicine, imaging probes may also need to develop into multidimensional and subcategory-specific direction. In addition to cancer field, many functional cells and immune cells have been the target of selective imaging probes so far [2]. The new probes with novel mechanisms may provide more and more colorful information of each cell in the context of the complex system.

The imaging probes may contribute not only for the diagnosis stage, but also for the prognosis purpose to monitor the effect of personalized treatment. Depending on the selectivity mechanism, some imaging probes (such as TiY [3] or TiNIR [4]) may offer therapeutic effects at high dose, as an example of theranostics. Even non-toxic fluorescent probes, if combined with photodynamic therapy, could be used for selective cell death as demonstrated with CDy1 [5]. Such a functional control of the stained cells will cast the universal tool for therapeutic or optogenetic research.

Finally, as the missing field, aging-related imaging probes and their applications are expected to emerge and blossom in the near future. Compared to the long history of cancer biomarker study, aging-related biomarkers are still in their infant stage of elucidation. With the power of imaging probes, development of young- or old-cell-selective probe would be straightforward and the probes will provide the clues for novel biomarker discovery. The functional control provided by theranostics technology may cast the possibility of slowing down the aging process or even reversing it. The fountain of youth may come true!

References

1 Su, D., Teoh, C.L., Wang, L. et al. (2017). *Chem. Soc. Rev.* 46: 4833–4844. https://doi.org/10.1039/C7CS00018A.

2 Liu, X. and Chang, Y.T. (2022). *Chem. Soc. Rev.* 51: 1573–1591. https://doi.org/10.1039/d1cs00388g.

3 Lee, Y.A., Kim, J.J., Lee, J. et al. (2018). *Angew. Chem. Int. Ed.* 57: 2851–2854. https://doi.org/10.1002/anie.201712920.

4 Kim, J.J., Lee, Y.A., Su, D. et al. (2019). *J. Am. Chem. Soc.* 141: 14673–14686. https://doi.org/10.1021/jacs.9b06068.

5 Cho, S.J., Kim, S.Y., Park, S.J. et al. (2016). *ACS Cent. Sci.* 2: 604–607. https://doi.org/10.1021/acscentsci.6b00099.

Appendix

No	Probe Company	Website
1	SenPro	https://senprobe.com/
2	Thermo Fisher	https://www.thermofisher.com/kr/ko/home/brands/invitrogen.html
3	Active Motif	https://www.activemotif.com/
4	Ebba Biotech	https://www.ebbabiotech.com/
5	Goryo Chemical	https://goryochemical.com/en/
6	BioActs	https://www.bioacts.com/main/main.php
7	Tocri	https://www.tocris.com/
8	BOC Sciences	https://probes.bocsci.com/
9	PerkinElmer	https://www.perkinelmer.com/
10	Spirochrome	https://spirochrome.com/
11	Enzo	https://www.enzolifesciences.com/platforms/cellular-analysis/dyes-stains-fluorescent-probes/
12	Dojindo	https://www.dojindo.com/ASIA/
13	Lumiprobe	https://www.lumiprobe.com/
14	Stemcell Technologies	https://www.stemcell.com/

Sensors and Probes for Bioimaging, First Edition. Young-Tae Chang and Nam-Young Kang.

Index

d